BURNHAM'S ARITHMETIC, REVISED.

A

NEW SYSTEM

OF

ARITHMETIC,

ON

AN IMPROVED PLAN:

EMBRACING THE

RULES OF THREE, SINGLE AND DOUBLE, DIRECT AND INVERSE; BARTER LOSS AND GAIN; REDUCTION; MULTIPLICATION AND DIVISION OF FRACTIONS: EXCHANGE OF CURRENCIES; INTEREST·

AND ALL

PROPORTIONAL QUESTIONS

IN ONE RULE APPLICABLE TO THE WHOLE

THE PROCESS GREATLY SIMPLIFIED AND ABRIDGED.

BY CHARLES G. BURNHAM, A. M.

KEENE, N. H.:
PUBLISHED BY S. & G. S. WOODWARD
1857.

PREFACE.

He who writes a book in this age merely for the sake of being a book-maker, will find that he has written for other times than these; and his fame will be like one of those "second sights," having existence only in the mind of him who sees it. Every invention, every thing new, every book, from the child's primer to the most profoundly scientific text-book, must be tested by a comparison with others, professing, each and all, to be the best extant. Nor will any production gain for its author that for which he labored, unless it finally proves to be what it professes. Improvement is the "charmed word" of the age. It rings hourly in the ear of the multitude. The strong wind bears it onward, and the gentle zephyr wafts its echo. He who has already written his name far above his competitors, now seeks to outdo himself; and the tyro fancies, that he can begin where the best have ended, and run his race alone.

The author of this series of Arithmetics cherishes none of these fancies, having already received satisfactory compensation for all his toil, in imparting, from time to time, to those he has had the pleasure to instruct, the improvements here embodied. But should it be found, when the decisive test has been applied, that he has said some things not before said, which may be of benefit to teachers, and the cause of education generally, his pleasure will not be less, because he had ventured to indulge some slight anticipation of the fact. If the experience of some twenty-five years in teaching has not failed to discover to him the real wants of our schools, then it will be found that his series of Arithmetics is adapted to meet those wants, and is in some measure suited to the spirit of the age in which we live.

The Cancelling Arithmetic, published in 1837, was the first work known to the author, which, to any considerable extent illustrated, and practically applied the principle of cancelling. Although it is true that the principle is coextensive with the science of numbers, for no question in Simple Division can be solved without employing it, still Division was not explained as embodying the whole of it, nor was the principle so applied and illustrated as to simplify Division. The mode of writing numbers for the convenience of cancelling, in connection with the ordinary mode, affords a variety of illustrations interesting and useful, both to teachers and scholars.

The application of the cancelling principle is not, however, the only peculiar characteristic of this work. It aims throughout, by the connection of its subjects, and illustration of principle, to impress upon the mind of the scholar the truth, that he will never discover nor need a new principle beyond the simple rules. Hence the first object is to make the scholar thoroughly acquainted with those rules. *One* thing at a time, and *in* its time, is the plan. The simple rules are presented in their order singly; then in contrast; then a review of the whole, to exercise the judgment of the scholar. Fractions are introduced as the result of division, or rather as division implied. They are made to occupy the same position, and are illustrated and solved the same as whole numbers. The same numbers are again written in the fractional form, and the scholar is enabled to perceive, at a single view, that a change of position, and of names, is a matter of convenience and not of necessity. In the ordinary mode of presenting fractions, the idea is not precluded from the mind of the scholar, that new positions and new names do not necessarily introduce new principles. The result is, that he perceives no connection between the present and the past, and consequently the subject is ever new, and new difficulties are constantly arising. A new form of notation, and new names being introduced, it is in vain to insist that no new principle is employed, so long as the subject is but imperfectly illustrated, and the scholar does not perceive that the change is not a matter of necessity. It is one thing to gain the assent of the pupil to a truth, and it is often quite another to give him a practical understanding of it.

It is a fact too little realized, that much time is consumed in going over ground, from which no practical knowledge is gained. Not that the studies themselves are not practical, but they are not pursued in a practical manner. The scholar may be often informed that a fraction is the result of division; that the fractional form of writing numbers is division implied; and that numerator is the same as dividend, and denominator is the same as divisor; and yet difficulties will arise which did not occur in whole numbers. Whereas, a practical knowledge of this fact would enable him to solve most questions, in fractions, with the same facility as in whole numbers; nor would he find any necessity for some half dozen rules, which he is usually required to commit to memory.

When the simple rules are thoroughly understood, the pupil may be introduced to the subject of fractions, in a manner similar to the following, at the *blackboard*. If we divide 2 by 2, the quotient is a unit or 1, 2|2=1, for the dividend is just equal to the divisor. Were we required to divide 1 by 2, we should meet with a difficulty, for the dividend is less than the divisor, and consequently will not contain it; we must therefore employ a new form of notation, 2|1=½. We write the divisor under the dividend, and give a new name to the expression; we call it a fraction, which means a part of a thing. The quotient usually shows how many times the dividend contains the divisor. If the quotient is 2, the dividend contains the divisor twice; if 3, three times. But here the quotient is a fraction, less than a unit, or 1, which shows that the dividend is only a part of the divisor. But what part? The same part the quotient is of a unit. But what part is the quotient of a unit?

It will now be convenient to introduce new names, in order to value the fraction. You perceive, that the number which we employed as divisor, we have written under the line, and the number employed as dividend, is above the line. If our divisor be 2, our quotient is one-half of the dividend; if our divisor be 3, the quotient is one-third of the dividend. Thus it is plain, that in whole numbers, the divisor gives name to the quotient. The same is true when we imply division and write the numbers in the form of a fraction. Our divisor in this example is 2, and our quotient is one-half of the dividend: it is also one-half of a unit. The unit is divided into two parts; our quotient is now denominated; we therefore call the figure below the line, denominator, or namer, because it gives name to the parts into which the unit is divided. Thus we have our fraction named, or denominated; but what is its value? It is halves, but how many halves does it contain? Evidently one, which the figure above the line shows. We have now the fraction denominated or named, and numbered. Its denomination is halves, and their number is one. Making use of the figure above and below the line in one expression, we call the fraction *one half*, or one-half. Thus you perceive that numerator is the same as dividend, and denominator the same as divisor. And, as in division multiplying the dividend increased the quotient, so in fractions, multiplying numerator increases the value of the fraction. Thus:

$$\begin{array}{r|l} 2 & 1 \\ & 2 \\ \hline 2 & 2=1 \end{array} \qquad \frac{1}{2}\times 2=\frac{2}{2}=1.$$

Let the scholar write numbers in this manner, side by side, and be exercised, as in division, by multiplying dividend and divisor, numerator and denominator, employing the language of division and the language of fractions, until he is practically familiar with the fact that the principle employed in fractions and whole numbers is the same.

Whenever new names are introduced, and new positions employed, let the different forms be written side by side, and extra exercises be given, until the scholar clearly perceives the unity of the principle. (See example under Art. 147.) In Decimal Fractions, also, the points in which they are like whole numbers and common fractions, and points in which they differ, are distinctly brought out as the scholar proceeds, and then, at the close, those points are presented in one general view. In Proportion, new names and new positions are again employed. Let the same pains be taken to contrast the new positions with the former, and to explain the new terms introduced.

TO TEACHERS.

It cannot be expected that a School Arithmetic, limited in size as it must be, should exhaust its subjects, or give all those illustrations which might be both interesting and useful. The most it can do upon any one subject is to give a single illustration of a principle, a formula of a particular mode of teaching. And that text-book is the best, which by its connection of thought and subjects, and illustration of principle, interests both teacher and scholar, and incites the teacher to invent new modes for himself. Teachers are here presented with an Arithmetic which is the result of much experience in teaching and effort at improvement.

It has been the purpose and aim of the author to prepare a work which should accord with the spirit of the age, and be adapted to the schoolroom. It is not expected, nor is it desirable, that the teacher should be confined to the forms laid down in the book. They are designed simply to open the subject—to serve as hints to something better. The peculiar mode of stating questions for the convenience of cancelling and for illustrating fractions as whole numbers, teachers can adopt, or apply the principle of cancelling to the ordinary mode of statement. It will be well to employ both modes, as together they open a wider field for illustration.

It is sometimes remarked of the cancelling system, that it is good as far as it goes. The same may be said of arithmetic; for the principle is inseparable from it. It is the only principle by which any question in division can be performed. Wherever it cannot be applied, the numbers must be written in the form of a fraction. When the question involves multiplication and division, it will generally be found to be a great saving of labor, to write down all those numbers which are to be factors of the dividend and divisor, before proceeding to the operation. The eye will then detect at a glance equal factors, and they can be excluded from the operation. The teacher will bear in mind the importance of giving general illustrations of arithmetical principles, whenever it can be done, as its tendency is to enlarge the views of the pupil and to give importance to the study. For example, let simple division be illustrated not only arithmetically, but on general principles. Let it be required to divide 16 by 8, and it may be done and illustrated in the following manner:—

$$8)16=\overline{\cancel{8}\times 2\div\cancel{8}}=2 \textit{ Ans.}$$

Now substitute the letter a for 8, and the letter b for 2, and read the question thus: divide ab by a.

$$\begin{array}{l}\cancel{a})\cancel{a}b\\ \overline{\ b\ } \textit{ Ans.}\end{array}$$

Here, as before, we exclude from the dividend a factor equal to the divisor. But this latter process is algebraic; hence the scholar's views are extended, and he perceives at once, and for the first time, the connection between arithmetic and algebra. Formulas are also given to aid the less experienced teacher, and also to bring out more prominently arithmetical principles.

MANNER OF RECITATION.

Promptness and dispatch are characteristics of our times, and young men must be educated in reference to them. There is no place, perhaps, better calculated to train a scholar to think and act with precision and energy, than at the *blackboard*. When a scholar is called out from his class to solve a question, let him quickly, and with gentlemanly mien endeavoring to be self-possessed, take his stand at the board, read his question distinctly, and with the same reference to rhetorical notation as though he were called out on purpose for the reading of the question. Then let him state his question, giving the reasons for each step as he proceeds; or let him state and solve his question, then return to the commencement, and illustrate the principle, and give the reason for each step in the solution. Then let him pause at the board a moment, for his teacher to propose such questions as he may think proper.

A brief view has now been given of the plan and mode of teaching arithmetic adopted in this system. It is confidently believed, from the long experience the author has had in teaching, that the mode here adopted for presenting the subject of arithmetic, will be found better calculated to induce a fondness for the study; that it unfolds more of the science, and brings out principles more clearly than any other system now before the public. With these views the author submits the work to the candid perusal of all who are interested in the progress of knowledge.

CHARLES G. BURNHAM.

DANVILLE, VT., *Oct.* 18, 1849.

DEFINITIONS, AXIOMS, AND SIGNS.

DEFINITIONS.

A definition is what is meant by a word or phrase. The language of a definition should be so plain as not to be capable of misapprehension.

1. Quantity is any thing which may be multiplied, divided, and measured.

2. Magnitude is that species of quantity which is extended; i. e. which has one or more of the three dimensions—length, breadth, and thickness. A line is a magnitude, because it has length.

3. Mathematics is the science of quantity.

4. Arithmetic is the science of numbers.

5. Algebra is a method of computing by letters and other symbols.

6. Geometry treats of lines, surfaces, and solids. Arithmetic, Algebra, and Geometry are those parts of mathematics, on which all the others are founded.

7. A Demonstration is a course of reasoning which establishes a truth.

8. A Proposition is any thing proposed: if to be proved or demonstrated, it is called a Theorem; if to be done, it is called a Problém.

9. A *plus* quantity is a quantity to be added, and has this sign + before it; thus, +6.

10. A *minus* quantity is a quantity to be subtracted, and has this sign − before it; thus, −6.

11. An Equation is a proposition expressing equality between one quantity, or set of quantities, and another, or between different expressions for the same quantity; thus, $5=3+2$.

12. A member of an equation is the quantity or quantities on one side of the sign of equality.

OBS.—For definitions of terms in more common use in this work, see Art. 54, or Part I.

AXIOMS

An axiom is a self-evident proposition.

1. Things which are equal to the same thing are equal to each other.
2. If equals be added to equals, the wholes will be equal.
3. If equals be taken from equals, the remainders will be equal.
4. If equals be added to unequals, the wholes will be unequal.
5. If equals be taken from unequals, the remainders will be unequal.
6. Things which are double of equal things are equal to each other.
7. Things which are halves of the same thing, are equal to each other
8. The whole is greater than any of its parts.
9. The whole is equal to the sum of all its parts.

SIGNS.

$=$ Equality is denoted by two horizontal lines.

$+$ Addition: as $4+3=7$; which signifies that 4 added to 3 equals 7.

$\times$ Multiplication: as $4\times3=12$; which signifies that 4 multiplied by 3 equals 12.

$-$ Subtraction: as $4-3=1$; which signifies that 3 taken from 4 leaves 1.

)(, $\div$, $\frac{4}{2}$, $^2|^4$, Division: as, 2)4(2, and $4\div2=2$, and $\frac{4}{2}=2$, and $^2|^4=2$. In either case it signifies that 4 divided by 2 equals 2.

: :: : Proportion: as, 2 : 4 :: 6 : 12; which is read, 2 is to 4 as 6 is to 12.

$\overline{\quad}$ Vinculum: as $\overline{4+3}=7$; which is read, the sum of 4 and 3 equals 7, and $\overline{4-3}=1$, is read, the difference of 4 and 3 equals 1.

$\sqrt{}$ Radical sign: placed before a number denotes that the square root is to be taken.

4^2 implies that 4 is to be raised to the second power.

4^3 implies that 4 is to be raised to the third power.

$\sqrt[3]{}$ implies the third root.

CONTENTS.

PAGE

Simple Numbers.
Notation 11
Numeration 12
Addition 19
Subtraction 22
Addition and Subtraction 26
Practical Questions in Addition and Subtraction 27
Multiplication 28
Multiplication and Division Table 29
Division 37
Short Division 39
Long Division 42
Multiplication and Division 49
Multiplication and Division by Cancelling 49
Supplement to the Four Fundamental Rules 52
Exercises in the use of the Signs 54
Ratio, or the Relation of Numbers 55
Fractions—their Origin 55
Common or Vulgar Fractions 58
Definitions 60
To reduce a Compound Fraction to a Simple 63
To change any given Fraction to an Equivalent, having a given Denominator 64
To reduce a Whole Number to an Equivalent Fraction, having a given Denominator 64
To reduce Improper Fractions to Mixed Numbers, and the reverse 65
To reduce a Fraction to its Lowest Terms 66
To reduce a Complex Fraction to a Simple, and the reverse 68
General Rule for the Multiplication and Division of Fractions by Whole Numbers, Whole Numbers by Fractions, Fractions by Fractions 69
Multiplication of Fractions by Whole Numbers 70
Division of Fractions by Whole Numbers 72
Multiplication and Division of Fractions by Whole Numbers 73

PAGE

Multiplication of Whole Numbers by Fractions 74
Division of Whole Numbers by Fractions 75
Multiplication and Division of Whole Numbers by Fractions 77
Multiplication of Fractions by Fractions 78
Division of Fractions by Fractions 79
Multiplication and Division of Fractions 81
Promiscuous Examples 82
Addition of Fractions 83
To find a Common Denominator 85
Subtraction of Fractions 86
To find the Least Common Multiple 87
Decimal Fractions 90
How Decimal Fractions are produced 92
Addition of Decimals 93
Subtraction of Decimals 96
Multiplication of Decimals 97
Division of Decimals 98
Federal Money 100
Addition of Federal Money 101
Subtraction of Federal Money 102
Multiplication of Federal Money 103
Division of Federal Money 104
Supplement to Decimal Fractions and Federal Money 104
Bills of Parcels 105
Compound Numbers 106
Tables of Compound Numbers 106
Reduction 112
Reduction Descending and Ascending 113
Supplement to Reduction of Whole Numbers 117
Reduction of Fractions 118
Reduction Descending and Ascending 121
Comparison of Numbers and Quantities 122
To reduce Fractions to Integers of lower Denominations, and the reverse 123
Reduction of Vulgar Fractions to Decimal 125
To reduce a Decimal Fraction to a Vulgar 126

	PAGE
To reduce Integers of different Denominations to a Decimal Fraction, and the reverse	127
Inspection	128
Compound Addition	129
Compound Subtraction	133
To reduce Longitude to Time	136
Compound Addition and Subtraction	137
Compound Multiplication and Division	138
Supplement to Compound Numbers	142
Supplement to Fractions	145
Circulating Decimals	147
Ratio and Proportion	149
Direct Proportion	156
Inverse Proportion	167
Compound Proportion	169
Supplement to Proportion	176
Exchange	178
Domestic Exchange	179
Foreign Exchange	181
Bills of Exchange	185
Reduction of Currencies	188
Percentage	190
Interest	191
Interest by Cancelling	196
Time, Rate, and Amount given, to find the Principal	197
Time, Rate, and Interest given, to find the Principal	198
Principal, Interest, and Time given, to find the Rate per cent	198
Principal, Rate, and Interest given, to find the Time	199
To find the Interest on Notes, etc., when Partial Payments have been made	199
Commission, Brokerage, and Insurance	202
Compound Interest	203
Discount	206
Bank Discount	208
Loss and Gain	210
Stock	213
Barter	214
Supplement to Interest, Discount, etc.	216
Equation of Payments	218
Fellowship	220
Assessment of Taxes	222
Double Fellowship	224
Involution	225
Evolution	227
Extraction of the Square Root	228
Extraction of the Cube Root	235
Extraction of Roots in General	242
Arithmetical Progression	243
Geometrical Progression	247
Compound Interest by Progression	250
Annuities at Compound Interest	252
Permutation	256
Single Position	257
Double Position	259
Alligation Medial	262
Alligation Alternate	263
Duodecimals	267
Miscellaneous Rules	271
Mensuration of Surfaces	271
Mensuration of Solids	282
Gauging	284
Measuring Grain	285
Tonnage of Vessels	286
Mechanical Powers	287
The Lever	287
The Wheel and Axle	288
The Pulley	289
The Screw	290
The Wedge	291
Mathematical Problems	292
Levelling	297
Philosophical Problems	298
Table of Specific Gravities	302
Astronomical Problems	303
Of Balls and Shells	304
Piling of Balls	305
To find the Weight of Cattle	306
Miscellaneous Questions	307
Book-keeping	317
Notes, Receipts, etc.	322

ARITHMETIC.

Art. 1.—Arithmetic is the science of numbers. It explains their properties, and teaches how to apply them to practical purposes.

Art. 2.—The principal, or fundamental rules, are, Notation, Numeration, Addition, Subtraction, Multiplication, and Division. These are called fundamental rules, because all questions in Arithmetic are solved by one or more of them.

Art. 3.—Notation is the expressing of any number or quantity by figures; thus, 1 one; 2 two; 3 three; 4 four; 5 five; 6 six; 7 seven; 8 eight; 9 nine; 0 cipher. The first nine figures are sometimes called digits, from the Latin word *digitus*, which means a finger. In the early stages of society people counted by their fingers; they were also formerly all called ciphers—hence the art of Arithmetic was called ciphering.

Art. 4.—There are two methods of Notation—the Arabic, as above, and the Roman, which is expressed by the following seven letters of the alphabet:

I, V, X, L, C, D, M.

1	2	3	4	5	6	7	8	9	10	20	30	40	50
I,	II,	III,	IV,	V,	VI,	VII,	VIII,	IX,	X,	XX,	XXX,	XL,	L,

60	70	80	90	100	500	1000.
LX,	LXX,	LXXX,	XC,	C,	D,	M.

Art. 5.—When a letter of less, is placed before one of a greater value, it diminishes the value of the greater, by the value of itself—thus, X signifies ten, but IX is only nine. When a letter of less, is placed *after* one of greater value, it increases the value of the greater by the value of itself.

This method is seldom used except in numbering chapters, sections, etc.

Questions.—1. What is Arithmetic? 2. What are the principal, or fundamental rules? 3. Why so called? 4. What is Notation? 5. What are the first nine figures sometimes called? 6. What were they all formerly called? 7. How many methods of Notation, and what are they? 8. How many are the Arabic characters, or figures? 9. By what is the Roman method expressed? 10. How is a letter affected when one of less value is placed before it? 11. How when one of less value is placed after it? 12. For what is the Roman method of Notation principally used?

NUMERATION.

Art. 6.—Numeration teaches to express in words the value of any number represented by figures. Thus, 365 is read, three hundred and sixty-five.

Art. 7.—Figures have a simple and relative value. When a figure stands alone its value is simply so many units, or ones; as, 2 two; 3 three; 4 four. Their relative value is derived from the place they occupy when joined together, or from their distance from the unit's place. Thus, 2 and 3 express their own value; simply so many units; but they are made to express either 23 or 32; that is, either three units and two tens, or two units and three tens. Hence it appears that the first, or right-hand place, always expresses so many units; it is therefore called the unit's place; the second, the place of tens, expressing always as many tens as the figure contains units. The third place is hundreds; the fourth, thousands, as may be seen by the following

TABLE.

Hundreds of millions	Tens of millions	Millions	Hundreds of thousands	Tens of thousands	Thousands	Hundreds	Tens	Units	
								1	One.
							2	1	Twenty-one.
						3	2	1	Three hundred and 21.
					4,	3	2	1	Four thousand and 321.
				5	4,	3	2	1	Fifty-four thousand and 321.
			6	5	4,	3	2	1	Six hundred and fifty-four thousand 321.
		7,	6	5	4,	3	2	1	Seven millions 654 thousand 321.
	8	7,	6	5	4,	3	2	1	87 millions 654 thousand 321.
9	8	7,	6	5	4,	3	2	1	987 millions 654 thousand 321.

Questions.—13. What is Numeration? 14. What is the value of a figure standing alone? 15. From what is their relative value derived? 16. What does the first, or right-hand figure, always express, and what is it called? 17. What are the second, third, and fourth places called? 18. What is the value of the cipher, when standing alone, or at the left hand of another figure? 19. What effect has it when placed at the right of another figure?

Art. 8.—The cipher, when standing alone, or at the left hand of another figure, signifies nothing, as 05, 005, is five in either case, because it still occupies the unit's place. But when placed at the right hand of another figure, it increases its value in a tenfold ratio, by removing the figure farther from the unit's place. This may be seen by the following—

TABLE II.

0	Nothing.
20	Twenty.
200	Two hundred.
2,000	Two thousand.
20,000	Twenty thousand.
200,000	Two hundred thousand.
2,000,000	Two millions.

Art. 9.—*To know the value of any number of figures.*

Rule.—1. Numerate from the right hand to the left, by saying units, tens, hundreds, &c., as in the Table.

2. To the simple value of each figure join the name of its place, reading from the left hand to the r'gh'.

TABLE III.

Duodecillions.	Undecillions.	Decillions.	Nonillions.	Octillions.	Septillions.	Sextillions.	Quintillions.	Quadrillions.	Trillions.	Billions.	Millions.	Thousands.	Units.
783	694	542	987	562	714	923	610	782	184	542	365	987	963
Period of Sextillions.		Period of Quintillions.		Period of Quadrillions.		Period of Trillions.		Period of Billions.		Period of Millions.		Period of Units.	

The first division of the foregoing Table is according to the French method, into periods of three figures each: the name of the period is superadded. The second division is according to the English method, into periods of six figures each. The name of each period is subjoined. The two divisions of the

Questions.—20. How may the value of any number be found? 21. What are the two methods of numeration in the third table? 22. In what respect do they differ?

Table agree for the first nine figures—beyond that they assume different names. The principles of Notation in both are the same. In the former method the names, units, tens, hundreds, are repeated in each period; in the latter method, thousands, tens of thousands, hundreds of thousands, are repeated with the name of the period. If the sum be not expressed in figures, it is necessary to know the method of notation employed.

Art. 10.—Let the scholar point the following numbers into periods, and read them.

3445
67891
983452
5437643
67821356
436543897
5678923412
96754329876
1234678901263

Art. 11.—Express the following numbers in figures.

1. Twenty-three.
2. Thirty-five.
3. One hundred and twenty.
4. One hundred and twenty-six.
5. Ten thousand three hundred and twenty.
6. Four millions four thousand and four.
7. One hundred and seventeen millions, one hundred and two.
8. Three billions, three millions, seventeen thousand and ten.
9. One hundred billions, one hundred thousand, two hundred and fifty.
10. Twenty billions and twenty.
11. Seven billions, seven thousand and seventeen.
12. One hundred and seven billions, twenty-seven thousand and one.
13. Five hundred and four trillions, two billions, ten millions, ten thousand and ten.
14. Forty-five trillions, forty billions and thirteen.
15. Two millions, two thousand, three hundred and three.
16. Thirty quadrillions, fifty millions, four thousand, three hundred and forty-eight.

17. Four hundred and foui quadrillions, seven hundred and seven thousand, two hundred and two.

18. Four quintillions, thirty-five quadrillions, three trillions, two billions, twenty-seven millions, three hundred and forty thousand, four hundred and seventeen.

ADDITION TABLE.

Art. 12.—*Signs.*—A cross + is the sign of addition. It shows that the numbers between which t is placed are to be added. Two parallel horizontal lines = signify equality. Thus: 3+4=7 is read, 3 added to 4, or 3 plus 4 (*plus* is a Latin word, which signifies *more*) is equal to 7.

The following TABLE *may be read thus:* 2 *and* 0 *are two;* 2 *and* 1 *are* 3, *&c.*

2+0= 2	3+0= 3	4+0= 4	5+0= 5
2+1= 3	3+1= 4	4+1= 5	5+1= 6
2+2= 4	3+2= 5	4+2= 6	5+2= 7
2+3= 5	3+3= 6	4+3= 7	5+3= 8
2+4= 6	3+4= 7	4+4= 8	5+4= 9
2+5= 7	3+5= 8	4+5= 9	5+5=10
2+6= 8	3+6= 9	4+6=10	5+6=11
2+7= 9	3+7=10	4+7=11	5+7=12
2+8=10	3+8=11	4+8=12	5+8=13
2+9=11	3+9=12	4+9=13	5+9=14
6+0= 6	7+0= 7	8+0= 8	9+0= 9
6+1= 7	7+1= 8	8+1= 9	9+1=10
6+2= 8	7+2= 9	8+2=10	9+2=11
6+3= 9	7+3=10	8+3=11	9+3=12
6+4=10	7+4=11	8+4=12	9+4=13
6+5=11	7+5=12	8+5=13	9+5=14
6+6=12	7+6=13	8+6=14	9+6=15
6+7=13	7+7=14	8+7=15	9+7=16
6+8=14	7+8=15	8+8=16	9+8=17
6+9=15	7+9=16	8+9=17	9+9=18

QUESTIONS.—Two and 0—how many? 2. Two and 1—how many? 3. Two and 2—how many?

The scholar should be questioned in this manner, until he is familiar with the above table.

The scholar should be well versed in Notation and Numeration, before proceeding to the following questions.

EXERCISES.

Art. 13.—1. If John has 6 apples, and his brother gives him 3 more, how many will he have?

2. James being on a visit at his uncle's, one of his cousins gave him 3 walnuts, another 4, and his uncle gave him 9; how many did he receive?

3. Samuel bought a book for 15 cents, and a slate for 17; how many cents did he give for both?

4. If a boy pay 15 cents for a book, 10 for a knife, and 6 for a dozen of apples, how many cents does he pay in all?

5. If an inkstand cost 10 cents, an orange 5, a lemon 3, and a dozen of quills 14 cents, what is the cost of the whole?

6. A man bought of a drover 3 sheep and a cow; for one of the sheep he paid 4 dollars, for the other two he paid 3 dollars apiece, for the cow he paid 20 dollars; how many dollars did he pay for the whole?

7. Joseph bought a sled for 25 cents, a yoke for 12 cents, and a whip for 6 cents; what did the whole cost him?

8. If I pay 6 dollars for a hat, 8 for a cap, 4 for a vest, and 14 for a coat, what do I pay for the whole?

9 If I owe one man 6 dollars, another 8, another 12, another 20, how much do I owe in all?

10. The scholars in a certain school are divided into 4 classes; in the first class there are 10 scholars, in the second 12, in the third 9, and in the fourth 14; how many in all?

11. If from my library I lend to one man 5 books, to another 10, to another 8, to another 12, to another 20, how many do I lend in all?

12. In my garden there are 6 apple-trees, 8 pear-trees, 10 peach-trees, 18 plum-trees; how many trees are there in all?

13. In a certain school 10 study music, 12 French, 14 Spanish; how many are there in all these studies?

14. Eliza had 4 finger-rings, Mary had 10, and Susan had 7; how many had they in all?

15. A certain man had 4 boarders; for two he received 3 dollars each per week, for one 2 dollars, for another 5; how much did he receive per week for the whole?

16. A young lady bought two dresses; for one she paid 7 dollars, for the other 9 dollars; how much did she pay for both?

ADDITION.

Art. 14.—ADDITION is the putting together of two or more numbers so as to make but one. The number thus obtained, is called their *sum* or *amount.*

Art. 15.—SIMPLE ADDITION is the putting together of two or more numbers of the same kind.

OBS.—It is called Simple Addition, because the numbers are all of one denomination; that is, all dollars, or all cents. When the numbers are pounds, shillings, pence, &c., the denominations are different.

If one man owe me 25 dollars, another 22,—to find the amount of what both owe, I write the sums in the following manner, units under units, tens under tens, and add them together, thus:

		Tens.		Units.	
25	or thus,	2	+	5	
22		2	+	2	
47 amount.		4	+	7	= 47.

Illustration.—Beginning at the right hand, or unit's place, I say 2 and 5 are 7; then, in the second place, or place of tens, I say 2 and 2 are 4—which is 4 tens, or 40.

2. A man has three fields; one contains 31 acres, another 25, another 42; how many acres are there in all?

Operation.

```
     31
     25
     42
     --
Ans. 98
```

Illustration.—Having written the numbers according to the directions, units under units, tens under tens, &c., we begin at the right hand to add, and find the amount to be 8 units, which we place under the column of units. The amount of the second column, or column of tens, we find to be 9 tens, or 90. The answer, then, is 9 tens and 8 units, or 98.

3. What will a carriage, horse and harness cost, if the carriage cost 102 dollars, the horse 80 dollars, and the harness 16 dollars? *Ans.* 198.

4. If a wagon cost 78 dollars, and a yoke of oxen 96 dollars, what will be the cost of both?

QUESTIONS.—1. What is Addition? 2. What is Simple Addition? 3. How are the numbers to be added written? 4. By what number do you carry? 5. Why? 6. What is the number called arising from the operation? 7. What is the sign of Addition? 8. Sign of Equality? 9. Sign of Subtraction? 10. What does *plus* signify?

In the preceding examples, the numbers, when added, have been less than 10, and, of course, have required but one figure to express them. In the last example it will be seen that the numbers in the unit column, when added, amount to more than 10, and in the column of tens, the amount is more than ten—that is, ten tens.

Let the student write the numbers to be added on the *black-board,* and illustrate in the following manner:

Seventy-eight equals seven tens plus eight units, and ninety-six equals nine tens plus six units.

Operation 1*st*,	2*d*,	3*d*.
Tens. Units.	Tens. Units.	
7 + 8 =	7 + 8 =	78
9 + 6 =	9 + 6 =	96
16 + 14 =	17 + 4 =	174 *Ans.*

Writing the numbers, units under units, and tens under tens, and adding, we have sixteen tens plus fourteen units, but fourteen units equal one ten plus four units; the left-hand figure of the units therefore belongs in the column of tens. It will be seen by this operation, that what is called carrying for ten, is simply adding numbers to the column where they belong. In practice, numbers are written as in operation 3d, and a part of the operation is carried on in the mind.

The same may be illustrated, thus:

```
 78
 96
 --
 14
16
---
174 Ans.
```

Placing the numbers as before directed, and adding the right-hand column, we find it amounts to 14 units, or 1 ten and 4 units. The next column amounts to 16 tens, or 100 and 6 tens, which, when added, make 4 units, 7 tens, and 100, or 174, the answer. From the foregoing it is evident, that one in the column of tens is equal to ten in the column of units, and one in the column of hundreds is equal to ten in the column of tens. This is the reason why we carry for 10 rather than any other number.

RULE.

Add each column, beginning at the right hand, and set down the amount directly under the column, if it be less than 10; *but if it be* 10 *or more, set down the right-hand figure, and add the left to the next column. Under the last left-hand column set down the whole amount. This is the same as carrying one for every ten.*

QUESTIONS.—11. What is the rule for Simple Addition? 12. How is Addition proved?

Proof.—Perform the addition downwards, and if this last amount corresponds with the sum total, the work is supposed to be right.

The following method may be adopted when the scholar has become acquainted with the rule of Division :

Add the digits in the top line, and from their sum reject the nines, and write down the excess at the right hand, directly even with the figures in the line. Proceed in the same manner with each proposed line of figures. Then add the several remainders, or excess of nines, and from their sum reject the nines and write down the excess. Add the digits in the sum total, and reject the nines, and the excess, if the work be right, will be the same as the excess last obtained. Thus,

		Excess of 9's.
The sum of the digits in the sum total is 16, and the excess above 9 is 7; the same as the last excess.	3497	5
	6512	5
	8295	6
	18304	7Ex. 9's.

This method of proof depends upon a property of the number nine, which belongs to no other digit but 3, which is a factor of 9; —namely, that any number divided by 9 will leave the same remainder as the sum of its digits divided by 9. This peculiar property of the number 9 grows out of the decimal relation of place.

Were the ratio of increase any other than ten fold, it would belong to the number next before the ratio. E. g. Were the ratio five fold, it would belong to 4; were it six, it would belong to 5, &c. It will be observed, that if we remove a digit from the units' place to the tens, we increase its value by as many 9's as there are units in the digit. Thus: 2 units, 20 units, or two 9's plus 2 units.— If we remove the digit still further to the left, we increase the number of 9's ten fold at each remove. Thus: 200=22 nines plus 2 units, and 2000=222 nines plus 2 units. Hence it appears that whatever be the number of 9's expressed by the digit, the digit itself is always a remainder; and since the proof depends not upon the number of 9's, but upon the remainders, it is plain that if we add any number of columns of units upwards, and then add the several lines from right to left, we shall have the sum of the same digits. If therefore we cast out the nines from each sum, the remainders, if the work is right, must be the same.

OBS.—This method of proof, although not a demonstration, is, nevertheless, very satisfactory, for it presents an entirely different combination of digits; nor should we be likely to drop a 9, or a multiple of 9.

EXERCISES.

Art. 16.—1. If a man pay 1496 dollars for a house, 734 dollars for a lot of land, 300 dollars for railroad stock, and 145 dollars for a share in a bridge, how much does he expend in the whole? *Ans.* 2675 dollars.

2. A man sold plank to the amount of 834 dollars; boards to the amount of 376; shingles to the amount of 400; timber 621; two masts, one for 30 and the other for 50 dollars; what was the amount of the whole? *Ans.* 2311 dollars.

3. A merchant, on settling his accounts, finds himself in debt to A. $100; to B. 60; to C. 78; to D. 80; to E. 447; how much does he owe in all? *Ans.* 765 dollars.

4. From the creation of the world to the Christian era was 4004 years; from that time to the Declaration of American Independence was 1776, and 64 years since that period. How many years since the Creation? *Ans.* 5844 years.

5. A man by his will left his two sons 1450 dollars each; his four daughters 1200 each; to his wife 1500; to various charitable objects, 1834; what was the value of his estate? *Ans.* 11034 dollars.

6. If 1889 figures cover one side of a slate, how many will it take to cover both sides of 4 slates? *Ans.* 15112.

7. Bonaparte was born in the year 1769; lived 52 years. In what year did he die? *Ans.* 1821.

8. General Jackson took the Presidential chair in 1829; occupied it 8 years. In what year did his course terminate? *Ans.* 1837.

9. George Washington was born in the year 1732. He lived 67 years. In what year did he die? *Ans.* 1799.

10. The distance from New York to Rahway, N. J., is 20 miles, from Rahway to New Brunswick 12 miles, from New Brunswick to Princeton 18 miles, from Princeton to Trenton 12 miles, from Trenton to Bristol 10 miles, from Bristol to Philadelphia 20 miles. What is the distance from New York to Philadelphia? *Ans.* 92 miles.

11. Lafayette was born in the year 1757. He died at the age of 78. In what year did he die? *Ans.* 1835.

12. A man sold five oxen, each weighing 864 pounds; how much did they all weigh? *Ans.* 4320.

13. How many times does a common clock strike in 24 hours? *Ans.* 156.

14. A gentleman left his two sons each 1480 dollars; his

only daughter 1500 dollars, and his wife 200 more than all his children; what was the wife's portion, and what was the value of the whole estate?

Ans. { Wife's portion, 4660 } dolls.
{ Whole estate, 9120 }

15. There are two numbers the less is 1768; their difference is 961; what is the larger number? *Ans.* 2729.

16. From Boston to Providence it is 40 miles; from Providence to New York 198 miles; from New York to Philadelphia 92 miles; from Philadelphia to Wilmington 28 miles; from Wilmington to Baltimore 72 miles; from Baltimore to Richmond 110 miles; from Richmond to Raleigh 155 miles; from Raleigh to Charleston 256 miles; from Charleston to Savannah 113 miles; from Savannah to New Orleans 713 miles. How many miles from Boston to New Orleans, passing through the above places? *Ans.* 1777 miles.

17. A man bought five firkins of butter; one firkin contained 150 pounds, another 60, another 75, another 98, another 125. How much did they all contain? *Ans.* 508.

18. There were five churches erected, one in ——, which cost 16,500 dollars, two in ——, which cost 18,350 dollars each, one in ——, which cost 19,386 dollars, and one in ——, which cost 12,640 dollars. How much was the expense of the whole? *Ans.* 85,226 dollars.

When the columns to be added are long, the following method will be found convenient. Begin to add with the unit figure, as usual; and for every ten, place a dot against that figure which makes ten, or more than ten, and add the excess to the figure above it; and thus proceed to the top of the column. Write the excess of ten at the foot of the column added; then count the dots, and as many as they are, so many carry to the next left-hand column.

27687	3978	3.4.9.5
78989	2129	6 7 8.9.
87896	9723	2.3.1 6.
98988	1320	7 6 3 4
65769	9621	2 3.1 6.
75645	8732	5.5 6.7.
	1256	2 1 4 8
		3 0 2 6 5

QUESTION.—When the columns to be added are long, how may you proceed?

SUBTRACTION.

Art. 17.—1. John's father gave him 6 apples. He gave his brother 4 of them. How many had he left?

2. Joseph bought sixpence worth of candies, and ninepence worth of hazel-nuts. How much more did he give for the hazel-nuts than for the candies?

3. Henry was 10 years old when his mother died; his sister was 6. How much older was Henry than his sister?

SUBTRACTION TABLE.

Art. 18.—*Signs.* A short horizontal line — signifies subtraction. Thus: 7—4=3, is read: 7 minus 4 (*minus* is a Latin word, which signifies *less*) equals 3.

2—2= 0	3—3= 0	4—4= 0	5—5= 0
3—2= 1	4—3= 1	5—4= 1	6—5= 1
4—2= 2	5—3= 2	6—4= 2	7—5= 2
5—2= 3	6—3= 3	7—4= 3	8—5= 3
6—2= 4	7—3= 4	8—4= 4	9—5= 4
7—2= 5	8—3= 5	9—4= 5	10—5= 5
8—2= 6	9—3= 6	10—4= 6	11—5= 6
9—2= 7	10—3= 7	11—4= 7	12—5= 7
10—2= 8	11—3= 8	12—4= 8	13—5= 8
11—2= 9	12—3= 9	13—4= 9	14—5= 9
12—2=10	13—3=10	14—4=10	15—5=10
6—6= 0	7—7= 0	8—8= 0	9—9= 0
7—6= 1	8—7= 1	9—8= 1	10—9= 1
8—6= 2	9—7= 2	10—8= 2	11—9= 2
9—6= 3	10—7= 3	11—8= 3	12—9= 3
10—6= 4	11—7= 4	12—8= 4	13—9= 4
11—6= 5	12—7= 5	13—8= 5	14—9= 5
12—6= 6	13—7= 6	14—8= 6	15—9= 6
13—6= 7	14—7= 7	15—8= 7	16—9= 7
14—6= 8	15—7= 8	16—8= 8	17—9= 8
15—6= 9	16—7= 9	17—8= 9	18—9= 9
16—6=10	17—7=10	18—8=10	19—9=10

QUESTIONS.—1. Two from 2—how many? 2. Two from 3—how many? 3. Two from 4—how many? 4. Two from 5—how many?

In this manner the scholar should be questioned, until he is familiar with the above Table.

Art. 19.—Susan had 6 frocks; 4 of them she burnt. How many had she left?

Four from 6, and 2 remain; 6, the larger number, is called the *Minuend*, because it is the number to be diminished, or made less; 4 is called the *Subtrahend*, because it is the number to be subtracted; 2, the difference, is called the *Remainder*, because it is the number left after subtraction. *The process of finding the difference between two numbers, is called Subtraction.*

Art. 20.—SIMPLE SUBTRACTION teaches to find the difference between two numbers of the same name or kind. (OBS. Art. 15.)

The object in subtraction is to take the whole subtrahend from the whole minuend. Whenever the numbers are small, the operation may be performed in the mind; but when they are large, it is better to write them down, and subtract a part at a time. Thus, from 252 subtract 161.

First Operation. *Second Operation.*

Hunds. Tens. Units. Hunds. Tens. Units.

$$\begin{array}{rcl} 252 = 2 + 5 + 2 & = & 1 + 15 + 2 \\ 161 = 1 + 6 + 1 & = & 1 + 6 + 1 \\ \hline \textit{Ans. } 91. \qquad & = & 9 + 1 \;\textit{ Ans.} \end{array}$$

We first subtract 1 unit from 2, and write down the remainder. Then, because 6 tens, the next left-hand figure of the subtrahend, cannot be taken from 5, the figure above it, we take one from the next left-hand figure of the minuend, or place of hundreds, equal to 10 tens, which, added to 5 tens, makes 15 tens. *Second Operation.* Then, 6 from 15, and 9, or 9 tens remain. Subtracting tens from tens, the remainder is tens. Now, because we have taken 1 from 2, in the place of hundreds, and added it to the place of tens, we call the 2,—as it really is,—1, and say, 1 from 1, and 0 remains; or, which is the same thing, we may add 1 to the lower figure, and say, 2 from 2 and 0 remains. Thus it appears that what is sometimes called borrowing ten is really making a new division of the minuend. This may be illustrated in the following man-

---

QUESTIONS.—1. What is Subtraction? 2. What does Simple Subtraction teach? 3. Why is it called simple? 4. How many numbers are required to perform the operation? 5. Which is the minuend? 6. Why called minuend? 7. Which is the subtrahend? 8. Why called subtrahend? 9. What is the remainder? 10. Why called remainder?

ner: Suppose a man have 252 bushels of grain in 3 boxes; in the first, 200 bushels; in the second, 50; in the third, 2. He sells to A. 100 bushels; to B. 60; to C. 1. How many has he left?

252=200+50+2
161=100+60+1

He may take one bushel from the smallest box, but the 60 bushels cannot be taken from the 50 in the second box; he therefore takes 100 bushels from the largest box, and adds it to the 50, in the smaller.

Thus: 100+150+2=252
100+ 60+1=161
90+1= 91 *Ans.*

He can now take 60 from 150, or 6 tens from 15 tens, and 9 tens remain. Then, because he has taken 100 bushels from the largest box, there remains but 100; therefore, 100 from 100, or 1 from 1, and nothing remains. Whenever, therefore, the lower figure exceeds the upper, we take 1 from the next left-hand column of the upper line, calling it 10, because 1 in the left-hand column is equal to 10 in the right, and add it to the upper figure.

OBS.—We take from the left-hand place, because the right can never contain enough. We take but 1, because 1 is always sufficient.

2. From seven thousand and five, take six thousand seven hundred and forty-six.

Operation.

7005=6000+990+15
6746=6000+740+ 6
Ans. 259= 250 + 9

In this example the 6 units of the subtrahend cannot be taken from the 5 units of the minuend. We must, therefore, borrow, or rather make another division of the minuend; but as the second and third places contain ciphers, we must go to the thousand's place. From the one thousand, which we borrow, we take ten units, and add them to the 5 units of the minuend. The remainder, nine hundreds and nine tens, now occupy the second and third places, instead of the ciphers, and we say 4 from 9, and 5 remain; 7 from 9, and 2 remain.

From the foregoing we derive the following

RULE.

Place the numbers, the less under the greater; units under units, tens under tens, hundreds under hundreds, etc. Begin at the right hand, or unit's place, and take each figure in the lower line from the one above it, and set down the remainder. If either of the lower figures be greater than the one above it, suppose ten to be added to the upper figure, subtract the lower figure from it, and set down the difference, observing to carry one to the next left-hand figure of the subtrahend,—or suppose the next left-hand figure of the minuend to be diminished by one. If the next figure of the minuend be a cipher, call it 9.

Proof.—Add the remainder and lower line together. If the work be right, the amount will correspond with the upper line.

EXAMPLES.

From 39070	From 506789	From 67023491
Take 28931	Take 467898	Take 57216532

EXERCISES.

Art. 21.—1. From four hundred and seventy-nine, take three hundred and seventy-five.

2. Take twenty-five thousand nine hundred and twenty-three, from forty-four thousand five hundred and twenty.

3. What number must be subtracted from 2081 that the remainder may be 1104? *Ans.* 977.

4. From thirty-four thousand, take seventeen thousand and ninety-one.

5. John's uncle gave him 20 cents. He lost 5 of them; how many had he left?

6. A western hunter met with 45 buffaloes in one drove, and killed all but 18; how many did he kill? *Ans.* 27.

7. The Arabian, or Indian method of notation, was first known in England about the year 1150. How long is it since to the present year, 1849? *Ans.* 699.

8. The mariner's compass was invented about the year 1302. How long before that period was the Arabian method of notation known in England? *Ans.* 152 years.

9. The first settlement in New England was made at Plymouth, by the Puritans, in the year 1620. How long is it since that time to the year 1837? *Ans.* 217 years.

QUESTIONS.—11. What is the rule for Simple Subtraction? 12. How do you prove Subtraction? 13. When the lower figure exceeds the upper, what is to be done? 14. What do you call it? 15, Why? 16. Suppose the next left-hand figure be a cipher, what is to be done?

10. Gunpowder was invented in the year 1320. How long was it after the invention of the mariner's compass?

Ans. 18 years.

11. Virginia contains 64000 square miles; New York contains 46000. What is the difference? *Ans.* 18000.

12. The library of Dartmouth College contains 12,800 volumes; Harvard University contains 34,600 volumes. How many does one contain more than the other? *Ans.* 21,800.

13. Dartmouth College was incorporated in the year 1769; Harvard University in the year 1638. What is the difference in time? *Ans.* 131 years.

14. The population of the state of New York in 1820 was 1,372,812; in the year 1835 it was 1,616,482. How many years between these two periods, and how much was the increase? *Ans.* { 15 years. 243,670 increase.

15. An officer, with a company of 102 soldiers, was met by a party of Indians, who killed all his army but 17 men. How many were killed? *Ans.* 85.

16. A merchant bought 40 tuns of wine, containing 10080 gallons, which cost him 2410 dollars. He sold 28 tuns, containing 7056 gallons, for 1814 dollars. How many gallons had he left, and how much money did he want to make up the first cost? *Ans.* { 3024 galls. 596 dolls

---

ADDITION.	SUBTRACTION.
Art. 22.—1. If a harness is worth 18 dollars, and the horse is worth 68 dollars more than the harness, what is the value of the horse?	**Art. 23.**—2. If a horse cost 86 dollars, and the harness 18 dollars, how much more than the harness did the horse cost?
3. If a merchant have 1734 yards of cloth, after selling 6588 yards, how many had he at first?	4. A merchant having 8322 yards, sells 6588 yards; how many has he left?
5. What is the amount of 2269+8625?	6. What number must be added to 8625 to make 10894?
7. Dr. Franklin was born in 1706, 93 years before the death of Washington. In what year did Washington die?	8. How many years before the death of Washington, in 1799, was the birth of Franklin?

9. The mariner's compass was invented in 1302; Sir Isaac Newton was born 340 years after. In what year was he born?

10. Sir Isaac Newton was born in 1642; the mariner's compass was invented 340 years before. In what year was it invented?

11. If a piece of land be bought for 550 dollars, and sold for 250 more than it cost, for how much is it sold?

12. If a piece of land sell for 800 dollars, which is 250 more than it cost, what was the first cost?

13. Peter the Great died in 1725, 112 years before the independence of Texas was acknowledged by the United States. In what year was the independence acknowledged?

14. Peter the Great died in the year 1725. How many years from that period to the acknowledging of the independence of Texas, in 1837?

15. Supposing a man to be born in the year 1738; lived 98 years; in what year did he die?

16. Supposing a man to be 98 years old in the year 1836, in what year was he born?

17. Columbus first sailed for America in the year 1492; the independence of America was declared 284 years after. In what year was it declared?

18. The independence of America was declared in the year 1776; 284 years before, Columbus first sailed for America; in what year did he sail?

19. Washington was born in 1732; was 67 years old when he died; in what year did he die?

20. Washington was born in 1732, died in 1799; how old was he when he died?

21. Noah's flood happened about the year of the world 1656; the birth of Christ was about 2348 years after; in what year was he born?

22. Noah's flood happened about the year of the world 1656; the birth of Christ was about 4004; how long was the flood before the birth of Christ?

PRACTICAL QUESTIONS IN ADDITION AND SUBTRACTION.

Art. 24.—1. Add 900, 400, and 752; subtract from their sum 647. *Ans.* 1405.

2. Charles had 18 peaches. He gave his mother 6 and his sister 4. How many had he left? *Ans.* 8.

3. A man buys at one store 84 eggs, at another 4 dozen, at another 3 dozen. As he returns, he sells 5 dozen. How many did he purchase, and how many had he left when he arrived at home? *Ans.* { 168. 108.

4. James owes A. 20 cents; B. 30; C. 40; D. 50. John owes A. 18 cents; B. 23; C. 35; D. 47. How much do they both owe, and which owes the most?

5. A merchant owes A. 1300 dollars; B. 1900 dollars; C. 2500 dollars. He is worth 3500 dollars. How much does he owe, and how much more than he is worth?

6. A man borrowed of his friend, at one time 100 dollars; at another, 150 dollars; at another, 175 dollars. He paid 275 dollars. How much did he borrow, and how much does he still owe?

Ans. { Borrowed 425.
Owes 150.

7. On a certain farm there are 750 apple-trees, 425 pear-trees, 1000 peach-trees, and 389 plum-trees. What is the amount of the whole, and how many more apple-trees than pear-trees, and how many more peach than plum?

8. A man left to his wife 2500 dollars; to his four sons 900 dollars each; to his three daughters 450 each. What was the amount of property left, and how much more was left to the mother than to the daughters, and to the sons more than to the mother?

9. There were five important events in the course of 215 years, viz:—1. The invention of the mariner's compass. 2. The invention of gunpowder. 3. The art of printing. 4. The discovery of America. 5. The reformation. The last was accomplished A. D. 1517; the third, 77 years before; the second, 18 years after the first, and the fourth 172 years after the second. The question is, In what year did each happen?

Ans. { Mariner's compass in 1302; gunpowder in 1320; printing, 1440; discovery of America, 1492.

MULTIPLICATION.

Art. 25.—1. If Mary give 4 cents for one picture-book, how much must she give for 2 books? how much for 4? how much for 5?

2. If one dozen of eggs cost 10 cents, how many cents will two dozen cost? how many will 4? 5? 6?

3. If one share in a library cost 5 dollars, how much will three shares cost? how much 4? how much 5? how much 6?

4. If a picture-frame cost 12 dollars, what will 4 cost? what will 6? what will 7? what will 8? what will 9?

5. Four men bought a piece of land, each paying 12 dollars. What did they all pay?

6. If a horse can trot 11 miles in one hour, how many miles can he travel in 8 hours? how many in 10? 12?

7. If a bushel of wheat cost 2 dollars, how many dollars will 8 bushels cost? how many will 9? how many will 10? how many will 11? how many will 12?

8. If a man receive 4 shillings for a day's work, how much will he receive for a week's work?

9. A pound of sugar is worth 8 cents. What are 6—7—8—9—10—11—12 pounds worth?

10. John bought a writing-book for 6 cents. What will 2 cost? what will 4? what will 6? what will 8?

Art. 26.—The scholar should commit to memory the following Table before proceeding any further.

MULTIPLICATION AND DIVISION TABLE

TWICE		3 TIMES		4 TIMES		5 TIMES		6 TIMES		7 TIMES	
1 make	2	1 make	3	1 make	4	1 make	5	1 make	6	1 make	7
2	4	2	6	2	8	2	10	2	12	2	14
3	6	3	9	3	12	3	15	3	18	3	21
4	8	4	12	4	16	4	20	4	24	4	28
5	10	5	15	5	20	5	25	5	30	5	35
6	12	6	18	6	24	6	30	6	36	6	42
7	14	7	21	7	28	7	35	7	42	7	49
8	16	8	24	8	32	8	40	8	48	8	56
9	18	9	27	9	36	9	45	9	54	9	63
10	20	10	30	10	40	10	50	10	60	10	70
11	22	11	33	11	44	11	55	11	66	11	77
12	24	12	36	12	48	12	60	12	72	12	84

8 TIMES		9 TIMES		10 TIMES		11 TIMES		12 TIMES	
1 make	8	1 make	9	1 make	10	1 make	11	1 make	12
2	16	2	18	2	20	2	22	2	24
3	24	3	27	3	30	3	33	3	36
4	32	4	36	4	40	4	44	4	48
5	40	5	45	5	50	5	55	5	60
6	48	6	54	6	60	6	66	6	72
7	56	7	63	7	70	7	77	7	84
8	64	8	72	8	80	8	88	8	96
9	72	9	81	9	90	9	99	9	108
10	80	10	90	10	100	10	110	10	120
11	88	11	99	11	110	11	121	11	132
12	96	12	108	12	120	12	132	12	144

Obs.—The student may be required to write out the table, as an exercise, up to 24 times 24, and commit it to memory.

11. If a man pay 85 dollars for a carriage, what must he pay for 5 carriages?

The answer may be obtained by setting down 85 five times, and adding them up, thus:

It will be seen, by examining this operation, that the product of five times five units is two tens and five units,—and five times eight tens is 40 tens. The answer, then, is 40 tens, 2 tens and 5 units, or 425.

```
 85
 85
 85
 85
 85
---
425
```

This method would be tedious when a number is to be many times repeated, and can be solved much easier by multiplication, thus:

```
 85
  5
---
425
```

Instead of setting down 85, five times, we write 5, the multiplier, under the unit figure of the number to be multiplied; then say, 5 times 5 are 25, setting down 5, the excess of tens; and reserving in the mind, 2, the number of tens, we say, 5 times 8 are 40; adding the two tens which we reserved from the unit column, we set down 42. The answer, then, is 42 tens and 5 units, or 425.

Art. 27.—From the above we derive the following definitions:

1st. Multiplication is the concise method of performing many additions.

2d. Multiplication consists in repeating a given number a required number of times.

Obs. 1.—It is always true of multiplication, that it can be performed by addition; but it is not always true that addition can be performed by multiplication: it is only the case when a number is to be repeated.

Obs. 2.—The word *factor* signifies an agent, or doer: it is derived from the Latin word *factum*, which signifies a deed, or thing done. A person employed to do business for another, is called an agent, or factor. Hence, when two numbers are employed as multipliers, or as the means of obtaining a product, they are called *factors*. (See def. Art. 54.)

12. If a share in a bridge is worth 142 dollars, how many dollars are 6 shares worth?

Operation.

```
 142            Hunds. Tens. Units. Hunds. Tens. Units.
   6     or thus, 1 +  4 +  2 = 1 + 4 + 2 = 142
                            6            6     6
  12=  2×6     ---------------------------------
 240= 40×6     6 + 24 + 12 = 8 + 5 + 2 = 852
 600=100×6
```

Ans. 852=142×6

13. If one man receive 164 dollars for a year's labor, what ought 32 men to receive, for the same time?

Operation.

```
 164
  32
 328=  2 times the mult.
492  =30 times the mult.
5248=32 times the mult.
```

Since we cannot conveniently multiply by a larger number than 12 collectively, it will be necessary, in this example, to adopt a new mode of operation. The multiplier consists of 2 units and 3 tens. We first multiply each figure of the multiplicand by the units of the multiplier. Units into units give units; units into tens give tens; units into hundreds give hundreds. We next multiply each figure of the multiplicand, beginning with the units, by the tens of the multiplier, observing to set 2, the first figure of the product, in the place of tens, because it is the product of tens. We next multiply the 6 tens of the multiplicand by the 3 tens of the multiplier, carrying one for every ten as in Addition, and set the product in the place of hundreds. The product of tens into tens is hundreds. Lastly, we multiply the hundreds of the multiplicand by the tens of the multiplier, and set the product in the place of thousands. The product of tens into hundreds is thousands. Adding together the several products, we have 5248, the answer.

The above illustration may be better understood by setting the product of each figure of the multiplier into each figure of the multiplicand down by itself; thus,

```
       Hunds. Tens. Units.   Hunds. Tens. Units.
        1 +  6 + 4     =   1 + 6 + 4
                3 + 2             3 + 2
        2 + 12 + 8         3 + 2 + 8        ┌5000
  3 + 18 + 12            4 + 9 + 2          │ 200
  3 + 20 + 24 + 8      5 + 2 + 4 + 8 =     {   40
                                            │   8
                                            └5248 Ans.
```

From the foregoing examples we derive the following

RULE.

I. *Place the multiplier directly under the multiplicand, units under units, tens under tens, etc., then draw a line underneath.*

II. *When the multiplier is* 12, *or less than* 12, *begin at the right hand of the multiplicand, and multiply each figure contained in it by the multiplier, setting down the numbers, and carrying as in Addition.*

III. *When the multiplier is greater than* 12, *write down the figures, as before directed, and multiply the multiplicand by each figure in the multiplier, commencing with the unit figure; observing to place each figure in the product directly under the figure by which you multiply. In this way proceed, and the sum of the products will be the answer.*

There are three methods of proving Multiplication.

First—Make the multiplicand and multiplier change places, and multiply the latter by the former, in the same manner as before; if the latter product be the same as the former, the work is supposed to be right.

Second—Cast the 9's out of the product, or answer, and set down the remainder. Cast the 9's out of the sum of the two factors; multiply the two remainders together, and cast the 9's out of the product. The last remainder, if the work is right, will be equal to the first.

OBS. 3.—The four remainders may be set within the four angular spaces of a cross, as in the following example.

Third.—Multiplication may be proved by Division. The product divided by either of the factors will give the other.

OBS. 4.—The second and third methods can be deferred until the scholar becomes acquainted with Division.

QUESTIONS.—1. What is Multiplication? 2. How many numbers are required to perform the operation? 3. What is the number to be multiplied, called? 4. What is the number by which you multiply, called? 5. Taken together, what are they called? 6. Why called factors? 7. What is the answer called? 8. How many figures are there in the multiplier of the 13th question? 9. By which do we multiply first? 10. What is the product of units multiplied into units? 11. Of tens into tens? 12. Of hundreds into hundreds? 13. How are the numbers placed in Multiplication? 14. How do you proceed when the multiplier is 12, or less than 12? 15. When the multiplier is greater than 12?

EXAMPLES.

(14)

3542	
96	
21252	
31878	
340032	

Proof.

3 / 5 × 6 / 3

(15)	(16)	(17)	(18)
3467	124567	54678901	34567892
6	8	341	4567

(19)	(20)	(21)	(22)
679834	678345	126789123	908764584768
12	23	27678	632976

Multiply 24 by 2; then double the multiplier; then double the multiplicand; then double the product.

1*st.*	2*d.*	3*d.*
24	$24 \times 2 =$	48
$2 \times 2 =$	4	2
$48 \times 2 =$	96 =	96

What effect upon the product has multiplying the multiplier?

What effect upon the product has multiplying the multiplicand?

23. What will 432 barrels of flour cost, at 14 dollars a barrel? *Ans.* 6048 dollars.

24. How many rods in 84 miles, there being 320 rods in a mile? *Ans.* 26880 rods.

25. What will be the cost of 6328 thousands of boards, at 18 dollars per thousand? *Ans.* 113904 dollars.

26. How many dollars would a man count in 12 days, if he count 42000 in one day? *Ans.* 504000.

How would you solve the above question by Addition?

27. What will 64 cows cost, at 16 dollars apiece? *Ans.* 1024 dollars.

The multiplier, 16, in the last example, is a number which can be formed by the multiplication of two numbers—thus:

$4 \times 4 = 16$: or $8 \times 2 = 16$. Any number thus produced is called a composite number. The numbers thus multiplied are called component parts. Sixteen, then, is a composite number, and 4 and 4, or 8 and 2 are the component parts of 16. Thus, taking the above question, $64 \times 4 = 256$, the price of 64 cows at 4 dollars each; and this product multiplied by 4 gives 1024, the price at 16 dollars each, because 4 times 4 are 16. The same result will be produced if we multiply 64 by 8 and 2.

Art. 28.—When the multiplier is a composite number.

RULE.

Multiply first by one of the component parts, and that product by the other, and so on, if there be more than two; the last product will be the answer.

28. What is the product of 78 multiplied by 25?

It will be seen that 5 and 5 are the component parts of 25.

```
  78
   5
 ----
 390
   5
 ----
1950 Ans.
```

29. There are 365 days in a year. If a man live 48 years, how many days does he live? *Ans.* 17520.

30. Multiply 7684 by 112. $8 \times 7 \times 2 = 112$.

31. Multiply 8410736 by 56.

32. Multiply 17548671 by 81.

33. Multiply 998673214 by 1864.

34. Multiply 99998887777 by 445566.

35. Multiply 88900236789456 by 77889123.

36. If it take 142 stones to build a rod of wall, how many will it take to build 10 rods?

It will be seen by this example, that the answer is obtained by annexing a cipher to the multiplicand; annexing a cipher, therefore, to any number multiplies that number by 10. Therefore,

```
 142
  10
----
1420
```

Art. 29.—To multiply by 10, 100, 1000, or by 1, with any number of ciphers, we have this

QUESTIONS.—16. What is a composite number? 17. What are the numbers called which form a composite number? 18. What is the rule for multiplying by a composite number?

RULE.

Annex as many ciphers to the multiplicand as there are ciphers at the right hand of the multiplier, and it will give the answer required.

37. Multiply 142 by 100. *Ans.* 14200.
38. Multiply 864 by 1000. *Ans.* 864000.
39. Multiply 999 by 100000. *Ans.* 99900000.
40. Multiply 2400 by 2200. *Ans.* 5280000.

OBS.—A significant figure is one which has value in itself. The nine digits are significant figures.

In this example, we multiply by the significant figures only, placing as many ciphers at the right hand of the product as there are ciphers in the multiplier and multiplicand.

Operation.

```
 2400
 2200
 ----
  48
 48
-------
5280000
```

41. What is the product of 68400 multiplied by 18000? *Ans.* 1231200000.

Art. 30.—When there are ciphers between the significant figures of the multiplier.

RULE.

Reject the ciphers and multiply by the significant figures, observing to place the first product of each figure directly under that by which you multiply.

42. Multiply 2008 by 604.

Operation.

```
   2008
    604
   ----
   8032
 12048
-------
1212832
```

43. Multiply 8624 by 108.
44. Multiply 340824 by 909.
45. Multiply 5678902 by 770901.
46. What will 412 hogsheads of molasses cost, at 31 dollars per hhd.? *Ans.* 12772 dollars.

QUESTIONS.—19. When the multiplier is 10, 100, 1000, &c., how may you proceed? 20. When there are ciphers at the right hand of the multiplier and multiplicand? 21. How do you proceed when there are ciphers between the significant figures? 22. What are significant figures?

47. What number is that of which 8, 9, 11, are factors? *Ans.* 792.

48. If 80 men dig a canal in 94 days, how many men could dig the same in one day? *Ans.* 7520.

49. How many shillings ought 7520 men to receive for one day's work, at 5 shillings each per day? *Ans.* 37600.

50. A merchant bought 28 boxes of sugar, each weighing 235 lbs., at 8 cents per lb. How many cents did they cost? *Ans.* 52640.

51. How many shillings will 89 cords of wood cost, at 15 shillings per cord? *Ans.* 1335.

52. A merchant bought 15 pieces of cloth, each piece containing 27 yards, at 7 dollars per yard. How much did he pay for the whole? *Ans.* 2835 dollars.

53. If a ship sail 12 miles per hour, how far will she sail in 12 days? *Ans.* 3456 miles.

54. If a man hoe 3 rows of corn, 28 hills each, in 1 hour, how many hills will he hoe in 12 days, working 8 hours in a day? *Ans.* 8064.

55. What will 50 firkins of butter cost, weighing 54 lbs. each, at 14 cents per lb.? *Ans.* 37800.

56. A man has 9 piles of wood, 16 cords in a pile. What is it worth at 7 dollars per cord? *Ans.* 1008 dollars.

Art. 31.—When the multiplier is 9, or any number of 9's.

RULE.

Annex as many ciphers to the multiplicand as there are 9's in the multiplier, and from it subtract the given multiplicand.

EXAMPLES.

57. Multiply 162 by 9.

Operation 1*st.*	*Operation* 2*d.*
1620	162
162	9
Ans. 1458	1458 *Ans.*

The reason of this process is evident; annexing a cipher to the multiplicand, multiplies it by 10, which is repeating it once more than is required.

QUESTIONS.—23. How do you proceed when the multiplier is 9? 24. How is this explained? 25. How is Multiplication proved?

On the same principle we may multiply by 8, by annexing a cipher, and subtracting twice the multiplicand; or by 98, by annexing two ciphers, and subtracting twice the multiplicand.

58. Multiply 3452 by 99.
59. Multiply 46784 by 999.
60. Multiply 576213 by 98.

Art. 32.—To multiply by any number between 10 and 20.

RULE.

Annex a cipher to the multiplicand, which is to multiply it by 10, *and to this result add the product of the given multiplicand into the right-hand figure of the multiplier; their sum will be the answer required.*

EXAMPLES.

61. Multiply 562 by 11.

Operation.
```
 5620
  562
 ----
 6182 Ans.
```

62. Multiply 789 by 12.

Operation.
```
 7890
 1578
 ----
 9468 Ans.
```

63. Multiply 843 by 19.

Operation.
```
  8430
  7587
 -----
 16017 Ans.
```

DIVISION.

Art. 33.—1. John has 5 oranges given him. He keeps one himself, and divides the others equally between his two sisters. How many did each receive?

2. Samuel divided 9 walnuts equally between three boys. How many did each receive?

3. James's father gave him 8 butternuts to divide equally between himself and his three brothers. How many did each receive?

4. If Harriet gave 24 cents for 6 pictures, what did she pay for 1?

5. Mary divided 36 cents equally between 6 poor children. What did each receive?

6. If 8 yards of cloth cost 56 cents, what did one cost?

7. How many yards of broadcloth can be bought for 72 dollars, at the rate of 8 dollars per yard?

8. How many bushels of apples can be bought for 100 cents, at 25 cents per bushel?

9. How many barrels of flour may be bought for 77 dollars, at 7 dollars per barrel?

10. In how many hours will a man travel 48 miles, at the rate of 4 miles per hour?

11. Eighty cords of wood are piled in 8 different piles. How many cords in each pile?

12. A farmer sold wool to the amount of 81 dollars, for 9 shillings a fleece. How many fleeces did he sell?

13. How many thousands of boards may be bought for 144 dollars, at 12 dollars per thousand?

14. How many books may be bought for 84 cents, at 7 cents apiece?

15. If 16 apples be divided equally between 4 boys, how many does each receive?

It is evident, that as many times as 4 is contained in 16, or, as many times as it can be subtracted from it, so many apples each boy will receive.

Operation

$$\begin{array}{r} 16 \\ \underline{4} \\ 12 \\ \underline{4} \\ 8 \\ \underline{4} \\ 4 \\ \underline{4} \\ 0 \end{array}$$

We find, by trial, that 4 is contained in 16 four times, which is the number of apples each boy is to receive.

16. If 4 boys receive 4 apples each, how many do they all receive?

It is plain that 4 boys will receive 4 times as many as one; therefore, if one boy receive 4 apples, 4 boys will receive $4\times4=16$.

Art. 34.—From the foregoing we derive the following definitions:

1. *Division is a concise method of performing many subtractions, or, the reverse of Multiplication.*

2. *Division consists in finding how many times one number contains another.*

As in Multiplication two numbers are required to perform the operation, so in Division. *The number to be divided is called the Dividend; the number by which you divide is called the Divisor. The Dividend is to be regarded as the product of two factors, of which the Divisor is one, and the other is sought, which is the Quotient after division. The Divisor and Quotient multiplied together produce the Dividend. Thus, it appears that Division and Multiplication mutually prove each other.*

OBS.—All questions in Division may be performed by Subtraction; but all questions in Subtraction cannot be performed by Division.—When a number is to be divided into equal parts, the operation may be performed by Division.

Art. 35.—*When the Divisor is not greater than* 12, *the process of operation may be carried on in the mind, and the Quotient only be written down. This process is called*

SHORT DIVISION.

17. If 336 dollars be divided equally among 3 men, how many dollars will each receive?

Illustration.—To subtract 3 from 336 as many times as would be necessary to give each man his share, would be long and tedious; but, by Short Division the operation becomes simple: 336 is 3 hundreds, 3 tens, and 6 units. It will be perceived, that if 300 be divided into 3 equal parts, one of these parts will be 100; and 3 tens divided in like manner become 1 ten; and 6 units divided by 3 become 2 units. The answer, then, is 1 hundred, 1 ten, and 2 units, or 112.

300 divided by 3 gives 100, 30 divided by 3 gives 10, and 6 divided by 3 gives 2: Then, $100+10+2=112$ *Ans.*

Operation.

$$\begin{aligned} 300 \div 3 &= 100 \\ 30 \div 3 &= 10 \\ 6 \div 3 &= 2 \\ \hline 336 \div 3 &= 112 \ \textit{Ans.} \end{aligned}$$

QUESTIONS.—1. What is Division? 2. How many numbers are required to perform the operation? 3. What are they called? 4. How is the dividend to be regarded? 5. What are the two factors, which, multiplied together, will produce the dividend? 6. How is Division proved?

hun. tens. units.

Or thus: 3)3+3+6

1+1+2=112 *Ans.*

By carrying on the process partly in the mind, the operation may be made still shorter, thus:

3)336 — 112 *Ans.* *Proof.* 112×3=336

18. If 4 shares of bank stock cost 456 dollars, what will 1 share cost?

4)456 — 114 *Ans.*

We first set down the number to be divided, or the dividend; at the left of this number, place the number by which we divide, or the divisor. Taking the first left-hand figure, or hundreds, we find how many times the divisor is contained in it. The number of times, or 1, we place directly under the divided figure. We next divide the tens of the dividend: 4 is contained in 5 once, and 1 ten over, which also must be divided. If this ten be added to the unit, it will make 1 ten and 6 units—equal to 16 units: 4, then, is contained in 16, four times, which we place in the column of units; and in 456, 114 times.

19. Six brothers received a legacy of 1512 dollars. What was the share of each?

6)1512 — 252 *Ans.*

In this question the divisor is not contained in the first left-hand figure of the dividend. We, therefore, take the next figure, 5, which with the 1 makes 15 hundred; 6 is contained in 1500, 200 times, and 30 tens, or 300 over. The 3 added to the 1 ten in the next column, is 31 tens; 6 is contained in 31, five times, or in 31 tens, 50 times and 1 ten over, which, with the two units, makes 12 units: this, divided by 6, is 2 units. The answer, then, is 200+50+2=252.

From the foregoing, we derive the following

RULE.

Write the divisor at the left-hand of the dividend, with a line drawn between them.

QUESTIONS.—7. How is the process partly carried on in Short Division? 8. When do you work by Short Division? 9. What is the method of procedure in the 18th question?

Find how many times the divisor is contained in the first left-hand figure or figures of the dividend, setting the result directly under the divided figure or figures. The remainder, if there be any, carry to the next figure, calling it so many tens.

Find how many times the divisor is contained in this dividend, and set it down as before; and so continue to do until the figures in the dividend are all divided.

EXERCISES.

Art. 36.—1. Paid 150 dollars for six tons of hay. How much was it a ton? *Ans.* 25 dollars.

2. In a certain town there are 1280 inhabitants. The average number in each family is 8. How many families are there? *Ans.* 160.

3. How many yards of cloth can be bought for 1155 dollars, at 7 dollars per yard? *Ans.* 165.

4. If a man labor one month for 12 dollars, how many months will he labor for 1008 dollars? and how many years, allowing 12 months to a year?

Ans. 84 months—7 years.

Art. 37.—*When the divisor is a composite number, and greater than* 12.

1. If 15 horses consume 2550 bushels of oats in one year, how many will one horse consume?

It will be seen that 15 is a composite number, produced by the multiplication of 5 and 3, thus: $5 \times 3 = 15$. As Division is the reverse of Multiplication, it is evident that when the divisor is a composite number, we may divide, first, by one of the component parts, and that quotient by the other. For example: Suppose 30 apples to be divided equally between 15 boys. In the first place, we divide the whole number by 5. If there were only 5 boys, they would receive 6 apples each; but as there are 3 times 5 boys, they can have only one-third as many as 5 boys would have.

Operation.

```
5)2550
  ----
3)510
  ----
  170 Ans.
```

2. How many days would it take a man to travel from Boston to New York, travelling at the rate of 30 miles a day, the distance being 240 miles? *Ans.* 8 days.

QUESTIONS.—10. What is the rule for Short Division? 11. When the divisor is a composite number, how may you proceed? 12. What is the first step in the 1st example? Second step?

3. If a horse travel 2184 miles in 42 days, how many miles will he travel in one day? *Ans.* 52 miles.

4. If 112 barrels of flour cost 672 dollars, what will 1 barrel cost? *Ans.* 6 dollars.

Art. 38.—*When the divisor is not a composite number, and is greater than* 12.

1. If 2624 bushels of corn be divided equally among 41 men, how many bushels will each receive?

As 41 is greater than 12, and not a composite number, the operation must be performed by the whole divisor at once. This process is called

LONG DIVISION.

Operation.

```
41)2624(64
   246
   ---
    164
    164
    ---
```

Setting down the numbers as before directed, and taking 41 for the divisor, we find that it is not contained in the 1st figure, nor in the 1st and 2d taken together, but it is contained in the first three figures, 6 times; that is, 41 is contained in 262 tens, 60 times and something over. To find what this remainder is, we find the product of 6 times 41, which we place under the three figures employed in the dividend, and, subtracting it therefrom, we find the remainder to be 16, which is 16 tens. We next bring down the 4 units of the dividend, and place them at the right hand of the 16 tens, which make 164 to be divided by 41, which is contained in it 4 times. The answer, then, is 64.

By examining the work of the last question, it will be seen, that it is the same as Short Division, only that the operation is all set down, instead of being carried on in the mind. For example—divide 868 by 7, Long Division.

Operation.

```
7)868(124
  7
  --
  16
  14
  --
   28
   28
   --
```

By Short Division, we say, 7 in 8, once, and 1 over; 7 in 16, twice, and 2 over; 7 in 28, 4 times, and no remainder.

In Long Division, we say, 7 in 8 once, and place 1 for the first figure in the quotient; we then multiply 7 by this quotient figure, and place the result under the 8, and subtracting it we find the difference to be 1, to which we bring down the next figure for a new dividend, and proceed as before.

From the preceding explanations is deduced the following

RULE.

Place the divisor at the left hand of the dividend. Draw a line at the right and left of the dividend, and take as many figures of the dividend as will contain the divisor one or more times. Place the number of times at the right hand of the dividend, for the first figure of the quotient. Multiply the divisor by this quotient figure, and place the result under the divided figures; find the difference between them, by subtraction, and to this difference bring down the next figure of the dividend, and divide as before; so continue to do until all the figures of the dividend are brought down. Should it be necessary to bring down more than one figure to contain the divisor, a cipher must be annexed to the quotient.

OBS.—The number of figures of the dividend we assume at any one tep is a matter of convenience.

EXERCISES.

Art. 39.—2. A man raised 6996 bushels of potatoes on 33 acres How many did he raise per acre?

Ans. 212 bushels.

3. How many years in 32485 days, if 365 days make a year? *Ans.* 89 years.

4. A legacy of 15808 dollars was left to a certain number of men, giving them 832 dollars each. How many men were there? *Ans.* 19 men.

5. How many pounds in 11520 farthings, there being 960 farthings in one pound? *Ans.* 12 pounds.

6. How many hogsheads in 49896 pints, if 504 pints make one hogshead? *Ans.* 99 hogsheads.

7. There are 8 furlongs in one mile. How many miles in 123 furlongs?

Operation.

$$8)\underline{123}$$

$15\frac{3}{8}$ *Ans.*

QUESTIONS.—13. What is the difference between Long and Short Division? 14. Rule for Long Division?

OBS.—For the illustrations of this and the following questions, see *Fractions.*

By dividing 123 by 8, we find the quotient to be 15 miles, and there is a remainder of 3 furlongs. As it takes 8 furlongs to make a mile, it is evident that 1 furlong is $\frac{1}{8}$ of a mile, and 3 furlongs are $\frac{3}{8}$ of a mile, and 8 furlongs are $\frac{8}{8}$, equal one mile. The answer, then, is 15 miles and $\frac{3}{8}$, which we place at the right of the quotient. We have, then, when there is a remainder in division, this

RULE.

Place it at the right hand of the quotient, as the numerator of a fraction, and under it place the divisor, as a denominator.

A number like this is called a mixed number: thus, $15\frac{3}{8}$ is a mixed number. To prove this last question, we multiply the quotient into the divisor, and add to the product the numerator of the fraction, or the remainder, thus:

$$\begin{array}{r} 15\frac{3}{8} \\ 8 \\ \hline 123 \end{array}$$

That the remainder, as the numerator of a fraction, is a part of the quotient, will appear from the following:

$\overline{120+3}$ is the whole dividend; 8, the divisor, is contained in 120 units 15 times; it is also contained in three units $\frac{3}{8}$ of a time. Since, therefore, $120+3$ is the whole dividend, it follows that $15+\frac{3}{8}$ is the whole quotient.

8. What is the quotient of 1832 divided by 16? *Ans.* $114\frac{8}{16}$.

9. There are 320 rods in a mile. How many miles in 66327 rods? *Ans.* $207\frac{87}{320}$.

10. How many miles from Boston to Providence, the distance being 12800 rods? *Ans.* 40 miles.

11. If 10 shares in a factory be worth 2220 dollars, what is one share worth?

Operation.

$$\begin{array}{r} 10)222|0 \\ \hline 222 \textit{ Ans.} \end{array}$$

We have seen, that annexing a cipher to any number is the same as multiplying by 10. To remove a cipher, therefore, from the right of any number, is dividing

QUESTIONS.—15. When there is a remainder, after dividing, what is to be done with it? 16. How do you know it is a part of the quotient? Illustrate. 17. What is a number like $15\frac{3}{8}$ called? 18. When there are ciphers at the right hand of the divisor how may you proceed? 19. How must the figures cut off from the right hand of the dividend be placed?

that number by 10. To remove a cipher from divisor and dividend, is dividing both by 10. Therefore,

Art. 40.—When the divisor is 10, 100, 1000, or 1, with any number of ciphers, we have the following

RULE.

Cut off those ciphers from the divisor, and a corresponding number of figures from the right of the dividend. The figures on the left will be the quotient, and those on the right, a remainder.

12. Divide 18986421 by 10000.

Operation.
1|0000)1898|6421
Ans. 1898,6421

In this example, the divisor is not a factor of the whole dividend, but of a part only. 10,000 will divide 18980000, but is not contained in 6421. The quotient, therefore, is 1898, and 6421 units are left undivided, which are a remainder.

13. Divide 3330 by 30.

Operation 1st.
30)3330(111 *Ans.*
30
33
30
30
30

Operation 2d.
3|0)333|0
Ans. 111

In *Operation 1st* we reject a factor from the dividend equal to the whole divisor; but the dividend may be separated into the the factors 3, 10, and 111; 3 and 10 are also factors of the divisor. By cutting off the cipher from the dividend, the process of dividing by 10 is performed.—*Operation 2d.* We have now only to divide by 3, the other factor of the divisor, and the factor 3 is rejected from the dividend. The remaining factor, 111, is the quotient.

14. Divide 342871 by 7000.

Operation.
7|000)342|871
48,6871

In this example, cutting off three figures from the dividend is dividing by 1000: 1000 is contained in 342,000, three hundred and forty-two times, and there is a remainder of 871 units. 7 is contained in 342, forty-eight times, and 6 remain, which is 6000, because it was taken from the place of thousands, and therefore must be prefixed to the first remainder.

Proof. $48 \times 7000 + 6000 + 871 = 342871.$

Art. 41.—To divide by any number whose right-hand figures are ciphers.

RULE.

Cut off the ciphers, and figures of the dividend, as before directed, and divide the remaining figures of the dividend by the remaining figures of the divisor. To the right hand of the remainder bring down the figures cut off from the dividend.

15. What is the quotient of 421998 divided by 8400?
16. What is the quotient of 406224 divided by 9600?
17. What is the quotient of 7864234 divided by 67200?

Art. 42.—From the foregoing it is manifest, that in the process of division, a factor is rejected from the dividend equal to the divisor. Upon the same principle—*If equal factors be rejected from divisor and dividend, the value of the quotient will not be altered.*

18. Divide 16 by 8.

Operation 1st.

$$8)\overline{16} = \not{8} \times 2 \div \not{8} = 2 \textit{ Ans.}$$
$$2 \textit{ Ans.}$$

For convenience' sake, we will draw the line between divisor and dividend straight instead of curved, and write the factors one over the other, those of the dividend on the right, and those of the divisor on the left, and draw a line through those factors which are rejected or cancelled. Thus:

Operation 2d.

$$\begin{array}{r|l} \not{4} & \not{4} \\ 2 & 4 = 2 \textit{ Ans.} \end{array}$$

The dividend, 16, is resolved into the factors 4 and 4; and 8, the divisor, into the factors 4 and 2. If we strike out the factor 4 from each, we have $4 \div 2 = 2$, as before. Again, we may separate the dividend into the factors 4, 2, and 2, and strike out the factors 4 and 2 from each side of the line. Thus:

Operation 3d.

$$\begin{array}{r|l} \not{4} & \not{4} \\ \not{2} & \not{2} \\ & 2 \textit{ Ans.} \end{array}$$

It is evident, since to reject equal factors from divisor and dividend does not affect the quotient, that *to multiply divisor and dividend by the same quantity would not affect the quotient.* If 16 contain 8 twice, the double of 16 would contain the double of 8 twice, - - - - - - $\overline{16 \times 2} \div \overline{8 \times 2} = 2$ answer.

QUESTIONS.—20. In the process of division, what factor is rejected from the dividend? 21. What effect upon the quotient has rejecting equal factors from divisor and dividend?

This expression is read, The product of 16 into 2, divided by the product of 8 into 2, equals 2.

OBS. 1.—Let the scholar now be called upon to illustrate, by a variety of examples, the principle employed in division, in the following manner taking the above question:—

Teacher. What do you infer from the process in operation 1st? *Scholar.* That in the process of division we reject from the dividend a factor equal to the divisor.——*T.* What do you infer from the process in operation 2d? *S.* That to reject equal factors from divisor and dividend does not affect the quotient.——*T.* What from operation 3d? *S.* The same as from operation 1st.

In the following questions let the student be required to separate divisor and dividend into their prime factors, and write them down, as before directed. Then let him reject an equal factor from each, and perform the operation with the remaining; then let him reject another, and perform the operation with the remaining, and so on until all the factors of the divisor are rejected.

3. Divide 72 by 12.

$$72 \div 12 = \overline{2 \times 2 \times 3 \times 2 \times 3} \div \overline{2 \times 2 \times 3} = 6 \quad \textit{Ans.}$$

Or thus:

Operation 1st.		*Operation 2d.*		*Operation 3d.*	
$\not{2}$	$\not{2}$	$\not{2}$	$\not{2}$	$\not{2}$	$\not{2}$
2	2	$\not{2}$	$\not{2}$	$\not{2}$	$\not{2}$
3	3	3	3	$\not{3}$	$\not{3}$
	2		2		2
	3		3		3
6	36=6 *Ans.*	3	18=6 *Ans.*		6 *Ans.*

OBS. 2.—In the following questions, let the teacher proceed thus:

Teacher. What are the factors of 84? *Scholar.* 7 and 12.——*T.* Which is prime? *S.* 7.——*T.* Write it on the right of the line.—What are the factors of 12? *S.* 2 and 6.——*T.* Which is prime? *S.* 2.——*T.* Write it under the 7.—What are the factors of 6? *S.* 3 and 2.——*T.* Are they prime? *S.* They are.——*T.* Write them down.

4. Divide 84 by 21. *Ans.* 4.
5. Divide 108 by 18. *Ans.* 6.
6. Divide 112 by 28. *Ans.* 4.
7. Divide 224 by 56. *Ans.* 4.
8. Divide 336 by 16. *Ans.* 21.
9. Divide 96 by 8. 96÷8=12 *Ans.*

In solving this question, into what factors must the dividend be separated? *Ans.* 8 and 12.

How do you know? *Ans.* Because 8, the divisor, is one factor, and therefore 12 must be the other; for $12 \times 8 = 96$.

10. Divide 72 by 6. $72 \div 6 = 12$ *Ans.*

Into what factors is the dividend separated in this example? *Ans.* 6 and 12.

Why not 8 and 9? They are also factors of 72.—*Ans.* Because neither is like the divisor.

Let the teacher propose similar inquiries in regard to the following exercises:

11. Divide 84 by 12.
12. Divide 108 by 9.
13. Divide 121 by 11.
14. Divide 132 by 12.

Let the student read the following forms of implying division, and write others similar:

$16 \div 8 = 2$	$56 \div 7 = 8$
$\frac{24}{6} = 4$	$12\vert 60 = 5$
$6)36 = 6$	$\frac{72}{9} = 8$
$8\vert 48 = 6$	$12)84 = 7$

Obs. 3.—For the reading of the following forms, see Art. 48.

Art. 43.—*Illustration of general principles.*—What effect upon the quotient has multiplying the dividend?

What effect upon the quotient has multiplying the divisor?

What would be the effect of multiplying both divisor and dividend? Illustrate.

What effect upon the quotient has dividing the dividend?

What effect upon the quotient has dividing the divisor?

What effect upon the quotient has dividing both the dividend and divisor? Illustrate.

$$16 \div 8 = 2$$

$$\overline{16 \times 2} \div 8 = 4$$

$$16 \div \overline{8 \times 2} = 1$$

$$\overline{16 \div 2} \div 8 = 1$$

$$16 \div \overline{8 \div 2} = 4$$

Art. 44.—From the foregoing illustrations, the following principles are manifest. *The larger the dividend, with a given divisor, the larger the quotient; and the less the dividend, with a given divisor, the less the quotient. Therefore,* To multiply the dividend is the same as to multiply the quotient, and to divide the dividend is the same as to divide the quotient. To divide the divisor is the same as to multiply the dividend, and to multiply the divisor is the same as to divide the dividend.

MULTIPLICATION.

Art. 45.—1. What will 1574 yards of cloth cost, at 12 dollars per yard?

3. How many inches in 56541 feet?

5. If a man travel 38 miles in a day, how many will he travel in 16 days?

7. If 60 minutes make 1 hour, how many minutes in 13070026 hours?

9. How many hours in 336 days?

11. The quotient of two numbers is 46; the divisor 14; what is the dividend?

13. The quotient of two numbers is 72; the divisor 84; what is the dividend?

15. How many pounds of flour may be put into 640 barrels, each containing 196 pounds?

17. What will 24 oxen cost, at 46 dollars each?

19. If a carriage wheel turn round 340 times in a mile, how many times will it turn in going from Boston to New York, it being 240 miles?

21. If 33 men do a piece of work in 24 days, in what time will 1 man do it?

23. In 7894 feet, how many barley-corns?

DIVISION.

Art. 46.—2. If 1574 yards of cloth cost 18888 dollars, what will 1 yard cost?

4. How many feet in 678492 inches, if 12 inches make 1 foot?

6. If a man travel 608 miles in 16 days, how many will he travel in 1 day?

8. In 784201560 minutes, how many hours?

10. In 8064 hours, how many days?

12. The product of two numbers is 644; the multiplier 14? what is the multiplicand?

14. The product of two numbers is 6048; the multiplicand 84; what is the multiplier?

16. A man has 125440 pounds of flour to be put into barrels, containing 196 pounds each. How many barrels must he have?

18. If 24 oxen cost 1104 dollars, what do they cost apiece?

20. If a carriage wheel turn round 81600 times between Boston and New York, and turn 340 times in a mile, what is the distance?

22. If 1 man do a piece of work in 792 days, in what time will 33 men do it?

24. How many feet in 284184 barley-corns?

MULTIPLICATION AND DIVISION, BY CANCELLING.

Art. 47.—The operation of questions, involving Multiplication and Division, may be greatly abridged by the following

RULE.

I. *Draw a perpendicular line, and place dividends and numbers to be multiplied for dividends, on the right, and divisors on the left hand.*

Obs. 1.—The perpendicular line is the same as the curve line in Division, separating divisors from dividends.

II. *If there be two equal numbers on each side of the line, cross them out, and omit them in the operation.*

Thus: Multiply 8 by 9, and divide by 8.

Operation.

| ~~8~~
~~8~~ | 9 *Ans.*

As 8 is found on both sides of the line, cross them both, and 9, remaining on the right, is the answer.

The principle upon which this Rule proceeds is that of cancelling, or rejecting equal factors from dividends and divisors. Thus, taking the above example, 8 and 9 are the factors of 72: $8\times9=72$. The quotient of 72 divided by 8, is 9, one of its factors; the other factor, 8, equal to the divisor, is rejected.

III. *If a number on one side of the line will divide a number on the other side, without a remainder, erase both numbers, and substitute for the larger the number of times it contains the smaller. Multiply the remainders together, on the right, for a dividend, and the remainders on the left, for a divisor.*

Thus: Multiply 6 by 3, and divide by 2.

| ~~6~~ 3
~~2~~ | 3
9 *Ans.*

In this example the divisor, 2, is not the same as either figure of the dividend, but it is a factor of one of them, $2\times3=6$. We may, therefore, cross 2 and 6, since the divisor, 2, cancels one of the factors of 6, the dividend, and write 3, the other factor, against 6 as the quotient. The remainders on the right multiply together, $3\times3=9$, and $18\div2=9$, the answer, as before.

When there is no remainder on either side of the line, and the numbers are all cancelled, the answer is 1: that is, the right-hand side contains the left-hand, once.

Obs. 2.—A stroke drawn through any number denotes its being cancelled; and any number which takes its place may be set alongside of it.

3. Multiply 8 by 5, and divide the product by 3; multiply the quotient by 18, and divide the product by 9; multiply again by 9, and divide the product by 6; multiply the quo-

---

Questions.—1. What is the rule for Multiplication and Division by cancelling? 2. First, second, and third steps? 3. Is the answer affected by striking out equals on each side of the line? 4. Why not? 5. What is done with remainders? 6 When there is no number left on either side of the line, what is the answer?

tient by 24, and divide the product by 12; multiply the quotient by 2, and divide the product by 4.

Operations.

	8
~~3~~	5
~~9~~	~~18~~
~~6~~	~~9~~
~~12~~	~~24~~
~~4~~	~~2~~

40 *Ans.*

Or thus:

	8 × 5 = 40
~~3~~	5
~~9~~	~~18~~ = 1
~~6~~	~~9~~ = 1
~~12~~	~~24~~ = 1
~~4~~	~~2~~

40 *Ans.*

Having stated the question, according to the foregoing RULE, we proceed to cancel, or cross equals on each side of the perpendicular line. In the first place, 9 is found on each side of the line. We therefore cross them both; for 9 is contained in 9 once, and multiplying any number by 1 does not alter its value. Secondly: 3 and 6, on the left hand of the line, multiplied together make 18—equal to 18, on the right hand of the line, which may be crossed out. Again: 4 and 12, on the left, multiplied together, are 48, equal to the numbers 2 and 24 on the right multiplied together, and may be crossed out. The numbers now are all cancelled, except the 5 and 8, on the right, which, multiplied together, give 40, the answer.

4. A boy gathered 16 nuts under each of 4 trees, and divided them equally between himself and 7 schoolmates. How many did each receive?

Operation.

	~~16~~ 2
~~8~~	4 × 2 = 8

8 *Ans.*

In this example, it is evident, that had the boy gathered but 16 nuts, there would have been but 2 apiece; but as he gathered the same number under each tree, the 16 must be multiplied by 4; and as there were 8 to share them, the product of 16 multiplied by 4 must be divided by 8.

5. Multiply 20 by 5, and divide by 6; multiply by 7 and divide by 14; multiply this again by 6, and divide by 10, and multiply by 12. *Ans.* 60.

6. Multiply 120 by 40, divide by 400, multiply by 20, divide by 30, multiply by 250, divide by 50, multiply by 300, divide by 500, and give the answer. *Ans.* 24.

QUESTION.—7. Give the reason for placing 16 and 4, in Example 4, on the right of the line, and 8 on the left.

SUPPLEMENT

TO THE FOUR FUNDAMENTAL RULES OF ARITHMETIC, VIZ:

ADDITION, SUBTRACTION, MULTIPLICATION, AND DIVISION.

EXERCISES.

1. A man purchased a farm for 6720 dollars; sold it for 199 dollars more than he gave. For how much did he sell it? *Ans.* 6919 dollars.

2. Suppose a tree broken by the wind 39 feet from the ground, and the part broken off to be 56 feet in length. How high was the tree? *Ans.* 95 feet.

3. A merchant having 784 bushels of salt, sold 99 bushels. How many had he left? *Ans.* 685 bushels.

4. A man left his estate, valued at 8956 dollars, to his wife and daughters, giving his wife 4688 dollars. How much did the daughters receive? *Ans.* 4268 dollars.

5. Sir Isaac Newton was born in the year 1642, and died in the year 1727. What was his age? *Ans.* 85 years.

6. The greater of two numbers is 624; their difference is 89. What is the less number? *Ans.* 535.

7. What will 58 yards of broadcloth cost, at 4 dollars per yard? *Ans.* 232 dollars.

8. Bought 122 bushels of wheat, at 2 dollars a bushel; 8 oxen for 27 dollars each; 4 cows, 16 dollars each, and a wagon for 60 dollars. How much was paid for the whole, and how much more for the wheat and oxen than for the cows and wagon? *Ans.* { 584. 336.

9. The factors of a certain number are the difference between 1632 and 1700, and between 94 and 5 dozen. What is that number? *Ans.* 2312.

10. How many barrels of flour may be bought for 6721 dollars, at 13 dollars per barrel? *Ans.* 517 barrels.

11. Paid 57600 cents for eggs, paying at the rate of 12 cents a dozen. How many dozen did I buy? *Ans.* 4800 dozen.

12. What will 168 firkins of butter cost, at 29 dollars a firkin? *Ans.* 4872 dollars.

13. A man bought at vendue the following articles, viz.:— A colt for 18 dollars; a horse for four times as much as the colt; a wagon for 8 dollars less than the cost of the horse; 4 cows for 4 dollars more than the cost of the wagon; 12 sheep, at 3 dollars each; a plough for 5 dollars; a ton of hay for 16 dollars; and a pair of oxen for four times the cost of the hay. Now, supposing he sells the whole for 527 dollars, how much does he gain; and if with the gain he pays 4 men, to whom he is in debt, equal sums, what does each receive?

Ans. 46 dollars.

14. How many square feet in a board 12 feet long, and 2 feet wide?

It is evident that a board 12 feet long and 1 foot wide would contain 12 square feet; then a board of the same length and 2 feet in width would contain twice as many feet. The answer, then, is $12 \times 2 = 24$ feet.

15. How many feet in length is a board which contains 24 square feet, and is 2 feet in width? *Ans.* 12 feet.

It is evident that this question is the reverse of the preceding. Then, $24 \div 2 = 12$.

16. How many square feet of boards in a log which will make 26 boards, 15 feet in length and 3 feet in width?

Ans. 1170.

17. How many square feet will it take for the floor of a hall, 40 feet long, 22 in width, allowing 24 feet for waste?

Ans. 904 feet.

18. What is the width of a house which is 42 feet long, and the length and width multiplied make 1260 feet?

Ans. 30 feet.

19. Supposing it take 60 yards of carpeting to cover the floor of a room 15 feet in width, what is the length of the room, and how much will be the cost of the carpeting, at 1 dollar 50 cents per yard?

Ans. { 12 yds. in length.
9000 cents.

20. How much money will a man lay up in a year of 52 weeks, if he lay up 25 cents a day, Sundays excepted?

Ans. 7800 cents.

21. What is the difference between 7 times 35, and 7 times 5 and 30? *Ans.* 180.

22. How many days, months and years will a man be in travelling around the globe, it being 25000 miles, at the rate of 5 miles per hour, 10 hours in a day?

23. The less of two numbers is 432; the difference between them is 175. What is the greater? *Ans.* 607.

24. The remainder of a sum in Division is 423; the quotient 423; the divisor is the sum of both and 19 more. What, then, was the number to be divided? *Ans.* 366318.

25. What number, multiplied by 72084, will produce 5190048? *Ans.* 72.

26. The remainder of a sum in Division is 244; the quotient 1269; the divisor is twice the sum of the remainder, less 32. What was the sum divided? *Ans.* 578908.

27. What is that number, which, being divided by 7, the quotient resulting multiplied by 3, that product divided by 5, from the quotient 20 be subtracted, to the remainder 30 added, and half the sum shall make 35? *Ans.* 700.

Art. 48.—*Exercises in the use of the signs.*

1. Write 9, plus 3, minus 7, plus 4.
$9+3-7+4=9$ *Ans.*

2. Write the sum of 9 plus 3, minus the sum of 7 plus 4.
$\overline{9+3}-\overline{7+4}=1$ *Ans.*

3. Write the sum of the products of 8 into 7, and 9 into 4.
$\overline{8\times7}+\overline{9\times4}=92$ *Ans.*

4. Write the product of the sum of 8 and 7 into the sum of 9 and 4. $\overline{8+7}\times\overline{9+4}=195$ *Ans.*

5. Write the difference of the products of 8 into 7, and 9 into 4. $\overline{8\times7}-\overline{9\times4}=20$ *Ans.*

6. Write the product of the difference of 8 and 7, and 9 and 4. $\overline{8-7}\times\overline{9-4}=5$ *Ans.*

7. Write the sum of the difference of 9 and 3, and 7 and 4.
$\overline{9-3}+\overline{7-4}=9$ *Ans.*

8. Write the product of 16 into 2, divided by 8.
$\overline{16\times2}\div8=4$ *Ans.*

9. Write 16 divided by the product of 8 into 2.
$16\div\overline{8\times2}=1$ *Ans.*

10. Write the quotient of 16 divided by 2, divided by 8.
$\overline{16\div2}\div8=1$ *Ans.*

11. Write 16 divided by the quotient of 8 divided by 2.
$16\div\overline{8\div2}=4$ *Ans.*

Art. 49.—*Let the scholar write and perform the following questions, as the preceding.*

1. What is the product of the sum of 16 and 12 into the sum of 9 and 10? *Ans.* 532.
2. What is the sum of the products of 7 into 11, and 5 into 8? *Ans.* 117.
3. What is the difference of the products of 9 into 12, and 7 into 9? *Ans.* 45.
4. Divide the sum of 5 and 19, by the sum of 3 and 5. *Ans.* 3.
5. Divide the product of 7 into 10, by the product of 5 into 7. *Ans.* 2.
6. Divide the product of 8 into 16, by the sum of 9 and 7. *Ans.* 8.
7. Divide the sum of 15 and 17, by the product of 4 into 2. *Ans.* 4.

RATIO, OR RELATION OF NUMBERS.

Art. 50.—The ratio, or relation of one number to another, is found by division. It is the quotient arising from dividing one number by another. Thus the ratio of 8 to 4 is 2; $8 \div 4 = 2$. The quotient shows that the dividend is twice as large as the divisor. Instead, therefore, of the word *quotient*, we might use the word *ratio*.

EXAMPLES.

1. What is the ratio of 25 to 5?

Operation.

$$5\overline{)25}$$

5 *Ans.*

2. What is the ratio of 30 to 6? *Ans.* 5.
3. What is the ratio of 56 to 7? *Ans.* 8.
4. What is the ratio of 144 to 12? *Ans.* 12.
5. What is the ratio of 6 to 7? *Ans.* $\frac{6}{7}$.
6. What is the ratio of 7 to 8? *Ans.* $\frac{7}{8}$.

When the dividend is less than the divisor, the ratio is expressed by writing the divisor under the dividend.

Art. 51.—Since no new principle is ever discovered or needed in arithmetical operations not embraced in the simple rules, it is important that the student should understand these

---

QUESTION.—1. What is ratio?

rules, in all their varied applications. New names, and a new mode of writing and solving questions, naturally suggest the idea of new principles. Hence the beginner, in Fractions, is generally perplexed; to avoid this, fractions are written, and the various operations are explained, in the following exercises, as in whole numbers.

1. Divide 2 by 2. *Operation.* 2)2 = 1 *Ans.* The quotient of 2 divided by 2, is a unit, or 1.

2. Divide 1 by 2. *Operation.* 2)1 = $\frac{1}{2}$ *Ans.* The quotient of 1 divided by 2, is something less than a unit, and is called a fraction. A fraction is, therefore, the result of division. The terms of the division, which were dividend and divisor, are now the terms of the fraction, and assume the new names, "Numerator and Denominator." The scholar will, therefore, bear in mind, that numerator is the same as dividend, and denominator the same as divisor. The fraction $\frac{1}{2}$, as a quotient, expresses the relation of dividend to divisor; that is, it shows that the dividend was half as large as the divisor. But it may still be regarded as division implied—the numerator may be considered as a whole number, and the expression read, 1 divided by 2.

3. Multiply $\frac{1}{2}$, or 1 divided by 2, by 2.

Operation.

$\not{2}|1$
$\ \ |\not{2}$
1 *Ans.*

It is evident that twice $\frac{1}{2}$, or twice 1 divided by 2, is equal to unity, or 1. To multiply the dividend is the same as to divide the divisor. Twice 1 divided by 2, is multiplication and division of whole numbers, and requires no new rule.

4. Divide $\frac{1}{2}$, or 1 divided by 2, by 2.

Operation.

2|1
2|
4|1 = $\frac{1}{4}$ *Ans.*

To obtain one half of any number, we divide it by 2. But our dividend is a number already divided; the operation, therefore, is a repetition of division, and consequently no new rule is necessary.

5. Multiply $\frac{3}{4}$ by $\frac{2}{3}$. That is, multiply 3 divided by 4, by 2 divided by 3.

Operation.

2 4̸|3
3̸|2̸
2|1 $=\frac{1}{2}$ *Ans.*

Were it required to multiply 3 by 2, we should write 3 and 2 under each other as now. But in this example, the numbers to be multiplied are divided numbers. The operation, therefore, involves multiplication and division of whole numbers—rules with which the student is already familiar.

It will be perceived that the only difference between multiplication of whole numbers by whole numbers, and the multiplication of fractions by fractions, is, that in the former case, we have no divisor, in the latter we have, viz., the denominators of the fractions.

6. Divide $\frac{3}{4}$ by $\frac{2}{3}$. That is, divide 3 divided by 4, by 2 divided by 3.

Operation.

4|3
2|3
8|9 $=1\frac{1}{8}$ *Ans.*

Were we required to divide 3 by 2, we should write the 3 and 2 as we have now done,—the 3 in the place of dividends, and the 2 in the place of divisors, thus, 2|3. But the 3 is already a divided number; therefore we have another divisor, or a factor to introduce into the *divisor*, written thus, $\frac{2}{4}|^{3}$. *Illustration.*—To multiply divisor is the same as to divide dividend. But 2, the divisor, is also a divided number. We have, therefore, another factor to introduce into the *dividend*, written thus, $\frac{2}{4}|\frac{3}{3}$. *Illustration.*—To multiply dividend is the same as to divide divisor.

It will be seen, by this mode of writing fractions, that the numerators occupy the same position, that whole numbers would occupy standing in the place of fractions. Having thus disposed of the numerators of fractions, it is easy to recollect that their denominators are not to occupy the same side of the line. The terms of the fraction thus disposed of, we may apply the language of whole numbers to the statement; thus—Divide 3 by 2: multiply the quotient by 3, and divide the product by 4. Thus it appears that whole numbers may be written as fractions, and fractions as whole numbers, and the same principle of illustration employed. Fractions will hereafter be written and illustrated in both forms.

QUESTIONS.—2. What is a fraction the result of? 3. Does division always result in a fraction? 4. When does it? 5. Can division be performed when the dividend is less than the divisor? 6. How, then, does division result in a fraction? 7. What is the value of a fraction?

FRACTIONS.

Art. 52.—A fraction is part of a thing. The word *fraction* is derived from the Latin word *frango,* which signifies to break. When, therefore, any thing is broken into parts, those parts are called *fractions.* If a stick be broken into parts, each part becomes a fraction of the whole.

The method of expressing whole numbers has been shown in Notation. Thus, the characters, 1, 2, when written alone express their own value; that is, 1 unit, 2 units, &c.; but, when taken together, they express either 12 or 21. To express one half of a unit, or 1, we make use of the same figures, thus: $\frac{1}{2}$. The unit is here divided into two parts, and one of those parts is here expressed. If a thing be divided into two equal parts, these parts are called halves; if into three equal parts, they are called thirds; if into four, fourths, or quarters, &c. The equal parts of a thing are expressed thus:

$\frac{1}{2}$ read, *one half,* or 1 divided by 2;
$\frac{1}{3}$ – *one third,* or 1 divided by 3;
$\frac{1}{4}$ – *one fourth,* or 1 divided by 4;
$\frac{2}{2}$ – *two halves,* equal 1, or 2 divided by 2;
$\frac{2}{3}$ – *two thirds,* or 2 divided by 3;
$\frac{3}{3}$ – *three thirds,* equal 1, or 3 divided by 3;
$\frac{3}{4}$ – *three fourths,* or 3 divided by 4;
$\frac{4}{4}$ – *four fourths,* equal 1, or 4 divided by 4.

From the nature of division, *The greater the dividend with a given divisor, the greater the quotient, and the less the dividend with a given divisor, the less the quotient.* If the quotient of 2 divided by 2, be 1, then the quotient of 1 divided by 2 must be one half of 1. Whenever, therefore, the dividend is less than the divisor, the quotient will be less than a unit. A man divides an acre of land into 4 equal parts; each part is one fourth of the whole. The quotient of 1 divided by 4 is one fourth, ($\frac{1}{4}$.) It is evident, since 4 will contain 4, one time, that 1 will contain 4 one fourth of a time, and 2, two fourths ($\frac{2}{4}$), and 3, three fourths ($\frac{3}{4}$), and 4, four fourths ($\frac{4}{4}$), equal to one time.

When a less number is to be divided by a greater, the division is performed by writing the divisor under the dividend, and drawing a line between them. Thus we have, at one view, in a fractional expression, the divisor, dividend, and quotient. As a quotient, it expresses the ratio of dividend to divisor. The divisor, or figure below the line, is called the denominator, because it gives name to the parts, or shows into how many parts the unit is divided. If the denominator be 2, the unit is divided into two parts; if 3, three parts; if 4, four parts, etc. The figure above the line is called the numerator, because it numbers the parts, or shows how many parts are contained in the fraction. If the numerator be 2, the fraction contains two parts; if 3, three parts; if 4, four parts, etc. A fraction is also the result of division, when the dividend is greater than the divisor, but will not contain it without a remainder.

OBS.—The idea connected with the numerator and denominator of a fraction, may be familiarly illustrated thus: Suppose I have a box of 12 oranges, labelled on the outside, thus:

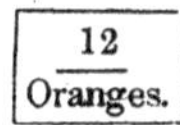

The *number* above the line shows how many things are contained in the box, and the *word* below the line shows what kind of things they are. If we write, instead of the word "Oranges," the figure 1, it would then show that the box contained 12 things, or units. Again, should we write the numbers, 2, 3, or 4, instead of the word "Oranges," they would show, not what kind of things were in the box, but into what parts the things were divided. Thus:

EXAMPLE.

Art. 53.—In the question, What is the quotient of 123 divided by 8? were the question, How many miles in 123 furlongs? (Art. 39,) 8 would still be the divisor. As 8 furlongs are equal to 1 mile, ($8 \div 8 = 1$,) 1 furlong is equal to one eighth of a mile, ($1 \div 8 = \frac{1}{8}$,) and 2 furlongs to $\frac{2}{8}$, and 3 furlongs to $\frac{3}{8}$. Increasing the numerator is only repeating the units to be divided; and as 123 units are to be divided by 8, we may write the whole in the form of a fraction, thus: $\frac{123}{8}$. The question now is, How many miles in 123 eighths of a mile? The same is required as at the first, viz., the quotient of 123 divided by 8.

QUESTIONS.—1. What are Fractions? 2. What is their origin? 3. If a thing be divided into two equal parts, what are those parts called? 4. Into four equal parts? To what does the dividing figure give name? 6. How are fractions expressed? 7. What is the figure below the line called? 8. What is the figure above the line called? What does it show? 10. What does the denominator show? 11. The numerator?

Operation.
8)123(15
120
3

We find by trial, that 15 is not the exact quotient of 123 divided by 8, and 16 would be too large; therefore, the true quotient is between those numbers, and must be expressed by a fraction. Having divided 120, the greatest number of units contained in the dividend of which 8 is a factor, we have 3 units of the dividend left, as a remainder, each of which must be divided by 8. $1 \div 8 = \frac{1}{8}$, and $3 \div 8 = \frac{3}{8}$, which evidently is the fraction required to express the exact quotient; for $120+3$ equals the dividend, and $\overline{120+3} \div 8 = 15 + \frac{3}{8} = 15\frac{3}{8}$. All questions in Division might be written in the form of a fraction, and to all fractional expressions the language employed in Simple Division might be applied, viz: What is the quotient of —— divided by ——? Hence it appears—

1. *That the value of a fraction is the quotient of the numerator divided by the denominator.*
2. *If the numerator be less than the denominator, the value of the fraction is less than a unit, or* 1.
3. *If the numerator be equal to the denominator, the value of the fraction is equal to a unit, or* 1.
4. *If the numerator be greater than the denominator, the value of the fraction is greater than a unit, or* 1

DEFINITIONS.

Art. 54.—Fractions are of two kinds; *Vulgar*, or common, and *Decimal*. They differ in the form of expression and mode of operation.

In *Decimal Fractions*, the unit, or integer, is divided into 10, 100, 1000, etc., equal parts; or the denominator is always 1, with as many ciphers annexed as the numerator has places.

In *Vulgar Fractions*, the integer may be divided into any number of parts; and the denominator being always expressed, may be any thing but 1 with a cipher or ciphers annexed.

Vulgar Fractions are either proper, improper, compound, or mixed.

Questions.—12. Fractions are of how many kinds, and what are they? 13. In what do they differ? 14. How is the unit divided in Decimal Fractions? 15. What is always the denominator? 16. In Vulgar Fractions, how is the integer divided? 17. What may the denominator be? 18. How are Vulgar Fractions subdivided?

1. A *Proper Fraction* is one whose numerator is less than the denominator, as $\frac{1}{2}$, $\frac{1}{3}$, $\frac{1}{4}$, etc.

2. An *Improper Fraction* is one whose numerator is equal to, or greater than the denominator, as $\frac{8}{8}$, $\frac{9}{8}$, $\frac{8}{7}$, etc.

3. A *Compound Fraction* is a fraction of a fraction, as $\frac{1}{2}$ of $\frac{3}{4}$ of $\frac{4}{5}$, etc.

4. A *Mixed Number* is a whole number and fraction written together, as $2\frac{1}{3}$, $14\frac{6}{7}$, 15.5, etc.

5. A *Complex Fraction* is one which has a fraction either in its numerator or denominator, or in both of them, thus:

$$\frac{6\frac{1}{2}}{8},\ \frac{8}{9\frac{1}{4}},\ \frac{3\frac{1}{4}}{5\frac{1}{2}},\ \frac{\frac{2}{3}}{5}.$$

6. A *Common Divisor*, or *Common Measure* of two or more numbers, is a number which will divide each of them without a remainder. 4 is the common measure of $\frac{12}{48}$.

7. The *Greatest Common Divisor* of two or more numbers, is the greatest number which will divide those numbers without a remainder. Thus, 12 is the greatest common measure of $\frac{12}{48}$.

8. Two or more fractions are said to have a *common denominator*, when the denominator of each is the same.

9. A *Common Multiple* of two or more numbers, is a number which may be divided by each of those numbers without a remainder.

10. The *Least Common Multiple* of two or more numbers, is the least number which may be divided by those numbers, without a remainder. Thus 8 is the least common multiple of 8, 4, and 2.

11. A *Prime Number* is that which can be measured only by itself or a unit.

12. Two numbers are prime to each other when a unit is the only number that will measure both of them. Thus, 3 and 5 are prime to each other.

13. A *Prime Factor* of a number is a prime number that will measure it; and all the prime factors of a number are all

---

QUESTIONS.—19. What is a proper fraction? 20. An improper fraction? 21. A compound fraction? 22. A mixed number? 23. What is a mixed fraction? 24. What is a common divisor, or common measure of two numbers? 25. The greatest common divisor? 26. When are fractions said to have a common denominator? 27. What is meant by a common multiple of two or more numbers? 28. The least common multiple? 29. What is a prime number? 30. What is a perfect number? 31. What is meant by the terms of a fraction? 32. When is a fraction said to be in its lowest terms?

the prime numbers that will measure it. Thus, 3 is a prime factor of 21, and 3 and 7 are all the prime factors of 21.

14. A *Component Factor* of a number is a composite number that will measure it; and all the component factors of a number are all the composite numbers that will measure it. Thus, 4 is a component factor of 12, and all the component factors of 12 are 4, 6, and 12.

15. An *Aliquot Part* of any number, is such a part of it as, being taken a certain number of times, will exactly make that number.

16. A *Perfect Number* is equal to the sum of all its aliquot parts.

The smallest perfect number is 6, whose aliquot parts are, 3, 2, 1; and $3+2+1=6$. The next perfect number is 28, the next 496, and the next 8128. Only ten perfect numbers are yet known.

17. The numerator and denominator of a fraction, taken together, are called the *terms* of the fraction.

18. A fraction is said to be in its lowest terms, when no number greater than 1, or unity, will divide the *terms* of the fraction without a remainder.

EXERCISES.

Art. 55.—1. If I divide an apple into 8 parts, by what fraction will one of those parts be expressed? 2 of those parts? 3, 4, 5, 6, 7, 8? *Ans.* $\frac{1}{8}, \frac{2}{8}, \frac{3}{8}, \frac{4}{8}, \frac{5}{8}, \frac{6}{8}, \frac{7}{8}, \frac{8}{8}$.

2. If 1 be divided by 4, what will be the quotient? if by 6, what? if by 7? if by 8? if by 9? if by 10? if by 11? if by 12? *Ans.* $\frac{1}{4}, \frac{1}{6}, \frac{1}{7}, \frac{1}{8}, \frac{1}{9}, \frac{1}{10}, \frac{1}{11}, \frac{1}{12}$.

3. If 2 be divided by 4, what will be the quotient? if by 6? by 7? by 8? by 13? *Ans.* $\frac{2}{4}, \frac{2}{6}, \frac{2}{7}, \frac{2}{8}, \frac{2}{13}$.

4. If a bushel be divided equally among 4 persons, what part of a bushel does each receive?

5. If 2 bushels of apples be divided equally among 4 persons, what will each receive?

6. If a bushel of corn be divided into four parts, what are those parts called? if into 5? into 6? into 7? 8? 9? 10?

7. If I give away 6 quarts of nuts, what part of a peck is it? if 7? if 8? if 9?

8. How many of the four last questions are proper fractions? Are any improper, and which are they?

9. How is the quotient of 4 divided by 3 expressed? What is the expression called?

10. If I divide an apple into halves, and give away $\frac{1}{2}$ of $\frac{1}{2}$, what part of the apple do I give way? What is the expression, $\frac{1}{2}$ of $\frac{1}{2}$ called?

In the foregoing question, the unit is divided into two equal parts, and each part is $\frac{1}{2}$ of the unit. A division is again made of one of these parts into two other equal parts, and each part is $\frac{1}{2}$ of $\frac{1}{2}$, or $\frac{1}{4}$ of the unit first divided. The expression, $\frac{1}{2}$, as it respects the unit of which it is a part, is a fraction, (*see definition*, Art. 52;) but as it respects itself, or a subsequent division, it is to be regarded as itself a unit, and may be divided into halves, or any number of parts. A quarter, or $\frac{1}{4}$ of a thing, is a whole quarter; and is made up of as many parts as the thing of which it is a part. It is, therefore, in relation to a division already made, that an expression is to be regarded as a fraction. As it respects itself, or a subsequent division, it is to be considered a unit. *Example.*—A yard may be divided into 3 equal parts, or feet. A foot, when spoken of in relation to the yard, is $\frac{1}{3}$; but $\frac{1}{3}$ of a yard is one foot, and may be divided into 12 equal parts, or inches, and each inch is $\frac{1}{12}$ of a foot, or $\frac{1}{12}$ of $\frac{1}{3}$ of a yard. The inch may be divided into 3 equal parts, or barley-corns, and each barley-corn is itself a unit of less value, and it is also a fraction of a unit of a higher value; that is, 1 barley-corn is $\frac{1}{3}$ of $\frac{1}{12}$ of $\frac{1}{3}$ of a yard. That the terms, *unit*, and *fraction*, are merely relative, may be seen by the following formula:

$$\overset{\text{yd.}}{1} \div 3 = \overset{\text{yd.}}{\tfrac{1}{3}} = \overset{\text{ft.}}{1}, \text{ and } \overset{\text{ft.}}{1} \div 12 = \overset{\text{ft.}}{\tfrac{1}{12}} = \overset{\text{in.}}{1}, \text{ and } \overset{\text{in.}}{1} \div 3 = \overset{\text{in.}}{\tfrac{1}{3}} = \overset{\text{bar.}}{1}.$$

Art. 56. To reduce a compound fraction to a simple one.

1. Reduce $\frac{2}{3}$ of $\frac{1}{4}$ to a simple fraction.

If we multiply the denominator of $\frac{1}{4}$ by 3, we obtain one-third of $\frac{1}{4}$. If we multiply this numerator by 2, we obtain two-thirds. Hence the

RULE.

Multiply the numerators together, and the denominators, having cancelled all the equal factors in the numerators and denominators.

Operation.

$$\frac{\not 2}{3} \times \frac{1}{\not 4\ 2} = \frac{1}{6} \quad \text{or} \quad \begin{array}{r|l} 3 & \not 2 \\ 2\ \not 4 & 1 \\ \hline 6 & 1 \end{array} = \tfrac{1}{6}\ \textit{Ans.}$$

2. Reduce $\frac{5}{6}$ of $\frac{3}{5}$ of $\frac{4}{7}$ to a simple fraction. *Ans.* $\frac{2}{7}$.

3. Reduce $\frac{1}{2}$ of $\frac{2}{7}$ of $\frac{3}{4}$ of $\frac{7}{8}$ to a simple fraction. *Ans.* $\frac{3}{32}$.

Art. 57.—To change any given fraction to an equivalent fraction, which shall have any required denominator.

Change $\frac{2}{3}$ to an equivalent fraction whose denominator shall be 6. In this example, the unit is already divided into thirds, and we wish to divide it into 6ths: We have, therefore, simply to reduce thirds to sixths. 2 sixths make a third, for the unit is divided into twice as many parts, and therefore the parts are one-half as large. Hence the

RULE.

Divide the required denominator by the denominator of the given fraction, and multiply the quotient by the numerator. The product will be the required numerator.

Art. 58.—To reduce a whole number to an equivalent fraction, having a given denominator.

1. Reduce 8 to a fraction whose denominator shall be 4.

As in 1 unit there are 4 fourths, so in 8 units there must be $8\times4=32$ fourths, expressed thus: $\frac{32}{4}$; therefore the

RULE.

Multiply the whole number by the given denominator, and set the product over the given denominator.

2. Reduce 16 to a fraction whose denominator shall be 7. *Ans.* $\frac{112}{7}$.

3. Reduce 40 to a fraction whose denominator shall be 9. *Ans.* $\frac{360}{9}$.

4. Reduce 129 to a fraction whose denominator shall be 21. *Ans.* $\frac{2709}{21}$.

5. Reduce 339 to a fraction whose denominator shall be 39. *Ans.* $\frac{13221}{39}$.

A whole number may be expressed fractionally, by writing 1 under it for a denominator.

Thus 2 may be written, $\frac{2}{1}$; and read 2 ones;
3 " " $\frac{3}{1}$ " 3 ones;
4 " " $\frac{4}{1}$ " 4 ones;

As the expression, $\frac{2}{1}$, is equal to 2, and $\frac{3}{1}$ to 3, the value

---

QUESTION.—How is a whole number reduced to an equivalent fraction, having a given denominator?

of a number is not affected by writing 1 under it, as a denominator.

To reduce improper fractions to mixed numbers, and mixed numbers to improper fractions.

IMPROPER FRACTIONS.

Art. 59.—1. Change $\frac{67}{4}$ to a whole, or mixed number.

$$4)\underline{67}$$
$$16\tfrac{3}{4}$$

As the denominator of a fraction denotes the number of parts into which the unit is divided, it is evident that $\frac{67}{4}$ contains as many units, or wholes, as 4 is contained times in 67, which we find, by trial, to be 16 times and $\frac{3}{4}$ of a time. Hence,

To reduce an improper fraction to a whole, or mixed number, we have this

RULE.

Divide the numerator by the denominator, and the quotient will be the whole number; the remainder, if any, written over the denominator, must be placed at the right hand of the quotient.

EXAMPLES.

3. Change $\frac{42}{8}$ to a whole or mixed number.

5. In $\frac{59}{9}$ how many wholes?

7. In $\frac{159}{15}$ of a week, how many weeks?

9. Change $\frac{2233}{14}$ to a whole or mixed number.

11. In $\frac{3247}{16}$ of a day, how many days?

13. In $\frac{49364}{39}$ of a year, how many years?

15. In $\frac{17}{5}$ of a cent, how many cents?

17. Change $\frac{229}{4}$ to a whole or mixed number.

19. In $\frac{504}{7}$ of a minute, how many minutes?

MIXED NUMBERS

Art. 60.—2. Change $16\frac{3}{4}$ to an improper fraction.

$$16\tfrac{3}{4}$$
$$\underline{4}$$
$$67$$
$$\overline{4}$$

The scholar will perceive, that the mixed number, $16\frac{3}{4}$, was the quotient, in the last question, of 67 divided by 4; and let it be remembered that a mixed number is the quotient of a division whose divisor is the denominator of the fraction; therefore,

To reduce a mixed number to an improper fraction, we have this

RULE.

Multiply the whole number by the denominator of the fraction, and to the product add the numerator; under the result, place the denominator of the fraction.

EXAMPLES.

4. Change $5\frac{3}{8}$ to an improper fraction.

6. In $6\frac{5}{9}$, how many ninths?

8. In $10\frac{9}{15}$ weeks, how many 15ths?

10. Change $159\frac{7}{14}$ to an improper fraction.

12. In $202\frac{15}{16}$ days, how many 16ths?

14. In $1265\frac{29}{39}$ years, how many 39ths?

16. Change $3\frac{2}{5}$ cents to the fraction of a cent.

18. Change $57\frac{1}{4}$ to an improper fraction.

20. In 72 minutes, how many 7ths?

Art. 61.—To reduce a fraction to its lowest terms.

1. Reduce $\frac{4}{8}$ to its lowest terms.

If 4 bushels were divided equally between two persons, it is evident that one person would receive $\frac{1}{2}$ of 4 bushels, or 2 bushels; so if $\frac{4}{8}$ of a bushel be divided equally between two persons, one person will receive one half of $\frac{4}{8}$, or $\frac{2}{8}$ of a bushel. Dividing the numerator by 2, we take one half of those parts which are contained in the fraction, while the value of each part remains the same. Therefore, *To divide the numerator diminishes the value of the fraction.*

If we divide the denominator of $\frac{4}{8}$ by 2, the fraction becomes $\frac{4}{4}$. In this expression the unit is divided into half as many parts as at the first, and consequently, these parts are twice as large. It is evident, therefore, that *To divide the denominator of a fraction, the numerator remaining the same, increases its value.*

If we divide the terms of the fraction by 2, it becomes $\frac{1}{2}$, which is equal to $\frac{2}{4}$, or $\frac{4}{8}$, for in either case the numerator is one half of the denominator. Hence it appears, that the value of a fraction is not affected by dividing or multiplying both the numerator and denominator by the same number. (See Art. 43.) To reducea fraction to its lowest terms, we have this

RULE.

Divide both the numerator and denominator by any number that will divide both without a remainder; and so continue to do until no number greater than 1 will divide them.

2. Reduce $\frac{84}{210}$ to its lowest terms.

$$7)\frac{84}{210}\left(\;{}^{(6)}\frac{12}{30}=\frac{2}{5}\right.\ \textit{Ans.}$$

3. Reduce $\frac{160}{760}$ to its lowest terms. *Ans.* $\frac{4}{19}$.

4. Reduce $\frac{371}{1232}$, $\frac{77}{171}$, $\frac{100}{180}$, $\frac{83}{249}$, to their lowest terms.

Art. 62.—Were the greatest number known which would divide the terms of the fraction, a simple division would at once reduce the fraction; but, as this is not the case, the greatest divisor may be found by the following

---

QUESTIONS.—1. What is the rule for reducing an improper fraction to a whole or mixed number? 2. For reducing a mixed number to an improper fraction?

RULE.

Divide the denominator by the numerator, or the larger number by the less, and if there be no remainder, the numerator, or the less number, will be that divisor; but if there be a remainder, divide the last divisor by the last remainder, and thus proceed until there be no remainder; and the last divisor will be the greatest common measure sought.

The above Rule may be illustrated in the following manner: Suppose I have two lines, and wish to obtain a third which shall be an exact measure of the two. I first apply the shorter line to the longer, and find it contains it twice, but not three times. The remainder I now apply to the shorter line, and find it contains it twice and no remainder. Therefore, this last divisor is the third line sought, and is the exact measure of the other two.

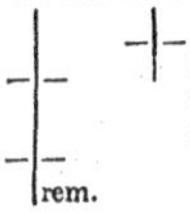

rem.

Suppose the longer line to be 40 feet and the shorter 16 feet, and we are required to find the greatest common measure, or divisor, of 16 and 40, or to reduce 16-40 to its lowest terms, we should proceed in the following manner:

Operation.

```
16)40(2
   32
   --
    8)16(2
      16
      --
```

It is evident that 16 is the greatest number that will divide 16 without a remainder; and would 16 divide 40 without a remainder, it would be the greatest common measure of the terms of the fraction. But we find, by trial, that 16 is contained in 40 twice and 8 remainder; hence 16 is not the common measure. Dividing 16, the last divisor, by 8, the remainder, we find it contains it twice and no remainder; therefore 8 is the greatest common divisor of the terms of the fraction. For, if 8 will divide 8, it will also twice 8, which is 16, and five times 8, which is 40.

6. Reduce $\frac{182}{196}$ to its lowest terms. *Ans.* $\frac{13}{14}$.
7. Reduce $\frac{468}{1184}$ to its lowest terms. *Ans.* $\frac{117}{296}$.
8. Reduce $\frac{6912}{32256}$ to its lowest terms. *Ans.* $\frac{3}{14}$.
9. Reduce $\frac{36}{1296}$ to its lowest terms. *Ans.* $\frac{1}{36}$.

If it be required to find the greatest common measure of more than two numbers, find the greatest common measure of two of them, as before; then, of that common measure, and of one of the other numbers, and so on through the whole. The common measure last found will be the one sought.

QUESTIONS.—3. What is the rule for reducing a fraction to its lowest terms? 4. Were the greatest number known which would divide the terms of the fraction, how might you proceed? 5. When this is not the case, how may the greatest divisor be found? 6. How is the common measure of more than two numbers found?

10. What is the greatest common measure of 48 and 192? *Ans.* 48.

11. Reduce $\frac{48}{192}$ to its lowest terms. *Ans.* $\frac{1}{4}$.

12. What is the greatest common measure of 35, 42, 63? *Ans.* 7.

Art. 63.—To reduce a complex fraction to a simple fraction.

1. Reduce $\frac{\frac{2}{5}}{\frac{3}{4}}$ to a simple fraction.

To multiply numerator is the same as to divide denominator, (Art. 44;) therefore, $\frac{\frac{2}{5}}{\frac{3}{4}}=\frac{2\times 4}{\frac{5}{3}}$; and to multiply denominator is the same as to divide numerator; therefore $\frac{\frac{2}{5}}{\frac{3}{4}}=\frac{2\times 4}{3\times 5}=\frac{8}{15}$. Hence the

RULE.

If the numerator be whole or mixed numbers, reduce them to improper fractions.

Then multiply the numerator of each fraction by the denominator of the other; the product will be the fraction required.

2. Reduce $\frac{5\frac{4}{9}}{7}$ to a simple fraction. *Ans.* $\frac{7}{9}$.

3. Reduce $\frac{\frac{5}{6}}{\frac{7}{8}}$ to a simple fraction. *Ans.* $\frac{20}{21}$.

4. Reduce $\frac{7\frac{1}{2}}{8\frac{1}{3}}$ to a simple fraction. *Ans.* $\frac{9}{10}$.

Art. 64.—To change a simple fraction to a complex.

1. Change $\frac{4}{5}$ to a complex fraction.

$\frac{4}{5}=\frac{2\times 2}{5}=\frac{\frac{2}{5}}{2}$ *Ans.* Hence the

RULE.

If the numerator or denominator, or both, be a composite number, separate them into factors, and transfer one or more from numerator to denominator, and from denominator to numerator, observing that a factor transferred, becomes a divisor.

2. Change $\frac{21}{40}$ to a complex fraction. *Ans.* $\frac{\frac{7}{5}}{\frac{8}{3}}$ or $\frac{\frac{3}{5}}{\frac{8}{7}}$.

OBS.—The answer depends upon the factor transferred.

3. Change $\frac{20}{35}$ to a complex fraction.
4. Change $\frac{15}{16}$ to a complex fraction.

Art. 65.—The following, if made familiar, will aid the scholar in cancelling.

1. Any number ending with an even number, or cipher, is divisible, or can be divided, by 2.
2. Any number ending with 5, or 0, is divisible by 5.
3. If the right-hand place of any number be 0, the whole is divisible by 10; if there be two ciphers, it is divisible by 100; if 3 ciphers, by 1000, and so on, which is only cutting off those ciphers.
4. If the two right-hand figures of any numbers be divisible by 4, the whole is divisible by 4; and if the three right-hand figures be divisible by 8, the whole is divisible by 8, and so on.
5. If the sum of the digits in any number be divisible by 3 or by 9, the whole is divisible by 3 or 9.
6. If the right-hand digit be even, and the sum of all the digits be divisible by 6, then the whole will be divisible by 6.
7. A number is divisible by 11, when the sum of the 1st, 3d, 5th, etc., or all the odd places, is equal to the sum of the 2d, 4th, 6th, etc., or of all the even places of digits.
8. If a number cannot be divided by some quantity less than itself, that number is a prime, and cannot be divided by any number whatever.

Art. 66.—To multiply and divide fractions by whole numbers, whole numbers by fractions, fractions by fractions.

RULE.

Draw a perpendicular line, and write numerators, in all cases, as you would a whole number standing in the place of the fraction; viz., the numerators of fractions to be multiplied or divided on the right of the line, and their denominators on the left. The question thus stated, equals on each side of the line may be crossed, as cancelling each other. (See Art. 47.) *When no two numbers remain, one on each side of the line, capable of being divided by any one figure, multiply the figures on the right of the line, for a numerator, or dividend, and those on the left, for a denominator, or divisor, and the result will be the answer in the lowest terms of the fraction.*

Multiplication of Fractions by Whole Numbers.

Art. 67.—1. If a man receive $\frac{1}{4}$ of a dollar for 1 day's work, what will he receive for 2 days' work?

It is evident, if a man receive $\frac{1}{4}$ of a dollar for 1 day's work, that he would receive, for 2 days' work, twice as much, or $\frac{2}{4}=\frac{1}{2}$. Multiplying the numerator by 2, the denominator remaining the same, we have twice the number of parts, while the value of each part remains the same. Dividing the denominator by 2, the numerator remaining the same, we have the same number of parts, while the value of each part is twice as great. Hence, to multiply the numerator of a fraction is the same, in effect, as to divide the denominator. If the numerator of $\frac{1}{2}$ be multiplied by 2, it becomes $\frac{2}{2}=1$. If the denominator be divided by 2, it becomes $\frac{1}{1}=1$. Therefore, to multiply a fraction by a whole number, we have the following

RULE.

Multiply the numerator, or divide the denominator, and the result will be the answer required.

2. If a pound of lead cost $\frac{1}{16}$ of a dollar, how much must 16 pounds cost?

Operation.

$$\begin{array}{r|l} \not{1}\not{6} & 1 \\ & \not{1}\not{6} \\ \hline & 1 \ \textit{Ans.} \end{array} \qquad \text{Or thus: } \frac{1}{\not{1}\not{6}} \times \frac{\not{1}\not{6}}{1} = \frac{1}{1} = \textit{Ans.}$$

It is evident, if one pound cost 1 dollar divided by 16, that 16 pounds would cost 16 dollars divided by 16, equal to 1 dollar. Therefore—*A fraction is multiplied into a quantity equal to its denominator, by cancelling or removing the denominator.*

3. If a pound of iron cost $\frac{1}{18}$ of a dollar, how much will 9 pounds cost?

Operation.

$$\begin{array}{r|l} 2 \ \not{1}\not{8} & 1 \\ & \not{9} \\ \hline 2 & 1=\frac{1}{2} \ \textit{Ans.} \end{array} \qquad \text{Or thus: } \frac{1}{\not{1}\not{8}} \times \frac{\not{9}}{1} = \frac{1}{2} \ \textit{Ans.}$$

QUESTIONS.—7. What is the rule for the multiplication and division of fractions, etc., by cancelling? 8. When the question is stated, what is the method of procedure? 9. When no two numbers are left, one on each side of the line, capable of being divided by any one figure, what is to be done? 10. How do you multiply a fraction by a whole number? 11. Why, in example 2, are 16 and the numerator of the fraction placed on the right of the line?

If 1 pound cost 1 dollar, 9 pounds would cost $1\times9=9$ dollars; but the cost of 1 pound is 1 dollar divided by 18: therefore, 9 dollars, the cost of 9 pounds, must be divided by 18. By the rule already given, the numerator of the fraction, with 9, its multiplier, is placed on the right of the line, and 18, the divisor, on the left. 9 and 2 are factors of 18; therefore, cross 9 and 18, and write 2, the remaining factor, in the place of 18. The answer, then, is, 1 divided by 2; or, $\frac{1}{2}$. (See Art. 66.) On the principle above stated, *A fraction may be multiplied into any factor in its denominator, by cancelling that factor.*

4. If a pound of lead cost $\frac{1}{16}$ of a dollar, how much will 8 pounds cost?

If the cost of 1 pound be $\frac{1}{16}$ of a dollar, 8 pounds will cost $\frac{1}{16}\times8=\frac{8}{16}=\frac{1}{2}$ of a dollar. Making the horizontal line, which separates the numerator of the fraction from the denominator, perpendicular, it will be seen that the numerator occupies the place of dividends, (the right of the line,) and the denominator the place of divisors, (the left of the line,) thus:

2 ~~16~~	1
	~~8~~
2	$1=\frac{1}{2}$ *Ans.*

In the latter mode the question is resolved into this. Multiply 1 by 8, and divide by 16; therefore, the numerator of the fraction and 8, its multiplier, occupy the right of the line, and 16, the divisor, the left. *It is to be remembered, that the numerator of a fraction, in all cases, is to be disposed of as a whole number, without regard to its denominator. On whichever side of the line the numerator falls, the denominator must be placed on the opposite side.*

5. What will 8 bushels of apples cost, at $\frac{1}{2}$ a dollar per bushel? *Ans.* \$4.

Obs.—This character (\$) placed before any number, shows that it is dollars.

6. If one man can plant $\frac{3}{4}$ of an acre in one day, how much could 12 men plant in the same time? *Ans.* 9 acres.

7. If 1 barrel of fish cost $6\frac{1}{4}$ dollars, what will 9 barrels cost? *Ans.* $56\frac{1}{4}$.

Obs.—Mixed numbers must be reduced to improper fractions.

8. If 1 chest of tea cost \$$25\frac{1}{4}$, what will 15 cost? *Ans.* \$$378\frac{3}{4}$.

9. If a man can walk $29\frac{1}{2}$ miles in 1 day, how far could he walk in 30 days? *Ans.* 885 miles.

10. What will 600 pounds of cotton cost, if 1 pound cost $9\frac{1}{2}$ cents? *Ans.* $57.

Obs.—Dividing any number of cents by 100, reduces them to dollars.

Division of Fractions by Whole Numbers.

Art. 68.—1. If a man receive $\frac{2}{4}$ of a dollar for two days' work, what does he receive per day?

We have seen, that a fraction is multiplied either by multiplying its numerator, or dividing its denominator; then, as Division is the reverse of Multiplication, the reverse of the rule for Multiplication will be the rule for Division. If I divide the dollar into 4 parts, or quarters, and pay a man $\frac{1}{4}$, or quarter, it is the same as though I should divide it into 8 parts, or half quarters, and pay him $\frac{2}{8}$, or 2 half quarters$=\frac{1}{4}$. Multiplying the denominator by 2, the numerator remaining the same, is dividing the unit into twice as many parts, and consequently the value of each part is diminished by one half. Dividing the numerator by 2, the denominator remaining the same, is taking half as many parts, while the value of each part is the same. Therefore, to divide a fraction by a whole number, we have this

RULE.

Divide the numerator of the fraction by the whole number, when it can be done without a remainder; otherwise, multiply the denominator.

2. If 8 pounds of lead cost $\frac{1}{2}$ of a dollar, what does it cost per pound?

Operation.

$$\begin{array}{r|l} 2 & 1 \\ 8 & \\ \hline 16 & 1=\frac{1}{16}\ \textit{Ans.} \end{array}$$

Were the cost of 8 pounds 1 dollar, the cost of 1 pound would be the quotient of 1 divided by 8. Regarding the numerator as expressing the cost of the lead, without reference to the denominator, we place it on the right, as a dividend, and 8, the number of pounds, on the left, as a divisor; but the cost of 8 pounds is 1 divided by 2; therefore, write 2 also on the left. As no

---

Questions.—12. How do you divide a fraction by a whole number? 13. Why, in example 2d, are 8 and 2, the denominator of a fraction, placed on the left? 14. Multiplying the denominator of a fraction, is the same as what?

reduction can be made, the 2 and 8 are to be multiplied together, for a divisor or denominator: (multiplying the denominator divides the fraction.) The answer, then, is 1 divided by 16, or $\frac{1}{16}$.

Or thus, $$\frac{1}{2\times 8}=\frac{1}{16}\ \textit{Ans.}$$

3. If a pair of oxen plough, in 4 days, $\frac{28}{31}$ of a field, what part do they plough in one day? *Ans.* $\frac{7}{31}$.

4. If 6 men earn $\frac{18}{28}$ of a guinea in 1 day, what part of a guinea does 1 man earn in the same time? *Ans.* $\frac{3}{28}$.

5. If 16 hats cost \$$64\frac{1}{2}$, what does 1 hat cost? *Ans.* \$$4\frac{1}{32}$.

6. What will 1 pipe of molasses cost, if 14 pipes cost \$$707\frac{1}{2}$? *Ans.* \$$50\frac{15}{28}$.

7. What will 1 pound of rice cost, if 50 pounds cost \$$150\frac{3}{4}$? *Ans.* \$$3\frac{3}{200}$.

Multiplication and Division of Fractions by Whole Numbers.

MULTIPLICATION.

Art. 69.—1. If a dollar will buy $\frac{1}{18}$ of an acre of land, how much will 9 dollars buy?

3. If a man travel $\frac{2}{27}$ of a mile in 1 minute, how far will he travel in 12 minutes?

5. If a man consume $\frac{5}{42}$ of a barrel of flour in 1 month, what will 7 men consume in the same time?

7. If $\frac{3}{14}$ of a box of glass cost 1 dollar, how many boxes will 21 dollars buy?

9. If a pound of chocolate cost $\frac{5}{28}$ of a dollar, how much will 7 pounds cost?

11. If a man can do $\frac{1}{12}$ of a piece of work in one day, how much could he do in 8 days?

13. What will 16 yards of cloth cost, at $\frac{3}{4}$ of a dollar per yard?

15. What will 40 yards of carpeting cost, at $\frac{7}{8}$ of a dollar per yard?

DIVISION.

Art. 70.—2. If 9 dollars will buy $\frac{1}{2}$ of an acre, how much will 1 dollar buy?

4. If a man travel $\frac{8}{9}$ of a mile in 12 minutes, how far will he travel in 1 minute?

6. If 7 men consume $\frac{5}{8}$ of a barrel of flour in 1 month, how much will 1 man consume?

8. If 21 dollars will buy $4\frac{1}{2}$ boxes of glass, how much will 1 dollar buy?

10. If 7 pounds of chocolate cost $\frac{5}{4}$ of a dollar, what will 1 pound cost?

12. If a man can do $\frac{2}{3}$ of a piece of work in 8 days, how much can he do in 1 day?

14. If 16 yards of cloth cost $\frac{12}{1}$ of a dollar, what will 1 yard cost?

16. If 40 yards of carpeting cost $\frac{35}{1}$ of a dollar, what will 1 yard cost?

17. If 1 pint of wine cost $\frac{7}{96}$ of a dollar, how much will 12 quarts cost?

18. If 12 quarts of wine cost $1\frac{3}{4}$ dollar, what is it per pint?

19. Multiply $\frac{7}{8}$ by 11.

20. Divide $9\frac{5}{8}$ by 11.

Art. 71.—Multiplication of whole numbers by fractions.

1. If a barrel of flour cost $9, how much will $\frac{2}{3}$ of a barrel cost?

Had the cost of 2 barrels been required, the price of 1 barrel being given, we should multiply the price of 1 barrel by the number of barrels. The same is now to be done; that is, the price of one barrel is to be multiplied by the part or parts of a barrel taken.

To multiply by 1, *is to repeat the units of the multiplicand once.*

To multiply by 2, *is to repeat the units of the multiplicand twice.*

To multiply by $\frac{1}{2}$ *of* 1, *is to repeat one-half of the units of the multiplicand once.*

The product of any number multiplied by a fraction is proportionally as much less than the multiplicand as the multiplier is less than the unit, or 1. Therefore, if we multiply 9 by $\frac{2}{3}$ of 1 the product will be $\frac{2}{3}$ of 9, or 6. Hence, it appears, that, to multiply by a fraction, is to repeat such a part of the multiplicand as the fraction is part of a unit. If 9 dollars be the cost of 1 barrel, then the quotient of 9 divided by 3 will be 3 dollars, the cost of $\frac{1}{3}$ of a barrel, and $3\times2=6$, the cost of $\frac{2}{3}$. Thus it appears, that the only difference between multiplying by a whole number and a fraction, is, that in the last case the multiplier is a number divided; now to divide the multiplicand or the product is the same as to divide the multiplier. Hence, *to multiply a whole number by a fraction,*

RULE.

Multiply the whole number by the numerator of the fraction, and divide the product by the denominator; or divide the whole number by the denominator of the fraction, and multiply the quotient by the numerator.

Obs.—As multiplying by a fraction is repeating a part only of the multiplicand, we divide by the denominator of the fraction, to obtain that part of the multiplicand to be repeated, and multiply by the numerator to repeat that part. Thus, to multiply by $\frac{2}{3}$, we divide by 3, and obtain one-third of the multiplicand, which is to be repeated twice, or multiplied by 2, the numerator. The same principle is applicable to the multiplication of fractions by fractions.

2. What will 16 yards of cloth cost, at $\frac{3}{4}$ of a dollar per yard?

Were the cost of 1 yard 3 dollars, the cost of 16 yards would be 16 times 3 dollars; but the cost of one yard is $\frac{1}{4}$ of 3 dollars; therefore the cost of 16 yards will be $\frac{1}{4}$ of 16 times 3 dollars.

Operation 1*st.*

4	3
	16 4
1	12 *Ans.*

Operation 2*d.*

$$\frac{3 \times \overset{4}{16}}{4} = \frac{12}{1} = 12 \text{ Ans.}$$

Excluding equal factors from divisor and dividend, or from numerator and denominator, does not affect the result. In this example we exclude the factor 4, and the remaining factors 3 and 4 multiplied together, being factors of the dividend or numerator, give 12, the answer.

3. What will 40 yards of carpeting cost, at $\frac{7}{8}$ of a dollar per yard? *Ans.* \$35.

4. What will 64 bushels of oats cost, at $\frac{3}{8}$ of a dollar per bushel? *Ans.* \$24.

5. What will 24 bushels of corn cost, at $\frac{5}{8}$ of a dollar per bushel? *Ans.* \$15.

6. Multiply 21 by $\frac{3}{7}$, by $\frac{6}{7}$, by $\frac{1}{3}$. *Ans.* 9, 18, 7.

7. What is the product of 324 multiplied by $\frac{11}{12}$?

8. What will 9 pounds of tea cost, at $\frac{27}{54}$ of a dollar per pound? *Ans.* \$4$\frac{1}{2}$.

9. What will 56 pounds of butter cost, at $\frac{2}{8}$ of a dollar per pound? *Ans.* \$14.

10. What will 124 pounds of sugar cost, at $\frac{2}{16}$ of a dollar per pound? *Ans.* \$15$\frac{1}{2}$.

11. Multiply 32 by $\frac{1}{4}$, by $\frac{2}{8}$, by $\frac{3}{8}$, by $\frac{7}{8}$. *Ans.* 8, 8, 12, 28.

12. Multiply 224 by $\frac{1}{56}$. *Ans.* 4.

Division of Whole Numbers by Fractions.

Art. 72.—1. If $\frac{2}{3}$ of a barrel of flour cost \$6, what will be the cost of 1 barrel?

QUESTIONS.—15. What is the product of any number multiplied by 1? 16. How many times greater than the multiplicand is the product of a multiplier greater than 1? 17. How much less than the multiplicand is the product, when the multiplier is less than 1? 18. What is the product of a whole number multiplied by a fraction? 19. Rule?

As this question is the reverse of question 1st, in the preceding section, the reverse of the rule there given will be the rule for solving this question. Had 6 dollars been the cost of 2 barrels, we should divide the price by the number of barrels. The same is now to be done; that is, the price is to be divided by the parts of a barrel taken.

The quotient of any number divided by a unit, or 1, is the same as the dividend.

The quotient of any number divided by a greater number than 1, is as many times less than the dividend, as the divisor is times greater than a unit, or 1.

On the same principle, if the divisor be less than a unit, or 1, the quotient will be greater than the dividend. If the divisor be $\frac{1}{2}$, the quotient will be $\frac{2}{1}$, or twice the dividend. If the divisor be $\frac{2}{3}$, the quotient will be $\frac{3}{2}$, or three halves of the dividend. Therefore, if we divide 6 by $\frac{2}{3}$ of 1, the quotient will be $\frac{3}{2}$ of 6, or 9. It is evident, that 6 will contain $\frac{1}{3}$ of 2 three times as often as it will contain 2. Again: if 6 dollars be the cost of $\frac{2}{3}$ of a barrel, the quotient of 6 divided by 2 will be 3 dollars, the cost of $\frac{1}{3}$ of a barrel, and $3\times3=9$ dollars, the cost of $\frac{3}{3}=1$ barrel. Thus it appears, that the difference between dividing by a whole number, or by a fraction, is, that in the latter case the divisor is a number divided; for, *To multiply the dividend or the quotient, is the same as to divide the divisor.* Hence the

RULE.

Divide the dividend by the numerator of the fraction, and multiply the quotient by the denominator; or multiply by the denominator, and divide the product by the numerator.

Operation.

~~2~~	~~6~~ $3\times3=9$ *Ans.*
	3

Regarding the numerator of the fraction as a whole number, we say, if 2 barrels cost 6 dollars, the quotient of 6 divided by 2 will be the cost of 1 barrel; therefore, place 6, the dividend, on the right, and 2, the divisor, on the left of the line. But the 2 barrels is 2 divided by 3; therefore, place 3, the divisor, or denominator, on the opposite side of the line, as a multiplier. *To multiply the dividend, or quotient, is the same as to divide the divisor.* We then say, 2 in 6, three times; cross 6 and 2, and multiply 3 into 3.

$3\times3=\$9$ *Ans.*

OBS.—Were the foregoing question, How many times will 6 bushels contain 2 pecks? we should multiply the dividend by 4, to bring it into pecks, or fourths of a bushel. 6 bushels equal 24 pecks, or $\frac{24}{4}$ of a bushel, and 2 pecks equal $\frac{2}{4}$ of a bushel. So, to divide 6 units by $\frac{2}{3}$, we multiply the dividend by 3, the denominator of the fraction, to bring it into thirds. The same is true in division of fractions by fractions.

2. If $\frac{5}{6}$ of a ton cost \$15, how much will 1 ton cost? *Ans.* \$18.

3. If $\frac{3}{8}$ of an acre be worth \$12, what is 1 acre worth? *Ans.* \$32.

4. If $\frac{3}{4}$ of the number of rows in a corn-field be 30, what is the whole number? *Ans.* 40.

5. Divide 20 by $\frac{1}{2}$, $\frac{1}{4}$, $\frac{1}{5}$. *Ans.* 40, 80, 100.

6. If a pound of tea cost $\frac{3}{5}$ of a dollar, how many lbs. may be bought for \$60? *Ans.* 100 lbs.

7. In what time can a man build 7 rods of wall, if he build $\frac{2}{5}$ of a rod in an hour? *Ans.* $17\frac{1}{3}$ hours.

8. At $\frac{6}{7}$ of a dollar for building one rod of stone wall, how many rods may be built for \$69? *Ans.* $80\frac{1}{2}$.

9. At $\$3\frac{3}{4}$ per yard, how many yards may be bought for \$80? *Ans.* $21\frac{1}{3}$ yards.

10. If $1\frac{1}{3}$ bushel of wheat sow an acre of land, how many acres will 12 bushels sow? *Ans.* 9 acres.

11. How many times is $\frac{7}{8}$ contained in 56? *Ans.* 64.

12. How many times is $\frac{24}{5}$ contained in 21? *Ans.* $4\frac{3}{8}$.

Multiplication and Division of Whole Numbers by Fractions.

MULTIPLICATION.	DIVISION.
Art. 73.—1. If a man can earn \$16 in a month, how much can he earn in $\frac{3}{8}$ of a month?	**Art. 74.**—2. If a man earn \$6 in $\frac{3}{8}$ of a month, how much can he earn in a month?
3. If a man lay up \$84 in a year, how much would he lay up in $\frac{5}{7}$ of a year?	4. If a man in $\frac{5}{7}$ of a year lay up \$60, how much would he lay up in a year?
5. If the price of a horse be \$75, what would be the price of a horse worth $\frac{4}{12}$ as much?	6. If $\frac{4}{12}$ of the value of a horse be \$25, what is the whole value?
7. If a house be worth \$672, how much is $\frac{9}{16}$ worth?	8. If $\frac{9}{16}$ of a house be worth \$378, what is the whole worth?
9. If a farm be worth \$840, what is $\frac{5}{8}$ of it worth?	10. If $\frac{5}{8}$ of a farm be worth \$525, what is the whole worth?

QUESTIONS—20. How much greater than the dividend is the quotient of a whole number divided by a fraction? 21. How do you divide a whole number by a fraction?

11. Multiply 156 by $\frac{3}{4}$.

12. Divide 117 by $\frac{3}{4}$.

13. A man has 360 apple-trees in two orchards; in the smaller, there is $\frac{1}{9}$ of the whole. How many are there in the smaller?

14. 40 is $\frac{1}{9}$ of what number?

15. A ship and cargo are valued at $100,000, the ship at $\frac{2}{40}$ of the whole. What is the value of the ship?

16. If $\frac{2}{40}$ of a ship and cargo are valued at $5,000, what is the value of the whole?

Multiplication of Fractions by Fractions.

Art. 75.—We have seen, that to multiply a whole number by a fraction, is to repeat such a part of the multiplicand as the multiplier is part of a unit. If the multiplicand be a fraction, the principle is the same.

1. If a bushel of corn be worth $\frac{2}{3}$ of a dollar, how much is $\frac{1}{2}$ of a bushel worth?

Operation.

$$\begin{array}{r|l} \not{2} & 1 \\ 3 & \not{2} \\ \hline 3 & 1=\frac{1}{3}\ \textit{Ans.} \end{array}$$

Were the cost of 2 bushels required, we should multiply the price by the quantity; but as the quantity is less than 1 bushel, we multiply by the parts taken; (*see* Art. 71.) The question then is, How much is $\frac{1}{2}$ of $\frac{2}{3}$? To multiply a whole number by $\frac{1}{2}$, we take $\frac{1}{2}$ of the multiplicand. The same is now to be done: $\frac{1}{2}$ of $\frac{2}{3}=\frac{1}{3}$. If the numerators be multiplied together, and also the denominators, we have the answer: Thus, $\frac{1}{2}\times\frac{2}{3}=\frac{2}{6}=\frac{1}{3}$. Multiplying the denominator of $\frac{2}{3}$ by 2, is dividing the fraction by 2. We thus obtain that part of the multiplicand to be repeated, or multiplied by the numerator of $\frac{1}{2}$. (*See* Art. 71, Obs.) Therefore, to multiply a fraction by a fraction, we have this

RULE.

Multiply the numerators together for a new numerator, and their denominators for a new denominator.

EXAMPLE.

1. A man owning $\frac{3}{4}$ of a ship, sold $\frac{2}{5}$ of his share: $\frac{2}{5}$ of $\frac{3}{4}$ is how much?

QUESTIONS.—1. How do you multiply a fraction by a fraction? 2. Why, in example 1st, are the numerators of the fractions placed on the right of the line, and the denominators on the left?

Operation.

$$\begin{array}{r|l} 5 & \not{2} \\ 2\ \not{4} & 3 \\ \hline 10 & 3 = \frac{3}{10}\ \textit{Ans.} \end{array}$$

As the numerators of the fractions are to be multiplied together for a new numerator, or dividend, they are placed on the right of the line, and the denominators, which are to be multiplied for a new denominator, or divisor, are placed on the left of the line. The numbers are cancelled and multiplied as in preceding examples.

Or thus: $\frac{3}{\underset{2}{\not{4}}} \times \frac{\not{2}}{5} = \frac{3}{10}$ *Ans.* The factor 2 in the numerator, cancels 2, one of the factors of 4, in the denominator.

2. What is the product of $\frac{7}{9} \times \frac{6}{7}$; of $\frac{7}{12} \times \frac{3}{9}$? *Ans.* $\frac{2}{3}$, $\frac{7}{36}$.

3. Multiply $\frac{7}{8}$ by $\frac{9}{2}$, and $\frac{6}{7}$ by $\frac{8}{8}$.

4. A boy having $\frac{1}{2}$ of a dollar, gave $\frac{1}{4}$ of it for toys; what did the toys cost him? *Ans.* $\frac{1}{8}$ of a dollar.

5. At $\frac{3}{8}$ of a dollar per yard, what will $\frac{8}{9}$ of a yard cost?

6. At $\frac{3}{4}$ of a dollar per pound, what will $\frac{7}{8}$ of a pound of tea cost? *Ans.* $\frac{21}{32}$ of a dollar.

7. At $\frac{1}{6}$ of a dollar a pound, what will $\frac{6}{7}$ of a pound of coffee cost? *Ans.* $\frac{1}{7}$ of a dollar.

8. At $2\frac{1}{4}$ dollars per bushel, what will $6\frac{3}{27}$ bushels of wheat cost? *Ans.* \$$13\frac{3}{4}$.

9. If a house lot be worth $100\frac{5}{75}$ dollars, what is $\frac{1}{25}$ of the lot worth? *Ans.* \$$4\frac{1}{375}$.

10. If a flock of sheep be worth $75\frac{1}{5}$ dollars, what is $\frac{1}{4}$ of the flock worth? *Ans.* \$$18\frac{4}{5}$.

Division of Fractions by Fractions.

1. If a bushel of corn cost $\frac{1}{3}$ of a dollar, how many bushels may be bought for $\frac{2}{3}$ of a dollar?

It is evident that $\frac{2}{3}$ will contain $\frac{1}{3}$ twice: $\frac{2}{3} \div \frac{1}{3} = \frac{2}{1} = 2$, *Ans.* In this example, both dividend and divisor are divided by 3. It has been shown, that to divide the dividend is the same as to divide the quotient, and to divide the divisor is the same as to multiply the quotient; and also, that to multiply and divide any number by the same quantity does not affect its value.

Art. 76.—Therefore, when the denominator of dividend and divisor are alike, *divide the numerator of the dividend by the numerator of the divisor, and the quotient will be the answer.*

Operation.

$$\begin{array}{c|l} \not{3} & 2 \\ 1 & \not{3} \\ \hline & 2 \end{array} \quad \textit{Ans.}$$

Or thus:

$$\frac{2\ \not{3}}{\not{3}\ 1}=\frac{2}{1}=2 \quad \textit{Ans.}$$

2. If a bushel of corn cost $\frac{3}{4}$ of a dollar, how much may be bought for $\frac{3}{7}$ of a dollar?

Were the money to be expended 3 dollars, and the price 3 dollars, the answer would be 1 bushel, $3 \div 3 = 1$. But suppose the price 3 dollars, and the money to be expended 3 dollars divided by 7, equal $\frac{3}{7}$. The answer would then be $\frac{3}{7} \div 3 = \frac{1}{7}$ of a bushel. To divide the quotient is the same as to divide the dividend. But the price is not 3 dollars, but 3 dollars divided by 4. To multiply the quotient is the same as to divide the divisor. Therefore, the true answer is, 4 divided by 7 equal $\frac{4}{7}$.

Again, $\frac{3}{7} = \frac{1}{7} \times 3$, and $\frac{3}{4} = \frac{1}{4} \times 3$. To multiply dividend and divisor by the same quantity does not affect the quotient.

Art. 77.—Therefore, when the numerators of divisor and dividend are alike,

RULE.

Divide the denominator of the divisor by the denominator of the dividend.

Art. 78.—3. If a bushel of oats cost $\frac{3}{8}$ of a dollar, how many bushels may be bought for $\frac{9}{16}$ of a dollar?

Operation: $\frac{3}{8})\frac{9}{16}(\frac{3}{2} = 1\frac{1}{2}$ *Ans.*

4. If a bushel of rye cost $\frac{5}{8}$ of a dollar, how many bushels may be bought for $\frac{7}{9}$ of a dollar?

In this example, the terms of the dividend cannot be divided by the corresponding terms of the divisor without a remainder; but to multiply the numerator is the same as to divide the denominator. Hence the

RULE.

Divide the terms of the dividend by the corresponding terms of the divisor, when it can be done without a remainder; otherwise, invert the divisor, and proceed as in Multiplication.

Operation.

$$\begin{array}{r|l} 9 & 7 \\ 5 & 8 \\ \hline 45 & 56 \end{array} = 1\tfrac{11}{45} \quad \textit{Ans.}$$

By the rule already given for placing the numerators of fractions as whole numbers, the numerator of $\frac{7}{9}$, the dividend, is placed on the right of the line, and the numerator of $\frac{5}{8}$, the divisor, on the left. This is the same as inverting the divisor. $\frac{7}{9}\div\frac{5}{8}=\frac{7}{9}\times\frac{8}{5}=\frac{56}{45}=1\frac{11}{45}$ *Ans.*

Division of fractions by fractions may be variously illustrated. 1. To multiply the dividend is the same as to divide the divisor, (*see Art.* 43.) To multiply the denominator divides the fraction. Therefore, $\frac{7}{9}\times{}^{8}\div 5=\frac{7}{9}\times\frac{8}{5}=\frac{56}{45}=1\frac{11}{45}$ *Ans.* 2. Multiplying the numerator of $\frac{7}{9}$ by 8 reduces it to eighths. Multiplying the denominator by 5 divides the fraction, (*see Art.* 72, *Obs.*)

1. If a bushel of potatoes cost $\frac{2}{8}$ of a dollar, how many bushels may be bought for $\frac{2}{9}$ of a dollar? *Ans.* $\frac{8}{9}$.

2. If $\frac{6}{7}$ of a bushel of apples cost $\frac{3}{5}$ of a dollar, how much will 1 bushel cost? *Ans.* $\frac{7}{10}$.

3. How many bushels of rye, at $\frac{8}{9}$ of a dollar per bushel, may be bought for $\frac{2}{3}$ of a dollar? *Ans.* $\frac{3}{4}$ of a bush.

4. If $\frac{2}{21}$ of a ton of hay cost $\frac{7}{8}$ of a dollar, what does it cost per ton? *Ans.* \$$9\frac{3}{16}$.

5. If $4\frac{1}{2}$ pounds of tea cost $3\frac{1}{4}$ dollars, what is it per lb.? *Ans.* $\frac{13}{18}$ of a dollar.

6. If $\frac{13}{18}$ of a dollar buy 1 pound of tea, how much will $3\frac{1}{4}$ dollars buy? *Ans.* $4\frac{1}{2}$ pounds.

7. Divide $17\frac{1}{2}$ by $7\frac{1}{2}$ and $18\frac{3}{4}$ by $\frac{25}{75}$. *Ans.* $2\frac{1}{3}$; $56\frac{1}{4}$.

Multiplication and Division of Fractions.

MULTIPLICATION.

Art. 79.—1. A man owning $\frac{7}{8}$ of a house, sold $\frac{5}{6}$ of his share. What part of the house did he sell?

3. If a bushel of salt cost $\frac{19}{21}$ of a dollar, what will $\frac{1}{5}$ of a bushel cost?

5. If a peck of coal cost $\frac{5}{61}$ of a dollar, what will $\frac{3}{4}$ of a peck cost?

7. If 1 cord of wood cost $\frac{22}{5}$ of a dollar, how much will $\frac{7}{9}$ of a cord cost?

9. If 1 foot of hammered stone cost $\frac{15}{98}$ of a dollar, what will $\frac{2}{15}$ of a foot cost?

DIVISION.

Art. 80.—2. A man sold $\frac{35}{48}$ of a house, which was $\frac{5}{6}$ of his share. What part of the house did he own?

4. If $\frac{1}{5}$ of a bushel of salt cost $\frac{19}{105}$ of a dollar, what does it cost per bushel?

6. If $\frac{3}{4}$ of a peck of coal cost $\frac{15}{244}$ of a dollar, what will one peck cost?

8. If $\frac{7}{9}$ of a cord of wood cost $3\frac{19}{45}$ dollars, how much is it per cord?

10. If $\frac{2}{15}$ of a foot of hammered stone cost $\frac{1}{49}$ of a dollar, what will one foot cost?

The simple rule may now be repeated for solving any question which may arise in Multiplication and Division of Fractions by Whole Numbers—Multiplication and Division of Whole Numbers by Fractions—Multiplication and Division of Fractions by Fractions.

RULE.

Place all those numbers which are to be multiplied together for a numerator, or dividend, on the right of the perpendicular line, and those numbers which are to be multiplied together for a denominator, or divisor, on the left of the line, and proceed to cancel, as before directed.

PROMISCUOUS EXAMPLES.

Art. 81.—1. A man owning $\frac{1}{2}$ of $\frac{2}{3}$ of $\frac{3}{4}$ of $\frac{4}{5}$ of $\frac{7}{8}$ of a ship, sold $\frac{3}{7}$ of $\frac{8}{3}$ of $\frac{2}{3}$ of his share. What part of the ship did he sell?

Thus:

$$\begin{array}{r|l} \not{2} & 1 \\ \not{3} & \not{2} \\ \not{4} & \not{3} \\ 5 & \not{4} \\ \not{8} & \not{7} \\ \not{7} & \not{3} \\ \not{3} & \not{8} \\ 3 & 2 \\ \hline 15 & 2 = \frac{2}{15} \textit{ Ans.} \end{array}$$

Or thus: $\frac{1}{\not{2}}\times\frac{\not{2}}{\not{3}}\times\frac{\not{3}}{\not{4}}\times\frac{\not{4}}{5}\times\frac{\not{7}}{\not{8}}\times\frac{\not{3}}{\not{7}}\times\frac{\not{8}}{\not{3}}\times\frac{2}{3}=\frac{2}{15}$ *Ans.*

Fractions connected by the word *of*, are called compound fractions. They are reduced to simple fractions, by multiplying all the numerators together for a new numerator, and all the denominators for a new denominator. By cancelling, the process of multiplying and reducing the fraction is performed at once.

2. Reduce $\frac{7}{8}$ of $\frac{3}{4}$ of $\frac{5}{7}$ of $\frac{2}{3}$ of $\frac{5}{6}$ of $\frac{8}{7}$ to a simple fraction. *Ans.* $\frac{25}{84}$.

3. A man owning $\frac{1}{2}$ of $\frac{5}{6}$ of $\frac{7}{9}$ of $\frac{9}{10}$ of a factory, sold $\frac{3}{4}$ of $\frac{3}{5}$ of $\frac{2}{3}$ of his share. What part of the factory did he sell? *Ans.* $\frac{7}{80}$.

4. What simple fraction is equivalent to $\frac{7}{8}$ of $\frac{22}{3}$ of $\frac{71}{8}$, of $\frac{5}{7}$ of $\frac{4}{18}$ of $\frac{3}{5}$ of 9, of $\frac{1}{36}$ of 18, of $\frac{4}{11}$ of 2. *Ans.* $17\frac{3}{4}$.

5. Multiply $\frac{3}{15}$ by 4.

6. Multiply $\frac{1}{3}$ by $\frac{1}{6}$.

7. Multiply $\frac{1}{2}$ by $\frac{1}{4}$.

8. Divide $4\frac{1}{2}$ by $3\frac{1}{2}$.

9. Divide $\frac{1}{4}$ of $\frac{4}{5}$ by $\frac{5}{8}$ of $\frac{8}{9}$ of 9. *Ans.* $\frac{1}{25}$.

10. Divide $\frac{1}{2}$ of 9 by $\frac{1}{3}$ of 7. *Ans.* $1\frac{13}{14}$.

11. Divide $\dfrac{16\frac{1}{3}}{18\frac{1}{4}}$ by $\dfrac{12\frac{2}{3}}{17\frac{1}{5}}$. Having reduced the terms of divisor and dividend to improper fractions, it will be found that the numerators and denominators themselves become fractions. Thus, the numerator of the dividend, $16\frac{1}{3}=\frac{49}{3}$, and the denominator, $18\frac{1}{4}=\frac{73}{4}$.

The dividend now assumes this form: $\dfrac{\frac{49}{3}}{\frac{73}{4}}$ The denominators may be removed from the terms of the fraction, and the process illustrated in the following manner. The numerator is 49 divided by 3. To multiply the denominator is the same as to divide the numerator.

$$\frac{\frac{49}{3}}{\frac{73}{4}}=\frac{49}{\frac{73\times3}{4}}$$

To multiply the numerator is the same as to divide the denominator. Therefore—

$$\frac{49}{\frac{73\times3}{4}}=\frac{49\times4}{73\times3}=\frac{196}{219}.$$

Let the scholar reduce the divisor, and illustrate in a similar mannner.

12. Divide $\frac{7}{8}\div\frac{6}{11}$ by $\frac{4}{5}\div\frac{3}{7}$. *Ans.* $\frac{55}{64}$.

13. Divide $\frac{1}{4}$ of $\frac{4}{5}$ of $\frac{5}{6}$ of $\frac{7}{8}$ of 18 by $\frac{4}{5}$ of $\frac{5}{6}$ of $\frac{7}{8}$ of 12; multiply by $\frac{1}{2}$ of $\frac{2}{3}$ of $\frac{3}{5}$ of 2; divide by $\frac{6}{7}$ of $\frac{7}{8}$ of $\frac{1}{2}$ of 6. *Ans.* $\frac{1}{15}$.

14. A man who owns $\frac{1}{2}$ of a farm, sells $\frac{1}{3}$ of $\frac{1}{4}$ of $\frac{3}{7}$ of $\frac{7}{8}$ of his half. What part of the farm does he sell? *Ans.* $\frac{1}{64}$.

15. Multiply 12 by $\frac{1}{2}$ of 3, divide by $\frac{1}{4}$ of 1, multiply by $\frac{1}{2}$ of 6, divide by $\frac{6}{7}$ of 14, multiply by $\frac{1}{2}$ of 18, divide by $\frac{2}{3}$ of 27. *Ans.* 9.

Addition of Fractions.

Fractions are added on the same principle as whole numbers. As tens can only be added to units, and pounds to shillings, by first reducing the higher denomination to the lower, so fractions of different denominations, or which have different denominators, can only be added by first reducing them to the same.

Art. 82.—Fractions which have a common denominator may be added by the following

RULE.

Add their numerators, and write their sum over the denominator.

Operation.

1. Add $\frac{3}{8}$ and $\frac{1}{8}$. $\frac{3}{8}+\frac{1}{8}=\frac{4}{8}$ *Ans.*
2. Add $\frac{1}{10}+\frac{2}{10}+\frac{3}{10}+\frac{5}{10}+\frac{7}{10}$. *Ans.* $\frac{18}{10}=1\frac{8}{10}$.

Art. 83.—Addition of fractions whose denominators are different, and one is a multiple of each of the others.

1. Add $\frac{2}{3}$ and $\frac{1}{6}$. In this example sixths is the lowest denomination mentioned; thirds must, therefore, be reduced to sixths. That is, $\frac{2}{3}$ must be reduced to an equivalent fraction, whose denominator is 6. (*See* Art. 58.)

OBS.—That fraction is of the lowest denomination whose denominator is the largest.

To ascertain how many of the smaller fractions make one of the larger, we divide the denominator of the smaller by the denominator of the larger; 6 contains 3 twice, or $\frac{1}{3}$ contains $\frac{1}{6}$ twice. That is, 2 sixths make $\frac{1}{3}$. Were the numerator a unity, we should now have the fraction required. But since it is greater, if we multiply it by 2, we have $\frac{4}{6}$, an equivalent fraction. Hence the

RULE

Divide the denominator of that fraction whose denominator is a multiple of each of the other denominators, first by the denominator of one of the other fractions; multiply its numerator into the quotient, and write the product over the denominator thus divided; and so continue to do, until the fractions are all reduced to the same denomination, or to a common denominator.

2. Add $\frac{1}{6}+\frac{2}{12}+\frac{1}{24}$.

Operation.

$$\left.\begin{array}{l}24\div12\times2=4\\24\div\ 6\times1=4\end{array}\right\}\text{ Then }\frac{4}{24}+\frac{4}{24}+\frac{1}{24}=\frac{9}{24}\ \textit{Ans.}$$

3. Add $\frac{1}{2}$, $\frac{1}{4}$, $\frac{1}{8}$, $\frac{1}{16}$. *Ans.* $\frac{15}{16}$.
4. Add $\frac{1}{3}$, $\frac{1}{15}$, $\frac{2}{5}$. *Ans.* $\frac{12}{15}$.

Art. 84.—Addition of fractions when no denominator is a multiple of each of the other denominators.

Add $\frac{2}{3}$, $\frac{1}{4}$, $\frac{2}{5}$. In this example we have no common denominator, or denomination given, to which each of the fractions may be reduced. If we multiply all the denominators together, we shall obtain a common multiple of all the denominators, or a denomination to which each of the fractions may be reduced. $3\times4\times5=60$. We have now to reduce each of the given fractions to 60ths. This may be done by the foregoing Rule. It will be observed, that the quotient of 60 divided by any one of the denominators is the product of all the other denominators. Hence the product of all the other denominators shows how many 60ths make *one* in the denomination of the fraction to be reduced; we may, therefore, adopt the common

RULE.

Multiply all the denominators together for a common denominator, and each numerator into all the denominators except its own for a new numerator.

Operation 1st.

Denominators, $3\times4\times5=60$ com. denominator.

Then, $\left.\begin{array}{l}60\div3\times2=40\\60\div4\times1=15\\60\div5\times2=24\end{array}\right\}$ new numerators.

Operation 2d.

Denominators, $3\times4\times5=60$ com. denominator.

$\left.\begin{array}{ll}\text{1st numerator,} & 2\times4\times5=40\\\text{2d \quad ``} & 1\times3\times5=15\\\text{3d \quad ``} & 2\times3\times4=24\end{array}\right\}$ new numerators.

Then $\frac{40}{60}+\frac{15}{60}+\frac{24}{60}=\frac{79}{60}=1\frac{19}{60}$ *Ans.*

EXAMPLES.

1. Reduce $\frac{1}{4}$, $\frac{1}{6}$, $\frac{2}{7}$, and $\frac{3}{8}$ to fractions having a common denominator. *Ans.* $\frac{336}{1344}$, $\frac{224}{1344}$, $\frac{384}{1344}$, $\frac{504}{1344}$.
2. Reduce $\frac{2}{9}$, $\frac{5}{12}$, $\frac{7}{14}$, and $\frac{2}{11}$ to fractions having a common denominator. *Ans.* $\frac{3696}{16632}$, $\frac{6930}{16632}$, $\frac{8316}{16632}$, $\frac{3024}{16632}$.
3. Add together $\frac{5}{6}$, $\frac{7}{8}$, $\frac{3}{4}$, and $\frac{2}{5}$. *Ans.* $\frac{2744}{960}=2\frac{103}{120}$.

Obs. 1.–Reduce the fractions to a common denominator, find new numerators, and add them together.

4. Add $\frac{1}{2}$, $\frac{3}{4}$, $\frac{3}{7}$, and $\frac{7}{8}$. *Ans.* $2\frac{31}{56}$.

5. Add together $\frac{6}{7}$ of $\frac{8}{9}$ and $\frac{9}{11}$ of $\frac{3}{13}$. *Ans.* $\frac{2855}{3003}$.

OBS. 2.—Compound fractions must be reduced to simple fractions.

6. Add $\frac{1}{3}$ of 96 and $\frac{7}{8}$ of $14\frac{1}{2}$ together. *Ans.* $44\frac{11}{16}$.

7. Add together $\frac{1}{2}$ of $\frac{7}{8}$ and $\frac{2}{3}$ of $\frac{19}{20}$. *Ans.* $1\frac{17}{240}$.

8. Add together 6 and $\frac{7}{8}$ of $\frac{9}{10}$ and $\frac{4}{7}$ of $\frac{1}{2}$ and $7\frac{1}{2}$. *Ans.* $14\frac{321}{560}$.

OBS. 3.—Mixed numbers may be reduced to improper fractions, or the fractional parts may be reduced to a common denominator, and added as in the foregoing examples. If their sum amount to an integer, add it to the whole numbers.

9. Add together $14\frac{3}{4}$ and $16\frac{2}{3}$.

Operation.

$14\frac{9}{12}$ $\frac{3}{4}=\frac{9}{12}$, and $\frac{2}{3}=\frac{8}{12}$; then $\frac{9}{12}+\frac{8}{12}=\frac{17}{12}=1\frac{5}{12}$.

$16\frac{8}{12}$

$31\frac{5}{12}$ *Ans.* We find the common denominator to be 12, and the new numerators to be 9 and 8, which when added are $\frac{17}{12}=1\frac{5}{12}$. Write the $\frac{5}{12}$ under the fractions, and carry 1 to the whole numbers.

10. Add together $17\frac{3}{4}$, $18\frac{1}{2}$, $19\frac{3}{5}$. *Ans.* $55\frac{17}{20}$.

11. A grocer sold the following parcels of sugar, viz: $16\frac{1}{2}$ lbs., $19\frac{1}{4}$, $13\frac{3}{4}$, $20\frac{1}{8}$, $25\frac{1}{16}$, $30\frac{5}{6}$, and $11\frac{1}{4}$ lbs. How many pounds did he sell in all? *Ans.* $136\frac{37}{48}$.

Subtraction of Fractions.

RULE.

Art. 85.—*Prepare the fractions as in Addition, and subtract the less numerator from the greater, and under the difference write the denominator.*

EXERCISES.

1. From $\frac{7}{8}$ take $\frac{3}{4}$. *Ans.* $\frac{1}{8}$.
2. From $\frac{3}{11}$ take $\frac{1}{22}$. *Ans.* $\frac{5}{22}$.
3. From $\frac{17}{21}$ take $\frac{9}{14}$. *Ans.* $\frac{1}{6}$.
4. From $\frac{209}{216}$ take $\frac{7}{144}$. *Ans.* $\frac{397}{432}$.

QUESTION.—What is the rule for the subtraction of fractions?

5. From $6\frac{5}{7}$ take $\frac{6}{7}$. *Ans.* $5\frac{6}{7}$.

6. From $\frac{5}{7}$ of $\frac{3}{4}$ take $\frac{3}{5}$ of $\frac{2}{3}$. *Ans.* $\frac{19}{140}$.

7. Add together $\frac{4}{5}$ and $\frac{5}{8}$, and from their sum subtract $\frac{1}{2}$ of $\frac{9}{11}$. *Ans.* $1\frac{7}{440}$.

8. A. owns $\frac{2}{3}$ of $\frac{3}{4}$ of a vessel; B. owns $\frac{3}{8}$ of $\frac{4}{5}$. How much greater is A.'s share than B.'s? *Ans.* $\frac{1}{5}$.

9. Subtract $13\frac{3}{4}$ from $15\frac{2}{3}$.

$15\frac{8}{12}$ $\quad \frac{3}{4}, \frac{2}{3}=\frac{9}{12}, \frac{8}{12}$.
$13\frac{9}{12}$
$1\frac{11}{12}$ *Ans.*

Having reduced the fractions to a common denominator, and found new numerators, as in Addition, we have $\frac{9}{12}$ to be taken from $\frac{8}{12}$. We therefore borrow a unit, and say $\frac{9}{12}$ from $\frac{12}{12}$, and add 3, the remainder, to 8, the numerator of the subtrahend. $3+8=\frac{11}{12}$, which we write under the fractions, and carry 1 to 13, the whole number, which makes 14; and 14 from 15, and 1 remains. The answer, then, is $1\frac{11}{12}$.

10. A man bought a horse for $\frac{1}{2}$ of $\frac{2}{3}$ of \$150, and sold him for $\frac{1}{4}$ of $\frac{6}{7}$ of $\frac{7}{1}$ of \$60. Did he gain or lose, and how much? *Ans.* \$40 gain.

To find the least Common Multiple.

Art. 86.—The common denominator found by the preceding rule, is a common multiple of the denominators of the given fractions; for every product must be divisible by all its factors; but it was not the *least* common multiple.

1. What is the least common multiple of 4, 6, 8, 10?

$4\times6\times8\times10=1920.$

Operation.

2)4, 6, 8, 10
2)2, 3, 4, 5
1, $3\times2\times5\times2\times2=120$.

1920 is evidently a common multiple of 4, 6, 8, and 10, because they are its factors; but it is not the *least* common multiple. We find, also, that each of these numbers is a multiple of 2, or that 2 is a prime factor of each. Dividing by 2, we find the other factors, which are 2, 3, 4, 5. Again: As the quotient, 4, is a multiple of 2, we may substitute for it 2, one of its factors; and as we employ the other factor for a divisor, we erase the other quotient, 2. We now have 3, 2, 5, undivided numbers, which are prime factors of the dividends 6, 8, 10; the other prime factors are the divisors. If now we multiply together these undivided numbers and the divisors, we shall have a combination of all the prime factors of each dividend, and consequently it must be divisible by them. Thus the divisors 2 and 2 are the factors of 4, the first dividend, $2\times2=4$. If 4 be divisible by 4, then 2, 3, and 5 times 4 must be divisible by 4; also 3, the first undivided number, is a factor of 6, the second dividend; and 2, the first divisor, is the other factor, $3\times2=6$. If 6 be divisible by 6, then 2 and 5 times 6 must be divisible by 6. The same may be said of 2 and 5, the other undivided numbers. Hence it appears, that the product of the continued multiplication of the remainders and divisors is divisible by the several dividends; and by examining the operation, it will be found to be the *least* number which can be divisible by them; for all repetition of prime factors beyond what is necessary to produce each dividend, is avoided. Therefore, to find the least common multiple of two or more numbers, we have the following

RULE.

Write the numbers in a horizontal line; divide them by the least prime number that will measure two or more of them; write the quotients and undivided numbers in a horizontal line under the given numbers; divide the numbers in this second line, in the same manner. Thus continue to divide until the quotients and undivided numbers are all prime to each other. The product of the continued multiplication of the divisors and undivided numbers will be the least common multiple required.

OBS. 1.–We divide by any number that will divide two or more of the numbers, to find first the least common measure of two or more.

EXERCISES.

2. What is the least common multiple of 3, 4, 9 and 12? *Ans.* 36.

3. What is the least number which can be divided by 7, 8, 10, and 12, without a remainder? *Ans.* 840.

4. What is the least common multiple of 7, 14, 28, 35? *Ans.* 140.

5. What is the least number which can be divided by the nine digits without a remainder? *Ans.* 2520.

QUESTIONS.—1. How is the least common multiple of two or more numbers found? 2. Why do you divide by any number that will divide two or more without a remainder?

6. Reduce $\frac{3}{4}$, $\frac{2}{5}$, $\frac{5}{6}$, to equivalent fractions having the least common denominator.

$$2\,|\,4 \quad 5 \quad 6$$
$$2\times5\times3\times2=60$$

60 being the least common multiple of 4, 5, and 6, it is, therefore, the least common denominator of the fractions $\frac{3}{4}$, $\frac{2}{5}$, $\frac{5}{6}$. The remaining part of the process is performed by the Rule under Art. 83. Still other illustrations may be given. The value of the fraction $\frac{3}{4}$, is three-fourths of a unit. That is, the unit is divided into 4 parts, and the fraction expresses $\frac{3}{4}$ of them. If we divide the unit into 60 parts, and wish to express the same part of a unit, we must take $\frac{3}{4}$ of 60. If we divide 60 by 4, we have one-fourth; if we multiply one-fourth by 3, we have three-fourths, $60\div4=15$, and $15\times3=45$. Now 3 is three-fourths of 4, and 45 is three-fourths of 60; therefore $\frac{45}{60}=\frac{3}{4}$, an equivalent fraction.

Operation.

$$\left.\begin{array}{l}60\div4\times3=45\\60\div5\times2=24\\60\div6\times5=50\end{array}\right\}\quad \text{Then } \frac{45}{60}, \frac{24}{60}, \frac{50}{60}\ \textit{Ans.}$$

OBS. 2.—By this process the fractions are all reduced to the same denomination.

7. What is the least common denominator of $\frac{1}{2}$, $\frac{1}{4}$, $\frac{3}{5}$, and $\frac{2}{8}$? *Ans.* $\frac{20}{40}$, $\frac{10}{40}$, $\frac{24}{40}$, $\frac{10}{40}$.

8. Reduce $\frac{8}{9}$, $\frac{9}{11}$, $\frac{7}{12}$, to fractions having the least common denominator. *Ans.* $\frac{352}{396}$, $\frac{324}{396}$, $\frac{231}{396}$.

9. Reduce $\frac{3}{7}$, $\frac{2}{8}$, $\frac{1}{9}$, and $\frac{11}{12}$, to fractions having the least common denominator. *Ans.* $\frac{216}{504}$, $\frac{126}{504}$, $\frac{56}{504}$, $\frac{462}{504}$.

10. A merchant buys 5 pieces of cloth. The first contains $40\frac{3}{4}$ yards; the second, $27\frac{1}{2}$; the third, $34\frac{7}{8}$; the fourth, $43\frac{3}{16}$; and the fifth, $39\frac{1}{4}$ yards. How many were there in the whole? *Ans.* $185\frac{9}{16}$.

11. Which is the greater fraction, $\frac{11}{16}$ or $\frac{13}{18}$. *Ans.* $\frac{13}{18}$ is greater by $\frac{5}{144}$.

OBS. 3.—If the denominator of either of the given fractions be a multiple of each of the other denominators, it will be the least common denominator.

QUESTION.—3. How are fractions of different denominators reduced to equivalent fractions having the same denominator?

12. Reduce $\frac{3}{8}$, $\frac{5}{16}$, and $\frac{7}{32}$ to equivalent fractions having the least common denominator.

Multiplying the first by 4 and the second by 2, we have the answer required.

$\frac{3}{8}\times 4=\frac{12}{32}$, $\frac{5}{16}\times 2=\frac{10}{32}$. *Ans.* $\frac{12}{32}$, $\frac{10}{32}$, $\frac{7}{32}$.

13. Which is the greater fraction, $\frac{5}{10}$ or $\frac{4}{20}$?

Dividing the terms of $\frac{4}{20}$ by 2, we have $\frac{2}{10}$, and $\frac{5}{10}-\frac{2}{10}=\frac{3}{10}$, the *Answer*.

Fractions whose denominators are 10, 100, or 1000, etc. form a very important class of fractions, and will be treated under a separate head, called

DECIMAL FRACTIONS.

Art. 87.—The term *decimal* signifies tenth. It is derived from the Latin word *decem*, which signifies ten. It is, therefore, applied to all fractions whose denominator is 10, or 1, with any number of ciphers. If a dollar be divided into ten parts, one of these parts, being worth ten cents, is one tenth of a dollar. If the dollar be divided into one hundred parts, one of these parts is the one hundredth part of a dollar. It is, nevertheless, a decimal fraction, because 100 is the product of 10's. The same may be said of a thousand, or ten thousand. A fraction is always known to be decimal, if its denominator be ten, a hundred, or a thousand. The denominator of a decimal fraction is not always expressed, but it can always be ascertained by the numerator. If it contains but one figure, the denominator is ten; if two, it is a hundred, etc. It is always one, with as many ciphers annexed as the numerator has places.

When the denominator is not expressed, the fraction is distinguished from a whole number by a period placed at the left of it. [The period is called the separatrix.] *Example:* .5, .50, is read five tenths, fifty hundredths, as though they were written $\frac{5}{10}$, $\frac{50}{100}$. If the numerator have not so many places as the denominator has ciphers, supply the defect by prefixing ciphers, thus: for $\frac{5}{100}$, write .05; $\frac{5}{1000}$, write .005. Ciphers placed at the right hand of a decimal do not affect its value,

QUESTIONS.—4. What does the term *decimal* signify? 5. From what derived? 6. To what applied? 7. How is a fraction known to be decimal? 8. Is the denominator always expressed? 9. How, then, can it be known?

as $\frac{5}{10}$, $\frac{50}{100}$ are the same in value; for while the addition of the cipher indicates a division into parts ten times smaller than the preceding, it makes the decimal express ten times as many parts. Thus, 5 tenths denotes 5 parts of a unit which is divided into 10 parts; and 50 hundredths denotes 50 parts of a unit which is divided into 100 parts. It is, therefore, plain, that the value is not altered, since 5 is half of 10, and 50 is half of 100.

The value of a decimal depends upon its distance from the unit's place. As whole numbers increase from the unit's place towards the left in a tenfold proportion, so decimals, in the same ratio, decrease from the unit's place towards the right hand; as will appear from the following

TABLE.

Millions.	Hundreds of thousands.	Tens of thousands.	Thousands.	Hundreds.	Tens.	Units.	Tenth parts.	Hundredth parts.	Thousandth parts.	Ten thousandth parts.	Hundred thousandth parts.	Millionth parts.
7	6	5	4	3	2	1	2	3	4	5	6	7

From the above Table it is evident that each figure, whether a whole number or decimal, takes its value from the unit's place. If it be in the first place on the right of units, it is tenths; if in the second, it is hundredths, etc. Consequently, every decimal will have for its denominator 1, with as many ciphers as the decimal is places distant from the unit's place; thus, 2 in the Table is $\frac{2}{10}$; 3 is $\frac{3}{100}$; 4 is $\frac{4}{1000}$, etc.

Art. 88.—The manner in which decimal fractions are produced, and the relation they bear to whole numbers, may be seen by the following formula:

QUESTIONS.—10. On what does the value of a decimal depend? 11. In what proportion do decimals decrease from the unit's place towards the right? 12. From what does each decimal figure take its value? 13. What is the value of the first figure on the right of units? 14. What effect have ciphers placed at the right hand of a decimal? 15. What effect at the left?

$1000 \div 10 = 100$

$100 \div 10 = 10$

$10 \div 10 = 1$

$1 \div 10 = \frac{1}{10} = .1$

$\frac{1}{10} \div 10 = \frac{1}{100} = .01$

$\frac{1}{100} \div 10 = \frac{1}{1000} = .001$

NOTE.—It will be noticed that the decimal point is the same as the straight line used Art. 40. Divide 1 by 10.

Operation. 1|0)|1=.1

Conceive the straight line employed to cut off figures from the right of the divisor and dividend to be contracted to a point, and you have the origin and use of the separatrix.

Thus it appears, that from any given place in whole numbers to any given place in decimals, is a regular descending series, formed by a uniform divisor. The right-hand place is the quotient of the left divided by 10. The first decimal place is the quotient of the unit's place divided by 10, and is called the tenth's place. The decimal point, therefore, occupies a position between the unit's place and its quotient. The second place is the quotient of the tenth's place divided by 10, or the unit's place divided by 100, and is called the hundredth's place. Thus decimal fractions, like whole numbers, have a local value, and are subject to the same law of increase from the right hand towards the left. As 1, in the place of tens, is equal to 10 in the unit's place, so 1 in the place of units is equal to 10 in the place of tenths. From this circumstance, we may know the value of those parts of the unit contained in the numerator, although the denominator be not expressed. This property of a decimal fraction also distinguishes it from a vulgar fraction, for there is no place on either side of the unit where the numerator of a vulgar fraction can be placed, which will give name to the fraction; its denominator must, therefore, always be expressed.

Although ciphers placed at the right hand of a decimal fraction do not affect its value, yet, placed at the left, they diminish it in a tenfold proportion, by removing the significant figure so much farther from the unit's place. Thus, .5 .05 .005 express different values, viz.—.5 is $\frac{5}{10}$, .05 is $\frac{05}{100}$, .005 is $\frac{005}{1000}$.

Write denominators to the following decimals: .5; .25; .026; .3245; .56783; .789024.

Write the following without their denominators.

1. Twenty-five hundredths. *Ans.* .25.
2. Four hundred and fifty-two thousandths. *Ans.* .452.
3. Five hundred and sixty ten thousandths.
4. Sixty-two hundred thousandths.
5. Forty-five millionths.
6. Eighty-seven billionths.

7. Ninety-eight trillionths.
8. Twenty-five, and four thousandths.

As whole numbers are written, units under units, tens under tens, from right to left, so decimals are written tenths under tenths, from left to right.

EXAMPLES.

1. Write 2 tenths; 3 hundredths; 4 thousandths; 6 ten thousandths.

.2
.03
.004
.0006

2. Write twenty-nine thousandths; three hundred and fourteen thousandths; five ten thousandths, and sixty-seven millionths.

.029
.314
.0005
.000067

3. Write five tenths; five hundredths; fifty thousandths, and forty-nine; one hundred thousandths, and sixteen thousandths.

4. Write forty-five and five tenths; six hundred and forty-five and four thousandths; twenty-nine and four thousandths; sixty-seven and forty-seven thousandths.

5. Write four hundred and fifty-three, and fifty-seven ten thousandths; five thousand and five hundredths; twenty-four and three millionths; thirty-six and eighty-two billionths.

ADDITION OF DECIMALS.

Art. 89.—1. Write one hundred and one tenth; twenty and two hundredths; five units and five thousandths, and add them together.

Operation.

100.1
20.02
5.005
125.125 *Ans.*

As whole numbers can only be added by writing them in their proper places and uniting those of the same name; so decimals, when written tenths in the place of tenths, hundredths in the place of hundredths, etc., are added by uniting those of the same name or denomination. The amount, both in decimals and whole numbers, takes its name from the lowest, or right-hand place of the numbers added: thus, 1 hundred, 2 tens and 5 units, when added, are read 125 units; and one tenth, 2 hundredths and five thousandths, when added, are read, 125 thousandths.

Decimal fractions may also be added and illustrated in the same manner as vulgar fractions.

2. Add two and five tenths; four and six hundredths; seven and three thousandths.

$$2.5=\frac{25}{10}, \text{ and } 4.06=\frac{406}{100}, \text{ and } 7.003=\frac{7003}{1000}.$$

Then $\frac{25}{10}\times 100=\frac{2500}{1000}$, and $\frac{406}{100}\times 10=\frac{4060}{1000}$.

OBS.—Multiplying the terms of a fraction by the same quantity does not alter its value. (See Art. 61.)

The fractions added:

$$\frac{2500}{1000}+\frac{4060}{1000}+\frac{7003}{1000}=\frac{13563}{1000}=13.563 \textit{ Ans.}$$

The same, added by decimal fractions:

$$\begin{array}{r@{\,=\,}l} 2.5 & 2.500 \\ 4.06 & 4.060 \\ 7.003 & 7.003 \\ \hline 13.563 & 13.563 \textit{ Ans.} \end{array}$$

From the foregoing it is evident, that decimal fractions are reduced to a common denominator by writing tenths in the place of tenths, and hundredths in the place of hundredths, and supposing those decimal places, which are deficient, to be supplied by ciphers.

Applying the decimal point to the amount, is equivalent to dividing it by its own denominator, which we have seen is the denominator of the lowest of the given decimals, or that decimal whose denominator is the largest. But the decimal places in the numerator of a decimal fraction, are equal to the number of ciphers in its denominator, the denominator being understood; therefore, addition of decimals may be performed by the following

---

QUESTIONS.—16. How is the first decimal place produced? 17. The second, third, &c.? 18. How are decimals to be added, written? 19. From what does the amount take its name? 20. Applying the decimal point is equal to what? 21. How are decimal fractions reduced to a common denominator?

RULE.

Place the numbers, tenths under tenths, hundredths under hundredths, etc.; or, so that the decimal points may stand directly under each other. Add as in whole numbers; observing to point off as many places for decimals in the amount, as will be equal to the greatest number of decimals in any of the given numbers.

EXAMPLES.

	(1)	(2)	(3)
			2345.6
Add	459.51	79.01	1987.51
	371.62	891.67	3456.712
	129.03	137.79	21098.6543
	1271.007	1239.812	16723.24567
	1090.215	2671.927	65431.002001
	3321.382		

4. Add thirty-five and four tenths; five hundred twenty-nine and seven millionths; sixty-nine, four hundred and sixty-three thousandths; two hundred, sixteen and two hundredths; seventy-seven, nine hundred and two tenths.

Ans. 1827.083007.

5. Add forty-nine and sixty-seven hundredths; six hundred seventy-nine, two hundred seventy-five thousandths; one thousand four hundred, fifty-five thousandths, nine hundred and ninety-nine millionths.

6. Add 249.39; 6712.9123; 6.3219; 2739.235; 5.671; 723.2674; 926.679; 72.601.

7. Add .7+9.2+.321+279.+4.67+349.2+3.956.

8. Purchased of one man 325.5 lbs. of beef; of another, 175.75; of another, 178.028; what was the amount?

9. I receive of A. $183.25; of B. $138.89; of C. $372.218; of D. $88.99; of E. $137.29; what is the amount of the whole?

10. Add $59.67; $158.355; $375.752; $167.375; $567.756.

SUBTRACTION OF DECIMALS.

Art. 90.—1. From three and two tenths, take one and five tenths.

Operation.

$$\begin{array}{r} 3.2 \\ 1.5 \\ \hline 1.7 \end{array} \textit{ Ans.}$$

Because five tenths cannot be taken from two tenths, we borrow 1 from the unit's place, which, reduced to tenths, equals 10 tenths; $\frac{10}{10}+\frac{2}{10}=\frac{12}{10}$, then $\frac{12}{10}-\frac{5}{10}=\frac{7}{10}=.7$. Lastly, 1 from 2, and 1 remains. Again, three and two tenths is the quotient of 32 divided by 10, (*see definition of a mixed number*, Art. 54;) therefore, $3.2=\frac{32}{10}$, and $1.5=\frac{15}{10}$; then $\frac{32}{10}-\frac{15}{10}=\frac{17}{10}=1.7$ *Ans.*, as before. Pointing off the remainder is dividing it by its own denominator. Hence the

RULE.

Write the numbers and point the result, as in Addition of Decimals, and subtract as in whole numbers.

	(2)	(3)	(4)	(5)
From	429.67	87.02	359.76	2029.5
Take	319.76	59.2	126.571	1718.279
	109.91			311.221

6. From two hundred and sixty-nine and three tenths, take fifty-seven and thirty-nine hundredths. *Ans.* 211.91.

7. Take twenty-four thousandths from nine hundredths. *Ans.* .066.

8. Take sixty-five millionths from five tenths.

9. From three hundred seventy-five thousand and three tenths, take two hundred forty-nine and thirty-nine one hundred thousandths. *Ans.* 374751.29961.

10. From 361.2 take 276.75.

11. From 456.35 take 27.356.

12. From 5678.0002 take 3980.96715.

QUESTIONS.—22. How are decimals to be subtracted, written? 23. How can five tenths be taken from two tenths? 24. What is done with the unit borrowed?

MULTIPLICATION OF DECIMALS.

Art. 91.—1. Multiply three and five tenths by five tenths.

Operation.

$$\begin{array}{r} 3.5 \\ .5 \\ \hline 1.75 \end{array} \text{ } \textit{Answer.}$$

The product of tenths into tenths is hundredths: $\frac{5}{10}\times\frac{5}{10}=\frac{25}{100}=.25$. The product of tenths into units is tenths: $3\times\frac{5}{10}=\frac{15}{10}=1.5$. The sum of the product, $.25+1.5=1.75$ *Ans.* Again, $3.5=\frac{35}{10}$ and $\frac{35}{10}\times\frac{5}{10}=\frac{175}{100}=1.75$ *Ans.*, as before. The value of the product is the quotient of its numerator divided by the denominator. Hence the figures cut off from the right of the numerator are equal to the ciphers in the denominator; but the ciphers in the denominator of the product, it will be perceived, are equal to the decimal places in both factors; therefore the multiplication of decimals may be performed by the following

RULE.

Multiply as in whole numbers, and point off as many places for decimals in the product as there are decimal places in both factors.

If there are not so many places, supply the defect by prefixing ciphers.

EXAMPLES.

2. Multiply five hundredths by five tenths.

Operation.

$$\begin{array}{r} .05 \\ .5 \\ \hline .025 \end{array} \text{ } \textit{Answer.}$$

The product of tenths into hundredths is thousandths. In this example, the tenth's place in the product is wanting; we must, therefore, supply it by prefixing a cipher.

QUESTIONS.—25. What is the product of tenths into units? 26. Of tenths into tenths? 27. What is the rule for the multiplication of fractions? 28. What is the value of the product?

3. Multiply 49.5 by 3.2.

```
   3.2
  ----
  99.0
 1485.
 ------
 158.40
```

4. Multiply 569.39 by 27.05.
5. Multiply 6.791 by 2.67.
6. Multiply 549.05 by 35.257.
7. Multiply six hundred and seventy-five by twenty-seven and three tenths.
8. Multiply sixty-seven thousand by three hundredths.
9. Multiply 34.56 by 1.3.
10. Multiply 674.49 by 37.16.
11. Multiply 5648 by 6.78.
12. Multiply 7864 by 467.
13. Multiply fifty-seven and three tenths by twenty-nine.
14. Multiply thirty-seven thousand by three hundredths.
15. Multiply fifty thousand and seven tenths by four hundredths.

DIVISION OF DECIMALS.

Art. 92.—1. Divide twenty-five hundredths by five tenths.

By Vulgar Fractions

Operation.

$\frac{5}{10})\frac{25}{100}(\frac{5}{10}=.5$ *Ans.*

When no remainder will arise from the division, the terms of the dividend may be divided by the corresponding terms of the divisor. (*See* Art. 78.) It will be seen that the decimal point implies a division of the numerator of the quotient by its own denominator.

By Decimal Fractions.

Operation.

```
.5).25
    .5 Ans.
```

Proof.

```
 .5
 .5
---
.25
```

We have seen that the decimal places in the product of any two factors are equal to the decimal places in both those factors. The divisor and quotient are factors of the dividend; therefore the decimal places in the quotient and divisor, taken together, must be equal to the decimal places in the dividend. Hence the

RULE.

Divide as in whole numbers, and point off so many places for ***decimals in*** *the quotient, that the decimal places in the quotient*

and divisor, taken together, shall equal the decimal places in the dividend; or, so many as the decimal places in the dividend exceed those of the divisor. If there are not so many, supply the deficiency by prefixing ciphers.

OBS. 1.–The above rule may be illustrated by reference to the operation of the preceding question by Vulgar Fractions, thus: the ciphers in the denominator of the divisor and quotient are equal to the ciphers in the denominator of the dividend; but the decimal places in the numerator of a decimal fraction are equal to the ciphers in its denominator; therefore the decimal places in the numerator of the quotient and divisor, taken together, must be equal to the decimal places in the numerator of the dividend.

2. Divide five tenths by twenty-five hundredths.

Operation.

$.5=\frac{5}{10}=\frac{50}{100}=.50$; then .25).50(2 *Answer.*
.50

OBS. 2.–Annexing a cipher to a decimal fraction multiplies the terms of the fraction by 10, and, therefore, does not alter the value. (See Art. 61) Whenever the decimal places in the divisor exceed those of the dividend, annex a cipher or ciphers to the dividend; this reduces it to the denomination of the divisor.

3. Divide three hundred and sixty-nine thousandths by nine.

Operation.

9).369
.041 *Answer.*

The necessity of prefixing a cipher to the quotient will be more readily seen by the following: $\frac{369}{1000}\div 9=\frac{41}{1000}$. If we remove the denominator of the quotient, and prefix the decimal point to the numerator, it will then be 41 hundredths, which is not its true value; but, by placing a cipher between the decimal point and the left-hand figure, the right-hand figure of the quotient will be made to occupy the thousandths' place, which will denominate the parts into which the unit is divided, or show their true value. Prefixing a cipher, therefore, divides the fraction, by multiplying its denominator.

4. Divide 36.72 by 18.

18)36.72(2.04
36
72
72

5. Divide 21.7 by 7.

7)21.7
3.1

QUESTIONS.—29. What is the rule for the division of decimals? 30. How is the quotient pointed off? 31. Illustrate the rule.

6. Divide 2.17 by 7. *Ans.* .31.
7. Divide .217 by .7. *Ans.* .31.
8. Divide .217 by 7. *Ans.* .031.
9. Divide one hundred and seventeen and nine tenths by nine tenths. *Ans.* 131.
10. Divide four hundred fifty-six and three hundred thirty-three thousandths by three hundredths.
11. If three hundred fifty pounds of beef cost twelve dollars twenty-five hundredths, what cost one pound? *Ans.* .035.
12. If 565.05 pounds cost 25.42725 dollars, what will one pound cost? *Ans.* .045.

Art. 93.—From the foregoing it appears that decimal fractions are like whole numbers in the following particulars:

1. The figures that compose them have an appropriate place to occupy, from which they take their value.
2. They take their name from the lowest right-hand place.
3. They increase in value from the right-hand place.
4. They can only be added by being first reduced to the lowest denomination.
5. They are reduced by writing them in their proper place.

They are unlike whole numbers in the following particulars:

1. They diminish in value from the unit's place.
2. A cipher, placed at the left hand, diminishes their value.
3. They may be written and treated as common fractions.

FEDERAL MONEY.

Art. 94.—Federal Money is the coin of the United States. Its denominations are eagles, dollars, dimes, cents, and mills.

From the above examples and illustrations in Decimal Fractions, we have seen that a decimal is the division of the unit into tens, and that from the unit's place towards the right hand it decreases in a tenfold proportion. If we examine the denominations of Federal Money, we shall find that all bear a decimal relation to the dollar, which is considered the unit. This will be seen by the following

TABLE.

10 Mills =1 Cent.
10 Cents =1 Dime.
10 Dimes =1 Dollar.
10 Dollars=1 Eagle.

OBS.—The eagle is a gold coin, the dollar and dime are silver coins, the cent is a copper coin. The mill is only imaginary, there being no coin of that denomination.

The dime being 1 tenth of a dollar, it occupies the first, or right-hand place from the dollar; thus, 0.1. The cent, being 1 tenth of a dime, and consequently 1 hundredth of a dollar, occupies the second place, or place of hundredths; thus, 0.01. The mill, being 1 tenth of a cent, and consequently 1 thousandth of a dollar, occupies the third place, or place of thousandths;

D. D. C. M.

thus, 0.001. Placing them together, 1 1 1 1. This may be read, one dollar, one dime, one cent, and one mill; or, one dollar, eleven cents, and one mill—as eleven cents is equal to one dime and one cent. The same may be said of eagles and dollars; thus, 25 dollars may be read, 2 eagles and 5 dollars, since 20 dollars are equal to 2 eagles. Write 4 eagles, 5 dol-

E. D. D. C. M.

lars, 8 dimes, 3 cents, 5 mills—4 5 8 3 5. This may be read, 4 eagles, 5 dollars, 8 dimes, 3 cents, and 5 mills; or, 45 dollars, 83 cents, and 5 mills. Hence, it is evident that the denominations in Federal Money are dollars and decimals of a dollar, and may be treated as Decimal Fractions. Federal Money is denoted by this character ($) placed before the figure.

ADDITION OF FEDERAL MONEY.

RULE.

Write the denominations, add and point the result as in Addition of Decimals.

EXAMPLES.

Art. 95.—1. If I buy a bushel of wheat for $2.25; a bushel of corn for $1.32; four yards of cloth for $14.285; how much do I pay for the whole?

QUESTIONS.—1. What is Federal Money? 2. What are its denominations?

OBS.—The scholar will do well to turn now to the rule for reducing a vulgar fraction to a decimal.

```
  2.25
  1.32
 14.285
$17.855 Ans.
```

2. Bought 8 yards of cloth for $16.25½; a pair of shoes for 87½ cents; a hat for $4.33; a whip for 42 cents; a knife for 37½ cents. How much did I pay for the whole?
Ans. $22.255.

3. Bought a cart for $17.62; a wagon, $62½; a plough, $7.48; 4 rakes, $1.26; 3 hoes, $2.15; a pitchfork, 87 cents. How much did the whole cost? *Ans.* $91.88.

4. Purchased a barrel of flour for $9.25; 4 pounds of tea, $2.08; 2 gallons of molasses, 64 cents; 3 pounds of raisins, 37½ cents; 9 pounds of sugar, $1.21½; 8 yards of calico, $2.23½. What is the amount of the whole?
Ans. $15.795.

5. Add forty dollars, sixty-seven cents and three mills; six hundred seventy-nine dollars, twenty-five cents and seven mills; one thousand and four dollars, five cents, and five mills; nine hundred, ninety-nine dollars, thirty-nine cents and nine mills.
Ans. $2723.384.

SUBTRACTION OF FEDERAL MONEY.

RULE.

Write the numbers, subtract and point the result as in Subtraction of Decimals.

EXAMPLES.

Art. 96.—1. A man bought 50 bushels of wheat for $125.50; sold it for $145.75. How much did he gain?
Ans. $20.25.

2. Bought 26 bushels of oats for $8.49; sold the same for $8.94. How much did I gain? *Ans.* $0.45.

3. Purchased a horse for $92; lost on the sale of him, $15.25. For how much did I sell him? *Ans.* $76.75.

4. Bought 2 barrels of flour for $22.50; but, it being damaged, I am willing to sell it at $4.25 less. What must I receive for it? *Ans.* $18.25.

5. Bought 8 yards of cloth for \$36; gave a \$50 bill. What must I receive in change? *Ans.* \$14.

6. Subtract 1 mill from \$333. *Ans.* \$332.999.

7. Subtract half of a cent from \$100,000.

8. Bought a wood lot for \$879; sold the same for \$1000.81. How much did I gain?

9. If a man's wages in a year amount to \$1434, and he spends \$928.45, how much does he save at the end of the year?

10. How much must be added to \$32.50 to make \$1000?

MULTIPLICATION OF FEDERAL MONEY.

RULE.

Write the numbers, and point the product as in Multiplication of Decimals.

EXAMPLES.

Art. 97.—1. How much will six pairs of shoes cost, at \$1.37½ a pair?

Operation.

```
       1.375
           6
Ans. $8.250
```

It will be seen that the operation is the same as in simple numbers. The product will always be in the lowest denomination of the given sum, until distinguished by points.

2. What will 9 sheep cost, at \$3.75 each? *Ans.* \$33.75.

3. How much must be paid for 45 bushels of corn, at \$1.37 per bushel?

4. What will 38 pounds of sugar cost, at 13½ cents per pound? *Ans.* \$5.13.

5. What will 3 dozen hats cost, at \$4.75 each? *Ans.* \$171.

6. What will 75 dozen eggs cost, at 15½ cents a dozen? *Ans.* \$11.625.

7. How much will a man spend in a year, if he spend 12½ cents a day?

8. What will 55 yards of broadcloth cost, at \$3.87½ per yard?

DIVISION OF FEDERAL MONEY.

RULE.

Write the numbers, and point the quotient as in Division of Decimals.

EXAMPLES.

Art. 98.—1. Bought 8 bushels of wheat for $17.92. How much was it per bushel? *Ans.* $2.24.

2. Bought 9 pounds of tea for $3.37½. What was it per pound?

Operation.

9)3.375

.375 *Ans.*

3. Bought tea to the amount of $3.37½, at 37½ cents per lb. What quantity did I buy? *Ans.* 9 lbs.

4. Bought 14½ bushels of corn for $21.75. How much was it per bushel? *Ans.* $1.50.

5. If a man pay $38.437½ for 20½ casks of lime, how much was it per cask? *Ans.* $1.87½.

6. Bought 6 yoke of oxen for $450. What was paid for each ox? *Ans.* $37.50.

SUPPLEMENT

TO DECIMAL FRACTIONS AND FEDERAL MONEY.

Art. 99.—1. Purchased 49.5 pounds of butter of A., at 12½ cents per pound; 37.51 pounds of B., at 18¾ cents per pound; 155.05 pounds of C., at 20 cents per pound. How many pounds did I buy, and what was the cost of the whole?

Ans. { $44.23+. 242.06 pounds.

2. When butter is worth 18 cents 4 mills per pound, how many pounds can be bought for $671.60?

Ans. 3650 pounds.

3. At 9 mills per yard, how many yards of tape can be bought for 45 dollars, 81 cents, 9 mills? *Ans.* 5091 yds.

4. If 5091 yards of tape be worth $45.819, what is 1 yard worth? *Ans.* 9 mills.

5. What will 629.21 feet of boards cost, at $20.18 per thousand?

6. What will 36 bushels 9 tenths of corn amount to, at 1 dollar 5 tenths per bushel? *Ans.* $55.35.

7. If corn be worth 3 and 5 tenths as much as potatoes, which are worth 25 hundredths of a dollar per bushel, and rye 5 tenths more than corn, and wheat 2 and 4 tenths more than rye, what is the value of wheat? *Ans.* $3.15.

8. Bought 4 cords of wood for $12.28; 15 pounds of beef for $1.25. How much do I pay for the whole, and how much more for the wood than for the beef?

9. Bought 28 bushels of potatoes, at 28 cents a bushel; 45½ bushels of apples at $1.12½ per bushel. How much did the whole cost, and how much more did the apples cost than the potatoes?

Ans. { $59.027½ whole cost. / $43.347½ difference. }

BILLS OF PARCELS.

Art. 100.—It is customary for the merchant, when he delivers goods, to give also a bill of the articles, and their prices, with the amount cast up. Such bills are called *Bills of Parcels.*

Concord, May 18, 1837.

Mr. JOHN WORTHY,

Bought of PETER TRUSTRUM,

5½ bushels of oats, at $0.63 per bushel	$3.465
12½ bushels of wheat, at $1.50 per bushel	18.750
7½ cords of wood, at $3.45 per cord	25.875
Received Payment,	$48.090

PETER TRUSTRUM.

Mr. BENJ. SAVAGE,

Bought of JOSEPH EASY,

12½ yards of broadcloth, at $3.87½ per yard	
5½ casks of nails, at $5.50 per cask	
112 pounds of iron, at 9½ cents per pound	
16 pounds of steel, at 18 cents per pound	
25 pounds of lead, at 9¼ cents per pound	
1 hogshead of sugar, (8½ cwt.) at $9.24 per cwt.	
2½ boxes of glass, at $7.50 per box	
	$191.810

COMPOUND NUMBERS.

Art. 101.—All preceding numbers have been simple; that is, numbers whose sum may be expressed by a certain number of units of one and the same kind, as 256. By reference to Notation, it will be seen, that this expression is 2 hundreds, 5 tens, and 6 units, which, instead of being written separately, are expressed as two hundred and fifty-six units, which are said to be of the same denomination. But if a man have 10 pounds and 2 shillings, he cannot add them so as to make 12 pounds, nor 12 shillings, but they must be expressed separately. So if a man travel 3 miles and 25 rods, the sum is neither 28 miles, nor 28 rods; but they must, in like manner, be expressed, the miles and the rods each by themselves. So of feet and inches, barrels, quarts, and pints. These are called different denominations. Hence, compound numbers are those which treat of quantities consisting of different denominations.

TABLES OF COMPOUND NUMBERS.

MONEY.

Art. 102.—*Federal Money.*

10 mills	make	........1 cent,	*ct.*
10 cents	"	........1 dime,	*d.*
10 dimes	"	........1 dollar,	*dol.*
10 dollars	"	........1 eagle,	*e.*

The above denominations of Federal Money are authorized by the laws of the United States; but, in the transaction of business in New England, we seldom hear any of them named but dollars and cents.

"A coin is a piece of money stamped, and having legal value. The coins of the United States are, three of gold; the eagle, half-eagle, and quarter-eagle; five of silver, the dollar, half-dollar, quarter-dollar, dime, and half-dime; and two of copper, the cent and half-cent. Of the small foreign coins current in the United States, the most common are the New England *four-pence-halfpenny*, or New York sixpence, worth 6¼ cents; and the New England *ninepence*, or New York shilling, worth 12½ cents. The value of the several denominations of English money is different in different places. A dollar is reckoned at 4*s.* 6*d.* in England; 5*s.* in Canada; 6*s.* in New England, Virginia, and Kentucky; 8*s.* in New York, Ohio, and North Carolina; 7*s.* 6*d.* in Pennsylvania, New Jersey, Delaware, and Maryland; and 4*s.* 8*d.* in South Carolina and Georgia."

Art. 103.—*English Money.*

4 farthings, *qrs.*,	make	........1 penny,	*d.*
12 pence	"	........1 shilling,	*s.*
20 shillings	"	........1 pound,	*l.* or £.

Art. 104.—*Time.*

Time is the measure of duration or existence.

60 seconds, *s.*, make	1 minute,	*m*
60 minutes "	1 hour,	*h.*
24 hours "	1 day,	*d.*
7 days "	1 week,	*w.*
365¼ days, or 365 *d.* 6 *h.*, or 52 *w.*, make	1 year.	*yr.*

"The year is commonly divided into 12 months, as in the following table, called calendar months:

Months.	Days.	M.	D.	M.	D.	M.	D.
1 January,	31	4 April,	30	7 July,	31	10 October,	31
2 February,	28	5 May,	31	8 August,	31	11 November,	30
3 March,	31	6 June,	30	9 September,	30	12 December,	31

Another day is added to February every fourth year, making 29 days in that month, and 366 in the year. Such years are called Bissextile, or Leap Year. To know whether any year is a common or leap year, divide it by 4; if nothing remain, it is leap year; but if 1, 2, or 3 remain, it is 1st, 2d, or 3d after leap year. The number of days in the several months may be called to mind by the following verse:

Thirty days hath September,
April, June, and November;
All the rest have thirty-one,
Excepting February alone,
Which hath twenty-eight, nay more,
Hath twenty-nine one year in four.

The true solar year consists of 365 days, 5*h.* 48*m.* 57*s.*, or nearly 365¼ days. A common year is 365 days, and one day is added in leap year to make up the loss of ¼ of a day in each of the preceding years. This method of reckoning was ordered by Julius Cæsar, 40 years before the birth of Christ, and is called the Julian Account, or Old Style. But, as the true year falls 11*m.* 3*s.* short of 365¼ days, the addition of a day every fourth year was too much by 44*m.* 12*s.* This amounted to one day in about 130 years. To correct this error, Pope Gregory, in 1582, ordered that 10 days should be struck out of the calendar, by calling the 5th of October the 15th; and to prevent its recurrence, he ordered that each succeeding century divisible by 4, as 16 hundred, 20 hundred, and 24 hundred, should be leap years, but that the centuries not divisible by 4, as 17 hundred, 18 hundred and 19 hundred, should be common years. This reckoning is called the Gregorian, or New Style. The New Style differs now 12 days from the Old Style."

Art. 105.—*Troy Weight.*

Troy weight is used in weighing gold, silver, platina, diamonds, and other precious stones. The standard Troy pound of the United States, is the weight of 22.794377 cubic inches of distilled water, weighed in air.

24 grains, *grs.*, make	1 pennyweight,	*pwt.*
20 pennyweights "	1 ounce,	*oz.*
12 ounces "	1 pound	*lb.*

Art. 106.—*Apothecaries' Weight.*

This weight is used only by apothecaries and physicians in compounding medicines.

20 grains, *grs.*	make	1 scruple,	℈.
3 scruples	"	1 dram,	ʒ.
8 drams	"	1 ounce	℥.
12 ounces	"	1 pound,	℔.

"The original standard of all our weights was a corn of wheat taken from the middle of the ear, and well dried. These were called grains, and 32 of them made one pennyweight. But it was afterwards thought sufficient to divide this same pennyweight into 24 equal parts, still calling the parts grains, and these are the basis of the table of Troy weight, by which are weighed gold, silver, and jewelry. Apothecaries' weight is the same as Troy weight, only having different divisions between grains and ounces. Apothecaries make use of this weight in compounding their medicines, but they buy and sell their drugs by Avoirdupois weight. In buying and selling coarse and drossy articles, it became customary to allow a greater weight than that used for small and precious articles, and this custom at length established Avoirdupois, or common weight, by which all articles are weighed, with the following exceptions. Avoirdupois weight is about one-sixth part more than Troy weight,—the former being 7000 grains, and the latter 5750 grains. In buying and selling by the hundred weight, 28 pounds have been called a quarter, 112 pounds a *cwt.*; but this practice of grossing, as it is called, is now pretty generally laid aside, and 25 pounds are considered a quarter, and 4 quarters (100 pounds) a hundred weight."

Art. 107.—*Avoirdupois, or Common Weight.*

This weight is used in weighing most kinds of merchandise, and all metals, except silver and gold.

16 drams	make	1 ounce,	*oz.*
16 ounces	"	1 pound,	*lb.*
28 pounds	"	1 quarter,	*qr.*
4 quarters,	"	1 hundred,	*cwt.*
20 hundred,	"	1 ton,	*ton.*

MEASURES.

Art. 108.—*Linear Measure.*

This measure is used in measuring distances, lengths, breadths, heighths, and depths.

3 barleycorns, *bar.*,	make	1 inch,	*in.*
12 inches	"	1 foot,	*ft.*
3 feet	"	1 yard,	*yd.*
5½ yards, or 16½ feet,	"	1 rod, or pole,	*rd.*
40 rods	"	1 furlong,	*fur.*

8 furlongs	make	1	mile,	*mi.*
3 miles	"	1	league,	*lea.*
69½ miles	"	1	degree,	*deg.*
360 degrees	"	1	circle of the earth.	
$7\frac{92}{100}$ inches	"	1	link,	*lk.*
25 links	"	1	rod,	*rd.*
4 rods, or 100 li.	"	1	chain,	*cha.*
80 chains	"	1	mile,	*mi.*

"The original standard of English Long Measure, was a barleycorn taken from the middle of the ear, and well dried. Three of these in length were called an inch, and then the others as in the table. Long Measure is employed for denoting the distance of places, and for measuring where length only is concerned. When measure is applied to surface where both length and breadth are concerned, it is called Square Measure. A square inch is a square measuring an inch on every side. The table of Square Measure is made from that of Long Measure, by multiplying the several numbers of the latter into themselves. Thus, 12 inches are a foot in length; a square foot, then, is a square which measures 1 foot, or 12 inches on every side, and contains 12×12=144 square inches. Three feet in length make a yard. A square yard is a square measuring 3 feet on each side; but such a square contains (see figure) nine (3×3=9) square feet; and when we say that a surface contains so many square feet, or square yards, we mean that the surface is equal to such a number of squares, measuring a foot, or a yard, on each side."

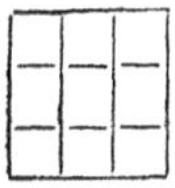

Art. 109.—*Cloth Measure.*

This measure is used for measuring cloth, and other goods which are sold by the yard or ell.

2¼ inches	make	1	nail,	*na.*
4 nails	"	1	quarter,	*qr.*
4 quarters	"	1	yard,	*yd.*
3 quarters	"	1	ell Flemish,	E. Fl.
5 quarters	"	1	ell English,	E. E.
6 quarters	"	1	ell French,	E. Fr.
37.2 inches	"	1	ell Scotch,	E. S.

Art. 110.—*Square Measure.*

This measure is used in measuring all kinds of surfaces, such as land, paving, flooring, plastering, and every thing which has length and breadth.

Gunter's chain, used by surveyors in measuring land, also in measuring distances, is 4 rods, or 66 feet in length, and is composed of 100 links.

144 inches	make	1	square foot,	*ft.*
9 feet	"	1	square yard,	*yd.*
30¼ yards	"	1	square rod,	*rd.*
272¼ feet	"	1	square rod,	*rd.*
40 rods	"	1	rood,	*ro.*
4 roods	"	1	acre,	*acr.*

640	acres	make	1 square mile,	*mi.*
10	square chains	"	1 acre,	*acr.*
6400	chains	"	1 square mile,	*mi.*

Art. 111.—*Solid, or Cubic Measure.*

Cubic Measure is used in measuring solids, or any thing that has the dimensions, length, breadth, and thickness.

1728 inches make	1 foot,	*ft.*
27 feet "	1 yard,	*yd.*
40 feet of round timber, or 50 feet of hewn timber,	1 ton,	*ton.*
128 feet	1 cord,	*cor.*

A perch of stone is equal to $24\frac{3}{4}$ cubic feet, used by masons in measuring stone walls. A square of earth is equal to 216 cubic feet.

Art. 112.—*Wine Measure.*

Wine Measure is used in measuring wine and all spirituous liquors, except porter, ale, and beer.

4	gills	make	1 pint,	*pt.*
2	pints	"	1 quart,	*qt.*
4	quarts	"	1 gallon,	*gal.*
$31\frac{1}{2}$	gallons	"	1 barrel,	*bar.*
63	gallons	"	1 hogshead,	*hhd.*
2	hogsheads	"	1 pipe,	*p.*
2	pipes	"	1 tun,	*t.*

Obs.—The wine gallon contains 231 cubic inches.

The hogshead is used only in estimating the contents of cisterns, wells or large bodies of water. The common gallon is 231 cubic inches. A gallon of milk, or malt liquor, is 282 cubic inches.

"Four pounds of Troy Weight of wheat, gathered from the middle of the ear, and well dried, were called one gallon; and this was the original standard of all English measures, both liquid and dry; and this was the same as the present wine gallon. But, in time, it became customary to use a larger measure in selling cheap liquors; and this custom at length established the Beer Measure, which bears about the same proportion to Wine Measure that Avoirdupois does to Troy Weight. The Dry Measure was also made larger than the Wine Measure, and was at length established at about a mean between Wine and Beer Measure. The statute bushel for measuring coal, ashes, and lime, in Vermont, contains 38 quarts, or 2553.6 cubic inches."

Remarks.—When measure is applied to magnitudes which have length, breadth, and thickness, it is called solid, or cubic measure. A solid inch is a body or block, having six sides, each of which is an inch square, and the number of inches in a solid foot is equal to the number of such blocks that would be required to make a pile a foot square and a foot high. Now it would require 144 blocks to cover a square foot one inch high. Hence to raise the pile 12 inches high would require 12 times 144=1728 blocks or inches. In like manner it would require 9 solid blocks, a foot each way, to cover a square yard to the height of one foot, and 3 times 9=27, to raise it 3 feet, or make one solid yard. A cord of wood is sometimes called 8 feet. In this case, four feet in length, four in breadth, and one in height=16 solid feet; or 8 feet in length, 4 in breadth, and 6 inches in height, a foot; that is, $\frac{1}{8}$ of a cord is called one foot, $\frac{2}{8}$, two feet, etc. In measuring lands, roads, etc., the distances are usually taken in chains and links. In ordinary business, feet and inches are the most common measures. By forty feet of round timber, in the table of solid measure, is meant so much round timber as will make forty feet after it is squared.

Art. 113.—*Ale, or Beer Measure.*

This measure is used in measuring porter, ale, beer, milk, and water.

2 pints, *pts.* make	1	quart,	*qt.*
4 quarts "	1	gallon,	*gal.*
36 gallons "	1	barrel,	*bar.*
54 gallons "	1	hogshead,	*hhd.*
2 hogsheads "	1	butt,	*butt.*
2 butts "	1	tun,	*tun.*

Obs.—The ale gallon contains 282 cubic or solid inches.

Art. 114.—*Dry Measure.*

This measure is used in measuring grain, fruit, seeds, roots, salt, sand, oysters, coal, etc.

2 pints, *pts.* make	1	quart,	*qt.*
4 quarts "	1	gallon,	*gal.*
8 quarts "	1	peck,	*pk.*
4 pecks "	1	bushel,	*bu.*
8 bushels "	1	quarter,	*qr.*
4 quarters "	1	chaldron,	*ch.*

Obs.—A gallon dry measure contains 268$\frac{4}{5}$ cubic inches.

A Winchester bushel is 18$\frac{1}{2}$ inches in diameter, 8 inches deep, and contains 2150$\frac{2}{5}$ cubic inches. The coal bushel must be 19$\frac{1}{2}$ inches in diameter; and 36 bushels, heaped up, make a London chaldron of coal, the weight of which is 3156 lbs. Avoirdupois.

Art. 115.—*Circular Measure.*

Circular Measure is used for measuring circles, latitude and longitude, and in computing the revolution of the earth and other planets round the sun.

60 seconds, ″ make	1	minute,	′
60 minutes "	1	degree,	°
30 degrees "	1	sign,	*s.*
12 signs, or 360°	1	circle.	

"Every circle, without regard to its size, is supposed to be divided into 360 equal parts, called degrees, and these again to be subdivided into minutes and seconds; so that the absolute quantity expressed by any of these denominations must always depend upon the size of the circle. A degree on the circumference of a circle, whose radius or semidiameter is 58 miles, is one mile; if the radius is 58 rods, the length of the degree is one rod. In this measure are reckoned latitude, longitude, and the planetary motions."

Art. 116.—*Miscellaneous.*

12	units, or things, make	1	dozen,	*doz.*
12	dozen "	1	gross,	*gro.*
12	gross, or 144 doz. "	1	great gross,	*G. gro*
20	things "	1	score.	
24	sheets of paper "	1	quire,	
20	quires "	1	ream.	
112	pounds "	1	quintal.	
6	points "	1	line.	
12	lines "	1	inch.	
4	inches "	1	hand.	
6	feet "	1	fathom.	

OBS.—Points and lines are applied to measuring pendulums of clocks; hands, to measuring horses; fathoms, to measuring depths of the sea.

Books.

When a sheet is folded into two leaves, it is called Folio.
When folded into 4 leaves, it is called Quarto.
When folded into 8 leaves, it is called Octavo.
When folded into 12 leaves, it is called Duodecimo, or 12*mo*
When folded into 18 leaves, it is called 18*mo*.
When folded into 24, it is called 24*mo*.
When folded into 48, it is called 48*mo*.

A gallon of train oil weighs	....	7½	pounds.
A stone of butcher's meat weighs	....	8	"
A gallon of molasses	"	11	"
A stone of iron	"	14	"
A firkin of butter	"	56	"
A fother of lead	"	19½	cwt.
A barrel of flour	"	196	pounds.
A " of pork or beef	"	200	"
A " of soap	"	256	"
A quintal of fish	"	112	"

REDUCTION.

Art. 117.—REDUCTION is the changing of numbers from one denomination to another, without altering their value.

1. In £6 7*s.* 8*d.* 2*qrs.*, how many farthings?

QUESTIONS.—1. What is Reduction? 2. By what numbers do you multiply in Example 1st? and why?

Operation.

£6 7*s.* 8*d.* 2*qrs.*
20
―――
127
12
―――
1532
4
―――
6130*qrs.* *Ans.*

It is plain, that if in one pound there are 20*s.*, in 6 pounds there are 6 times as many, or 120*s.*; and in £6 7*s.* there are 127*s.* Again, if in one shilling there are 12*d.*, in 127*s.* there are 127 times 12, which, with the 8*d.* added, equals 1532*d.* Lastly, it is evident that, if in 1 penny there are 4 farthings, in 1532*d.* there are 4 times the number of farthings, or 6130, the 2*qrs.* in the given question being added. Hence it appears, that in any given number of pounds there are 20 times as many shillings as pounds, 12 times as many pence as shillings, and 4 times as many farthings as pence. This process is called Reduction Descending, because higher denominations are brought into lower.

2. In 6130 farthings, how many pounds?

Operation.

4)6130
12)1532 2*qrs.*
2|0)12|7 8*d.*
£6 7*s.* 8*d.* 2*qrs.* *Ans.*

It will be seen that this question is the reverse of the former; and as 4 farthings are equal to 1 penny, so the number of pence in 6130*qrs.* will equal the number of times it contains 4, or 1532*d.* and 2*qrs.* over. Again, as 12*d.* are equal to 1 shilling, so in 1532*d.* the number of shillings will equal the number of times it contains 12, or 127*s.* and 8*d.* over. Lastly, since 20*s.* is equal to 1 pound, 127*s.* must equal 6 pounds and 7*s.*, because 20 is contained in 127*s.* 6 times, and 7*s.* over. It is always to be remembered, that the remainder is of the same denomination as the dividend, whatever may be the divisor. This latter process is called Reduction Ascending, because lower denominations are brought into higher. By these examples, it appears that Reduction Ascending and Descending mutually prove each other. From the preceding operations we derive the following Rules:

Art. 118.

Reduction Descending.

RULE.

Multiply the highest denomination given by that number which expresses how many it takes of the

Art. 119.

Reduction Ascending.

RULE.

Divide the denomination given by that number which expresses how many of that denomination it next lower to make one of the higher, observing to add the next lower denomination to the product. Multiply pounds by what makes a pound; shillings by what makes a shilling, and so on, until you have reduced it to the denomination sought in the question.

takes to make one of the next higher. Divide farthings by as many farthings as it takes to make a penny; pence by as many pence as it takes to make a shilling, and so on, until you have reduced it to the denomination required.

English Money.

EXAMPLES.

3. In 624 pounds, how many farthings?

5. If 6 shillings make a dollar, how many dollars in £780 18*s.*?

7. How many pence in £24 16*s.* 11*d.*?

9. How many guineas, 28*s.* each, in £49?

EXAMPLES.

4. In 599040 farthings, how many pounds?

6. In 2603 dollars, how many pounds and shillings?

8. How many pounds in 5963 pence?

10. How many pounds in 35 guineas?

Troy Weight.

11. In 18 lbs. 11 oz. 6 pwt. 18 grs., how many grains?

13. Bought jewelry weighing 1 lb. 10 oz. 15 pwt. 20 grains. Paid $0.04 per grain. What did I pay?

12. In 109122 grains, how many pounds?

14. If I pay $437.60 for jewelry, at the rate of $0.04 per grain, what was the weight?

Avoirdupois Weight.

15. In 5 tons, 16 cwt. 3 qrs. 12 lbs. 14 oz. 10 drs., how many drams?

17. What will be the cost of 40 tons of lead, at $0.12 per pound?

19. What must I pay for 20 tons, 18 cwt. of hay, at the rate of 5 cts. per pound?

16. Reduce 3350762 drs. to tons.

18. Bought lead to the amount of $10752, at $0.12 per pound. What was the weight?

20. How much hay can be bought for $2340.80, at 5 cts. per pound?

Apothecaries' Weight.

21. Reduce 6℔ 10℥ 7ʒ 2℈ 16 grs. to grains.

23. Reduce 21℔ 11℥ 3ʒ 1℈ 13 grs. to grains.

22. In 39836 grains, how many pounds?

24. In 126453 grains, how many pounds?

QUESTIONS.—3. What is the process called? 4. By what numbers do you divide in question 2? and why? 5. What is this process called? 6. What is the rule for Reduction Ascending? 7. For Reduction Descending?

Long Measure.

25. Reduce 640 degrees to feet.

To reduce degrees to statute miles, multiply first by 69, and for the $\frac{1}{2}$ in the multiplier take $\frac{1}{2}$ of the multiplicand. Thus:

```
2)640 degrees.
   69½
  ----
  5760
 3840
   320
 -----
 44480 miles.
```

26. In 234854400 feet, how many degrees?

To divide by $16\frac{1}{2}$, first reduce $16\frac{1}{2}$ to halves, and the dividend also to halves. Thus, $16\frac{1}{2}=33$ half feet; and $234854400=469708800$ half feet, which, divided by 33 gives 142338 rods.

27. How many barleycorns will reach across the Indian Ocean, it being 45 degrees?

28. How many degrees in 594475200 barleycorns?

29. A teamster, after travelling 20 miles, met a man who offered him as many 5 cent pieces for his load, as his larger wheel had turned round times since he commenced his journey, the wheel being 20 feet in circumference. How much did he get for his load?

30. If a teamster receive for his load $264.00, being paid 5 cents for each revolution of the larger wagon-wheel, the circumference being 20 feet, how far had he travelled?

31. How many barleycorns will reach round the globe, it being 360 degrees?

32. In 4755801600 barleycorns, how many degrees?

Land, or Square Measure.

33. If in a county there are 1200 square miles, how many square rods are there in this county?

34. How many square miles in 122880000 square rods?

35. How many square inches in 430 square acres?

36. In 2697235200 square inches, how many square acres?

Solid, or Cubic Measure.

37. How many solid inches in 10 cords of wood?

38. Reduce 2211840 solid inches to cords.

39. In 40 cords, how many cord feet?

40. In 320 cord feet, how many cords?

41. How many solid inches in a pile of timber containing 54 tons?

42. In 3732480 solid inches of round timber, how many tons?

Cloth Measure.

43. Reduce 6324 yds. 3 qrs. 3 na. to nails.

44. How many yards in 101199 nails?

45. What will 54 yds. 3 qrs. 0 na. of cloth cost, at $12\frac{1}{2}$ cts. an inch?

46. If I buy cloth to the amount of \$246.37$\frac{1}{2}$, at the rate of $12\frac{1}{2}$ cts. per inch, how many yards do I buy?

Wine Measure.

47. In 5 tuns, how many quarts?

48. In 5040 quarts, how many tuns?

49. What will be the cost of 8 hhds. of wine, at \$0.05 per pint?

50. How much wine can be bought for \$201.60, at 5 cents per pint?

Ale, or Beer Measure.

51. What will 40 hhds. of beer cost, at \$0.02 per pint?

52. How much beer can be bought for \$345.60, at 2 cents per pint?

Dry Measure.

53. Reduce 8 ch. 10 bush. 3 pks. 7 qts. 1 pt. to pints.

54. In 19135 pints, how many chaldrons?

55. How many pints in 30 bushels?

56. In 1920 pints, how many bushels?

Circular Measure.

57. In 6 s. 28 deg. 10 m. 8 sec., how many seconds?

58. Reduce 749408 seconds to signs.

59. Reduce 7 s. $21° 30' 29''$ to seconds.

60. In 833429 seconds, how many signs?

Time.

61. How many minutes in 10 yrs. 30 da. 18 hrs. 45 m.?

62. In 5303925 minutes, how many years?

63. How many seconds from May 1, 1830, to April 12, 1837, inclusive?

64. In 219326400 seconds, how many years?

65. Alexander the Great ascended the throne 332 years before the Christian era. How many minutes from that period to 1837?

66. In 1140807240 minutes, how many years?

SUPPLEMENT TO REDUCTION OF WHOLE NUMBERS.

Art. 120.—1. How many dollars in £480 18*s.*? *Ans.* $1603.

2. In £332 16*s.* 8*d.* how many ninepences? *Ans.* 8875 and 5*d.*

3. How many times will a regular clock strike in 400 years? *Ans.* 22776000.

4. A man sold four trees standing in the forest, measuring as follows: 6 tons, 5 tons, 4½ tons, 3 tons, at 12½ cts. per foot. What was their value? *Ans.* $92.50.

5. A man buys 20 tons of hay, at 45 cts. per cwt. He pays a man 62½ cts. a day for himself, and 50 cts. for his team. It takes him 6 days to cart it. How much does the hay cost him? *Ans.* $186.75.

6. How many plank one foot wide will it take to cover a bridge 60 rods in length, and 2 rods wide, and what will it cost, at 20 cts. per hundred feet? *Ans.* { 990 plank. $65.34 cost. }

7. If a boy be paid for wheeling a bushel of apples over said bridge 1 mill for every revolution of the wheel, which is 5 feet in circumference, how much does he receive? *Ans.* 19 cts. 8 m.

8. What will 2 tons of molasses amount to, at 6¼ cents a pint? *Ans.* $252.

9. A Vermonter, being in Boston, 80 miles from home, sold his dog and returned. At 6 o'clock the following night, the dog left his new master, and at 6 o'clock the next morning stood at the door of his former master. How many steps did he take of 8 inches each?

10. What will the plastering of a room, 15 feet square, the walls 9 feet high, amount to at 23 cents a square yard, deducting for 2 doors, 7 feet by 3, and 2 windows, 5 feet by 3? *Ans.* $17.71.

11. How much time would a person lose in 20 years, by lying in bed half an hour later every day than he ought? *Ans.* 152 days, 4 hours, 30 minutes.

12. How many cords of wood would a man draw in 6 weeks, drawing 4 loads a day, and 6½ cord feet at a load? *Ans.* 117 cords.

13. A merchant failing in trade, owes A. £15 7*s.* 9*d.*; B. £69 11*s.* 6*d.*; C. £102 16*s.* 11*d.*; D. £41 19*s.* 10*d.*;

E. £139 17*s.* 5*d.* His whole estate is valued at £300. How much does he owe more than he is worth?

Ans. £69 13*s.* 5*d.*

14. How many shingles will cover the roof of a factory 100 feet in length, one side of the roof being 40 feet in width, if 4 shingles in width cover 2 feet in length, and 2 courses make a foot? *Ans.* 32,000.

15. How many boxes, each 12 lbs., can be filled from a hogshead of sugar containing $7\frac{1}{2}$ cwt.? *Ans.* 70.

16. In 46 bales of cloth, each containing 24 pieces, and each piece 42 ells Flemish, how many yards?

Ans. 34,776 yards.

17. The sun travels through 6 signs of the zodiac in half a year. How many degrees, minutes, and seconds?

Ans. 180 deg., 10,800 m., 648,000 sec.

18. How many English crowns, at 6*s.* 8*d.* each, in 10 English guineas, at 28*s.* each, and 24 pistoles, at 22*s.* each?

Ans. 121*c.* 1*s.* 4*d.*

19. The forward wheels of a wagon are $14\frac{1}{2}$ feet in circumference, and the hind wheels $15\frac{3}{4}$ feet. How many more times will the forward wheels turn round than the hind wheels, in running from Concord to Boston, the distance being 60 miles?

Ans. 1734, rejecting fractions.

REDUCTION OF FRACTIONS.

Art. 121.—1. Reduce $\frac{1}{288}$ of a pound to the fraction of a penny.

We have seen that integers of a higher denomination are brought into integers of lower, by multiplication, (see Art. 117;) and also that fractions are multiplied either by multiplying the numerator, or dividing the denominator. (See Art. 67.) Pounds are reduced to shillings by multiplying by 20, and shillings to pence by multiplying by 12. *Therefore, to reduce $\frac{1}{288}$ of a pound to the fraction of a penny, multiply the fraction by 20 and 12, thus:*

$\frac{1}{288} \times 20 \times 12 = \frac{240}{288} = \frac{5}{6}$ *Ans.*

Or thus:

	1
6 ~~24~~ ~~288~~	~~20~~ 5
	~~12~~
6	$5 = \frac{5}{6}$ *Ans.*

As the numerator of the fraction is to be multiplied, place it with its multipliers on the right of the line, and 288, the divisor, on the left. Cross 288 and 12, and write 24 in the place of the larger number; 4 is contained in 20 five times, and in 24 six times. The answer, then, is 5 divided by 6, or $\frac{5}{6}$, in the lowest terms of the fraction.

Obs.—It will be seen, that the only difference between reducing 1 pound to pence, and $\frac{1}{288}$ of a pound, is, that in the latter case the multiplicand is a number divided; consequently, having multiplied the numerator as we should a whole number, we divide the product by the denominator. To divide the product is the same as to divide the multiplicand.

Art. 122.—To change fractions of a higher denomination into fractions of a lower denomination, we have the following

RULE.

Multiply the numerator of the fraction, or divide the denominator by all the denominations between it and that denomination into which it is to be reduced, including the lower denomination.

EXAMPLES.

1. Reduce $\frac{1}{320}$ of a pound to the fraction of a penny. *Ans.* $\frac{3}{4}$.
2. Reduce $\frac{1}{3360}$ of a pound to the fraction of a farthing. *Ans.* $\frac{2}{7}$.
3. What part of a pound is $\frac{1}{504}$ cwt.? *Ans.* $\frac{2}{9}$.
4. Reduce $\frac{1}{24}$ of a yard to the fraction of a nail. *Ans.* $\frac{2}{3}$.
5. Reduce $\frac{1}{1920}$ of a pound to the fraction of a farthing.

Art. 123.—It has been shown (see Art. 117) that Reduction Ascending is the reverse of Reduction Descending, and also, (Art. 68,) that a fraction is divided, either by dividing its numerator or multiplying its denominator. Farthings are reduced to pence by dividing by 4, and pence to shillings by dividing by 12; shillings to pounds by dividing by 20. Therefore—

Operation.

	1
~~1920~~	~~20~~
2	~~12~~
	~~4~~
2	$1 = \frac{1}{2}$ *Ans.*

To reduce $\frac{1}{2}$ *of a farthing to the fraction of a pound, divide the fraction by* 4, 12, *and* 20.

1 What part of a pound is $\frac{1}{2}$ of a farthing?

Operations.

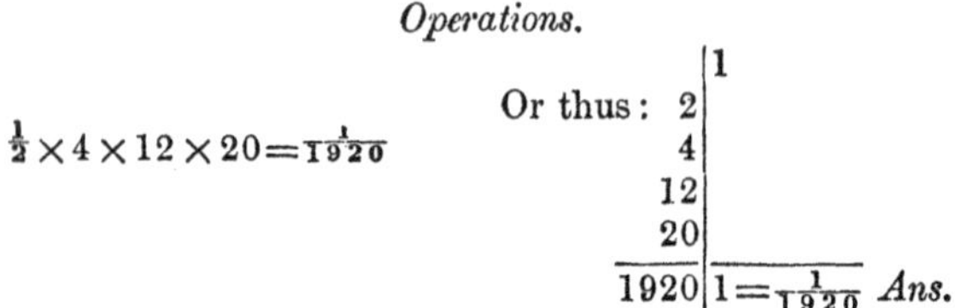

$\frac{1}{2} \times 4 \times 12 \times 20 = \frac{1}{1920}$

Or thus: 2, 4, 12, 20 | 1; 1920 | $1 = \frac{1}{1920}$ *Ans.*

Art. 124.—To change fractions of lower, into fractions of higher denominations, we have, then, this

RULE.

Multiply the denominator of the fraction by all the denominations between it and that into which it is to be reduced, and write the product under the numerator of the given fraction.

2. Reduce $\frac{5}{6}$ of a penny to the fraction of a pound.

Operations.

$\frac{5}{6} \times 12 \times 20 = \frac{5}{1440} = \frac{1}{288}$.

Or thus: 6, 12, 4 ~~20~~ | ~~5~~; 288 | $1 = \frac{1}{288}$ *Ans.*

As the denominator of the fraction is to be multiplied for a divisor, place it with its multipliers on the left of the line. Then, by cancelling, 5 on the right of the line, and 20 on the left, equal $\frac{1}{4}$: therefore cross 5 and 20, and write 4 on the left. Multiply the remaining numbers on the left together, for a divisor. We have, then, the answer in the lowest terms of the fraction, $\frac{1}{288}$.

3. Reduce $\frac{3}{4}$ of a penny to the fraction of a pound. *Ans.* £$\frac{1}{320}$.

4. Reduce $\frac{2}{7}$ of a farthing to the fraction of a pound. *Ans.* $\frac{1}{3360}$.

5. What part of a cwt. is $\frac{2}{9}$ of a pound? *Ans.* $\frac{1}{504}$.

6. Reduce $\frac{2}{3}$ of a nail to the fraction of a yard. *Ans.* $\frac{1}{24}$.

QUESTION.—1. What would be your multipliers in reducing the fraction of a pound to the fraction of a penny?

REDUCTION ASCENDING AND DESCENDING.

EXAMPLES.

DESCENDING.

Art. 125.—1. Reduce $\frac{1}{1440}$ of a pound to the fraction of a farthing.

3. Reduce $\frac{1}{480}$ of a pound to the fraction of a penny.

5. Reduce $\frac{5}{2688}$ of a guinea to the fraction of a penny.

7. Reduce $\frac{4}{7}$ of a guinea to the fraction of a pound.

	7	4
Guinea	1	28*s.* 4
5	*s.*20	1£
	5	4 $= \frac{4}{5}$ *Ans.*

9. Reduce $\frac{1}{360}$ of a guinea to the fraction of a penny.

11. Reduce $\frac{1}{84}$ of a shilling to the fraction of a farthing.

13. Reduce $\frac{5}{108}$ of a pound Troy to the fraction of an ounce.

15. Reduce $\frac{8}{6432}$ of a hhd. of wine to the fraction of a quart.

17. What fraction of a rod is $\frac{1}{7684}$ of an acre?

19. Reduce $\frac{1}{9216}$ of a mile to the fraction of a rod.

21. Reduce $\frac{1}{365040}$ of a degree to the fraction of a foot.

23. Reduce $\frac{1}{512}$ of a bushel to the fraction of a gill.

25. Reduce $\frac{1}{32256}$ of a tun to the fraction of a gill.

27. Reduce $\frac{1}{12}$ of $\frac{1}{3}$ of 4 pounds to the fraction of a penny.

29. $\frac{7}{60}$ of a pound is $\frac{1}{6}$ of what fraction of 7 guineas?

$\frac{7}{60}$ of a pound is $\frac{1}{14}$ of what fraction of 7 guineas?

$\frac{7}{60}$ of a pound is $\frac{1}{6}$ of $\frac{1}{14}$ of how many guineas?

ASCENDING.

Art. 126.—2. Reduce $\frac{2}{3}$ of a farthing to the fraction of a pound.

4. Reduce $\frac{3}{5}$ of a penny to the fraction of a pound.

6. Reduce $\frac{5}{8}$ of a penny to the fraction of a guinea.

8. What fraction of a guinea is $\frac{4}{5}$ of a pound?

	5	4
£	1	20*s.* 4
7	*s.*28	1 guinea.
	7	4 $= \frac{4}{7}$ *Ans.*

10. What part of a guinea is $\frac{14}{15}$ of a penny?

12. What part of a shilling is $\frac{4}{7}$ of a farthing?

14. What part of a pound Troy is $\frac{5}{9}$ of an ounce?

16. What fraction of a hhd. is $\frac{21}{67}$ of a quart?

18. What fraction of an acre is $\frac{40}{1921}$ of a rod?

20. Reduce $\frac{5}{144}$ of a rod to the fraction of a mile.

22. Reduce $\frac{2}{3}$ of a foot to the fraction of a degree.

24. Reduce $\frac{1}{2}$ of a gill to the fraction of a bushel.

26. Reduce $\frac{1}{4}$ of a gill to the fraction of a tun.

28. $\frac{80}{3}$ of a penny is $\frac{1}{12}$ of $\frac{1}{3}$ of how many pounds?

$\frac{80}{3}$ of 1 penny is $\frac{1}{12}$ of what fraction of 4 pounds?

$\frac{80}{3}$ of a penny is $\frac{1}{3}$ of what fraction of 4 pounds?

30. Reduce $\frac{1}{6}$ of $\frac{1}{14}$ of 7 guineas to the fraction of a pound.

31. Reduce $\frac{2}{4233600}$ of a week to the fraction of a second.

32. Reduce $\frac{2}{7}$ of a second to the fraction of a week.

33. Reduce $\frac{1}{43830}$ of a year to the fraction of an hour.

34. Reduce $\frac{1}{5}$ of an hour to the fraction of a year.

Comparison of Numbers and Quantities.

Art. 127.—We compare quantities and numbers, to ascertain what part the one is of the other.

Things compared must be of the same kind, or those properties compared must be alike. We do not compare rods with hours, nor minutes with days, but rods with rods, and minutes with minutes. (See Art. 166.)

The terms quotient, ratio, value of the fraction, each expresses what part the dividend is of the divisor.

What is the quotient of 6 divided by 3 ?
What is the ratio of 6 to 3 ?
What is the value of the fraction $\frac{6}{3}$?
What is one-third of 6 ?
How many times greater than 3 is 6 ?
What part of 3 is 6 ?
} *Ans.* 2.

The answer to each question is the same, and obtained by the same process. The quotient expresses what part the dividend is of the divisor. The quotient, with unity over it, expresses what part the divisor is of the dividend, $\frac{6}{3}=\frac{2}{1}$, and $\frac{3}{6}=\frac{1}{2}$. That is, 6 is twice as large as 3, and 3 is half as large as 6. The numerator of a fraction is the same part of the denominator, that the fraction is of unity.

1. What part of 4 is 3 ? *Ans.* $\frac{3}{4}$.

2. What part of 3 is 4 ? *Ans.* $\frac{4}{3}$.

3. What part of 1 dollar is 1 shilling ? *Ans.* $\frac{1}{6}$.

4. What part of £1 is 1*s.* ? *Ans.* $\frac{1}{20}$.

5. What part of 1 yard is 1 quarter ? *Ans.* $\frac{1}{4}$.

6. What part of 2 yards is 1 quarter ? *Ans.* $\frac{1}{8}$.

RULE

If the numbers consist of different denominations, reduce them to the same, and write that number which the question requires to be a part, as the numerator of a fraction, and the other as a denominator.

7. What part of 2 yards and 2 qrs. is 1 qr.?

yds. qrs.
2+2=10 qrs. *Ans.* $\frac{1}{10}$.

9. What part of 8 dollars is 2 dollars and 25 cts.? *Ans.* $\frac{225}{800}$.

11. When wood is worth 4 dollars a cord, what part of a cord can be purchased for 2 dollars? *Ans.* $\frac{2}{4}=\frac{1}{2}$.

13. What part of $5\frac{1}{2}$ is $3\frac{1}{2}$? *Ans.* $\frac{7}{11}$.

14. What part of $7\frac{3}{4}$ is $5\frac{2}{3}$? *Ans.* $\frac{68}{93}$.

8. What part of 2 yards and 2 qrs. is 1 yard and 1 qr.?

yds. qrs.
1+1= 5 qrs.
2+2=10 qrs. } *Ans.* $\frac{5}{10}=\frac{1}{2}$.

10. What part of 20 miles is 8 mi. 6 furlongs? *Ans.* $\frac{7}{16}$.

12. When corn is worth $\frac{3}{4}$ of a dollar a bushel, what part of a bushel can be bought for $\frac{7}{8}$ of a dollar?

$\frac{\frac{7}{8}}{\frac{3}{4}}=\frac{7}{8}\div\frac{3}{4}=\frac{7}{\cancel{8}\,2}\times\frac{\cancel{4}}{3}=\frac{7}{6}=1\frac{1}{6}$ *A.*

15. What part of 7*s.* 6*d.* is 4*s.* 2*d.*? *Ans.* $\frac{5}{9}$.

For a fuller illustration of the subject, see Art. 129.

To reduce Fractions to integers of lower denominations, and the reverse.

Art. 128.—1. What is the value of $\frac{2}{3}$ of a pound?

$\frac{2}{3}$ of a pound reduced to the fraction of a shilling is $\frac{2}{3}\times 20=\frac{40}{3}$ of a shilling—which, reduced to a mixed number, (Art. 59,) is $13\frac{1}{3}$*s.* The $\frac{1}{3}$ of a shilling reduced to the fraction of a penny, is $\frac{1}{3}\times 12=\frac{12}{3}=4$*d.* Hence, to reduce fractions of one denomination to integers of a lower, we have this

RULE.

Multiply the numerator of the given fraction by that number which expresses how many of the next lower denomination make 1 *of that denomination in which the fraction is given, and divide the product by the denominator of the fraction. If there be a remainder, proceed as before, until it is reduced to the lowest denomination. If there be still a remainder, place it at the right of the last answer.*

Art. 129.—2. Reduce 13*s.* 4*d.* to the fraction of a pound.

In £1 there are 240*d.* 1 penny then is $\frac{1}{240}$ of a pound, and 160*d.*, the number of pence in 13*s.* 4*d.*, is $\frac{160}{240}$ of a pound, or 160 times as much as 1 penny. Therefore, to reduce integers of lower, to fractions of a higher denomination, we have this

RULE.

Reduce the given numbers to the lowest denomination mentioned, for a numerator, and an integer of the denomination required to the same denomination for a denominator, and they will form the fraction required.

Obs.—Let question 2d be written upon the *blackboard*, in the following manner, and illustrated.

Teacher. Express by writing upon the *blackboard*, what part of a pound is $13\frac{1}{3}s$.

Scholar. $13\frac{1}{3}s.$ is $\frac{13\frac{1}{3}}{20}$ of a pound, which equals $\frac{\frac{40}{3}}{20}$.

T. What kind of fractions are those you have written?

S. Complex.

T. Remove the denominator of the numerator, and illustrate.

S. $\frac{\frac{40}{3}}{20}=\frac{40}{60}$. To multiply the denominator is the same as to divide the numerator, for, to multiply the divisor is the same as to divide the dividend.

T. Remove the denominator of the fraction, and illustrate.

S. $\frac{\frac{40}{3}}{20}=\frac{2}{3}$. To divide the numerator divides the fraction, for, to divide the dividend divides the quotient.

Let other questions be written and illustrated in a similar manner.

3. What is the value of $\frac{5}{8}$ of a shilling?

Operation.

$$\begin{array}{r|l} 2\ \cancel{8} & 5 \\ 1 & \cancel{12}\ 3 \\ \hline 2 & 15=7\frac{1}{2}\ \textit{Ans.} \end{array}$$

Or thus:

$$\frac{5}{2\,\cancel{8}}\times\frac{\cancel{12}\ 3}{1}=\frac{15}{2}=7\frac{1}{2}\ \textit{Ans.}$$

4. Reduce $7d.\ 2qrs.$ to the fraction of a shilling.

Operation.

Integer.	*d.*	*qrs.*
1*s.*	7	2
12	4	
12	6)30(5	
4	48(8 *Ans.*	
48		

Or, we may reduce the farthings to the fractions of a penny, and reduce the whole to an improper fraction. Thus:

$$7\frac{1}{2}=\frac{15}{2}.\qquad \begin{array}{r|l} 2 & \cancel{15}\ 5 \\ 4\ \cancel{12} & 1 \\ \hline 8 & 5=\frac{5}{8}\ \textit{Ans.} \end{array}$$

Obs.—It will be recollected that it was said (Art. 60) that a mixed number is the quotient of a division, whose divisor was the denominator of the fraction. Consequently, example 10 is the quotient of a division whose divisor was 11. We have therefore only to multiply quotient and divisor together, or to reduce a mixed number to an improper fraction, and we have the fraction of a cwt., which we divide by 20, to reduce it to the fraction of a ton. Thus:—

Questions.—1. Rule for reducing fractions to integers of lower denominations? 2. Rule for reducing integers of lower denominations to a fraction of a higher?

Operation.

	cwt.	qr.	lbs.	oz.	dr.
	16	1	12	11	$10\frac{2}{11}$
					11
11	180	0	0	0	0
20	9				
11	$9=\frac{9}{11}$ *Ans.*				

If there should be no fraction in the question, the lowest denomination may be reduced to a fraction of the higher. In example 12, the 36 minutes may be reduced to the fraction of an hour; thus, $\frac{36}{60}=\frac{3}{5}$. We then have $9\frac{3}{5}$ hours, a mixed number.

5. What is the value of $\frac{5}{8}$ of a mile?

6. Reduce 6 furlongs, 26 po. 11 ft. to the fraction of a mile.

7. What is the value of $\frac{1}{8}$ of a degree?

8. Reduce 8 mi. 5 fur. 20 po. to the fraction of a degree.

9. What is the value of $\frac{9}{11}$ of a ton?

10. Reduce 16 cwt. 1 qr. 12 lbs. 11 oz. $10\frac{2}{11}$ dr. to the fraction of a ton.

11. What is the value of $\frac{4}{5}$ of a month?

12. Reduce 3 w. 1 da. 9 hr. 36 m. to the fraction of a month.

13. What is the value of $\frac{5}{7}$ of a pound Troy?

14. Reduce 8 oz. 11 pwt. $10\frac{2}{7}$ grs. to the fraction of a pound.

15. What is the value of $\frac{5}{6}$ of an acre?

16. Reduce 3 roods, 13 rods, 90 feet, 108 in. to the fraction of an acre.

17. What is the value of $\frac{7}{8}$ of a yard of cloth?

18. Reduce 3 qrs. 2 na. to the fraction of a yard.

19. What is the value of $\frac{3}{4}$ of a dollar in shillings?

20. Reduce 4*s.* 6*d.* to the fraction of a dollar.

21. What is the value of $\frac{5}{9}$ of a ton?

22. Reduce 11 cwt. 0 qr. 12 lbs. 7 oz. $1\frac{7}{9}$ drs. to the fraction of a ton.

23. What is the value of $\frac{7}{9}$ of a hogshead?

24. Reduce 49 gals. to the fraction of a hogshead.

Reduction of Vulgar Fractions to Decimal.

Art. 130.—1. Reduce $\frac{1}{2}$ to a decimal fraction.

In this example, $\frac{1}{2}$ being a proper fraction, the numerator will not contain the denominator; but, by annexing a cipher, which reduces it to tenths, we can divide by the denominator. In 1 unit there are 10 tenths, but the example is one half of a unit; therefore, one half of 10 tenths, which is 5 tenths, will be the answer. Hence the

RULE.

I. *Annex a cipher, or ciphers, to the numerator, and divide by the denominator.*

11*

II. *If there be a remainder, a cipher, or ciphers, may be annexed, and the process of division carried on until there be no remainder, or the quotient is sufficiently exact.*

The decimal places in the quotient must be equal to the number of ciphers annexed to the numerator.

If, after division, the quotient does not contain so many, supply the deficiency by prefixing ciphers.

2. Reduce $\frac{3}{4}$ to a decimal fraction. *Ans.* .75.

3. Reduce $\frac{5}{8}$, $\frac{1}{8}$, and $\frac{3}{5}$, to decimal fractions.
Ans. .625, .125, .6.

4. Reduce $\frac{7}{8}$, $\frac{3}{8}$, $\frac{3}{4}$, $\frac{1}{4}$, and $\frac{1}{2}$, to decimal fractions.
Ans. .875, .375, .75, .25, .5.

5. Reduce $\frac{1}{2}$ of $\frac{2}{3}$ of $\frac{3}{4}$ to a decimal fraction.

Operation.

~~2~~	1
~~3~~	~~2~~
4	~~3~~
4	1.00
	.25 *Ans.*

6. Reduce $\frac{3}{4}$ of $\frac{5}{6}$ of $\frac{1}{8}$, divided by $\frac{3}{8}$ of $\frac{5}{9}$ of $\frac{1}{6}$, to a decimal fraction. *Ans.* 2.25.

7. Reduce $\frac{1}{25}$ to a decimal fraction. *Ans.* .04.

8. Reduce $\frac{162}{2563}$ to a decimal fraction. *Ans.* .632071+.

Art. 131.—To reduce a decimal fraction to a vulgar.

RULE.

Write down the given decimal, as a numerator, and for a denominator, write 1, *with as many ciphers annexed as there are figures in the numerator, and then reduce the fraction to its lowest terms.* (See Art. 61.)

1. Reduce .25 to a vulgar fraction.

Operation.

$$25 \left| \frac{25}{100} \right| \frac{1}{4} \text{ Ans.}$$

QUESTIONS.—1. Rule for reducing a vulgar fraction to a decimal? 2. How many decimal places must there be in the quotient? 3. If the quotient does not contain a sufficient number of figures, what is to be done?

2. Reduce .125 to a vulgar fraction. *Ans.* $\frac{1}{8}$.
3. Reduce .45 to a vulgar fraction. *Ans.* $\frac{9}{20}$.
4. Reduce .24 to a vulgar fraction. *Ans.* $\frac{6}{25}$.
5. Reduce .945 to a vulgar fraction. *Ans.* $\frac{189}{200}$

To reduce Integers of different denominations to a Decimal Fraction of a higher denomination, and the reverse.

Art. 132.—1. Reduce 4 pence 2 farthings to the decimal of a shilling.

Operation.

$$\begin{array}{r|l} 4 & 2.0 \\ 12 & 4.500 \\ \hline & .375 \end{array}$$

2 farthings is $\frac{2}{4}$ of a penny; then, by the rule for reducing vulgar fractions to decimal, we have $\frac{2}{4}$=.5, or $\frac{5}{10}$ of a penny. This, placed at the right of 4 pence, 4.5, and divided by 12, the number of pence in a shilling, or because 4 pence is $\frac{4}{12}$ of a shilling, gives .375 of a shilling. Hence the

RULE.

Place the numbers one above another, the highest denomination at the bottom. Divide the lowest denomination by that number which expresses how many of that it takes to make 1 of the next higher denomination, writing the quotient at the right of the next higher denomination; and so proceed until the whole shall be reduced to the required decimal.

Obs.—Integers of different denominations may be reduced to a decimal of a higher, by reducing the given numbers to the lowest denomination mentioned for a numerator, and the integer, to which the given numbers are to be reduced, to the same denomination for a denominator, and dividing the numerator by the denominator.

Art. 133.—2. Reduce .375 of a shilling to integers of lower denominations.

As this question is the reverse of the former, and as the decimal, .375, was obtained by dividing the integers, it is plain, that the integers may be obtained by multiplying the decimal by the same numbers.

Operation.

$$\begin{array}{r} .375 \\ 12 \\ \hline 4.500 \\ 4 \\ \hline 2.000 \end{array}$$

Hence the

RULE.

Multiply the given decimal by that number which expresses how many of the next lower denomination it takes to make one of that in which the decimal is given; observing to point off as many places in the product, for decimals, as there are figures in the given decimal; and so proceed through all the denominations; and the several numbers at the left of the decimal points will be the answer required.

Obs.—Pointing off the product is the same as dividing by the denominator of the decimal.

3. Reduce 8*s.* 4*d.* 2*qrs.* to the decimal of a pound.

Operation.

```
 4|2.0
12|4.500
2|0|8.375|0
   |.41875
```

4. Reduce .41875 to integers of lower denominations.

Operation.

```
      .41875
           2
  s. 8.375000
          12
  d. 4.50000
           4
qrs. 2.00000
```

5. Reduce 4 oz. 4 pwts. to the decimal of a pound.

6. What is the value of .35 of a pound Troy?

7. Reduce 4 oz. 8 drs. to the decimal of a pound.

8. What is the value of .28125 of a pound?

9. Reduce 2 cwt. 2 qrs. to the decimal of a ton.

10. What is the value of .125 of a ton?

11. Reduce 3 qrs. 2 na. to the decimal of a yard.

12. What is the value of .875 of a yard?

13. Reduce 20 h. 16 m. 48 sec. to the decimal of a day.

14. What is the value of .845 of a day?

15. Reduce 21*s.* 6*d.* to the decimal of a guinea.

16. What is the value of .875 of a guinea?

Art. 134.—To reduce shillings, pence, and farthings to the decimal of a pound, by inspection.

1. Reduce 7*s.* 8*d.* 2*qrs.* to the decimal of a pound?

One shilling is $\frac{1}{20}$ of a pound: therefore, two shillings is $\frac{2}{20}$, or $\frac{1}{10}$. Having, therefore, any number of shillings given, if we take one half the even number, they will be reduced at once to the decimal of a pound. If there is an odd shilling, it is the same as $\frac{5}{100}$ of a pound: $\frac{1}{20}$=.05. The farthing, which is $\frac{1}{960}$ of a pound, is made to occupy the 1000ths place. But $\frac{1}{960}$ is greater than $\frac{1}{1000}$ by $\frac{1}{24000}$; there will, therefore, be a loss of $\frac{1}{24000}$ on every farthing; but if we add one to the number, when they exceed 12 and do not exceed 36, and two when they exceed 36, the expression will be nearly so many 1000ths of a pound.

2. Reduce 4*s.* 6*d.* to the decimal of a pound.

Operation.

```
 .2   half of the even shillings.
 .024 farthings in 6d.
 .001 for excess of 12.
 ----
 .225 Ans.
```

If we call the farthings in 6*d.* $\frac{24}{1000}$, there will be a loss of $\frac{24}{24000}=\frac{1}{1000}$; if we add 1 to the 1000ths place, we have, in this instance, precisely the decimal required.

3. Reduce 7*s.* 8½*d.* to the decimal of a pound.

Operation.

.3 half of the even shillings.
.05 for the odd shilling.
.034 farthings in 8½*d.*
.001 for excess of 12.
.385 *Ans.*

4. Find by inspection the decimal expression of 18*s.* 3¼*d.*, and 17*s.* 8½*d.* *Ans.* £.914 and £.885.

5. Reduce to a decimal by inspection the following sums, and add them together, viz:—15*s.* 3*d.*; 8*s.* 11½*d.*; 10*s.* 6¼*d.* 1*s.* 8½*d.*; 2¾*d.* *Ans.* 1.832.

Decimals may be reduced back to shillings, pence, and farthings, by reversing the above process. Double the left-hand figure, or tenths, for the shillings; if the second figure be 5, or greater than 5, deduct 5 from it, and add 1 to the shillings. Then consider the second and third figures so many farthings; if they exceed 12, deduct 1; if they exceed 36, deduct 2.

6. Find by inspection the value of £.385.

7. Find by inspection the value of £.927. *Ans.* 18*s.* 6*d.* 2*qrs.*

8. Find by inspection the value of £.491, and £.984. *Ans.* 9*s.* 9*d.* 3*qrs.*; 19*s.* 8*d.* 1*qr.*

COMPOUND ADDITION.

Art. 135.—1. A boy bought a slate for 4*d.* and a book for 8*d.* What did both cost? *Ans.* 1*s.*

2. If I buy a book for 2*s.* 4*d.*, another for 4*s.* 8*d.*, what do I pay for both? *Ans.* 7*s.*

3. If a boy pay 4*s.* 8*d.* for a sled, and 5*s.* for a wagon, what does he pay for both? *Ans.* 9*s.* 8*d.*

4. How many shillings in 4*d.* 8*d.* 9*d.* 3*d.* 6*d.*? *Ans.* 2*s.* 6*d.*

5. How many pounds are 8*s.* 7*s.* 4*s.* 3*s.* 9*s.* 5*s.*?

Ans. £1 16*s.*

6. How many yards are 3 feet, 4 feet, 5 feet, 6 feet?

Ans. 6 yards.

7. Bought two pieces of cloth; one 10 yards, 1 foot; the other 12 yards, 2 feet. What was the length of both pieces?

Ans. 23 yards.

8. What is the amount of £1 4*s.* 2*d.* 3*qrs.*, and £10 8*s.* 3*d.* and 2*qrs.*? *Ans.* £11 12*s.* 6*d.* 1*qr.*

9. Add £4 5*s.* 6*d.* 3*qrs.*, and £5 17*s.* 7*d.* 2*qrs.*

£	*s.*	*d.*	*qrs.*	
4	5	6	3	
5	17	7	2	
10	3	2	1	*Ans.*

In adding the first column, or column of farthings, we find the amount to be 5 farthings. Now as 4 farthings are equal to 1 penny, we write the 1 farthing over, in the line of farthings, and carry the 1 penny to the column of pence. One to 7 is 8, and 6 are 14*d.* In 14*d.* there is 1 shilling and 2*d.* over, which we write in the column of pence, carrying the 1*s.* to the column of shillings. One added to 17 is 18, and 5 are 23. In 23*s.* there is £1 and 3*s.* over, which we write in the column of shillings, and carry 1 to the column of pounds.

Had the numbers to be added in the question been simple numbers, we should have had none to carry, because 5, in the column of units, is not equal to 1 in the column of tens. Again, had 14 been in the column of tens, we should have written 4 and carried 1. Lastly, had 23 been in the column of hundreds, we should have written 3, and carried 2, because 23 in the right-hand column, is equal to 2 in the left, and 3 remain; or, 23 hundred is equal to 2 thousand, and 3 hundreds remain.

Art. 136.—From the foregoing questions and illustration we derive the following definition and rules.

Compound Addition is the adding of numbers of different denominations. By different denominations is meant a different name—as shillings, pence, farthings, etc. Were the numbers given to be added, all pence, or all farthings, there would be but one denomination.

RULE.

I. *Write numbers of the same denomination directly under each other, pounds under pounds, shillings under shillings, etc.*

II. *Begin to add at the right-hand column, observing to carry one for as many in that column as make one in the next left-hand column.*

Proof—The same as in addition of simple numbers.

EXAMPLES.

1. Bought 4 books at the following prices, viz., £1 4*s.* 6*d.*; £2 3*s.* 8*d.*; £2 19*s.* 11*d.*; 2*s.* 3*d.* 2*qrs.* To what did they amount? *Ans.* £6 10*s.* 4*d.* 2*qrs.*

2. Add the following numbers: £46 26*s.* 7*d.* 3*qrs.*; £49 18*s.* 5*d.* 1*qr.*; £57 17*s.* 9*d.* 2*qrs.*; £102 19*s.* 10*d.* 1*qr.* *Ans.* £258 2*s.* 8*d.* 3*qrs.*

3. Add $286 12 cts. 6 m.; $347 20 cts. 4 m.; $119 18 cts. 7 m.; $542 93 cts. 9 m.; $314 89 cts. 1 m. *Ans.* $1610 34 cts. 7 m.

4. Add 45 lbs. 9 oz. 15 pwt. 18 grs.; 90 lbs. 6 oz. 16 pwt. 23 grs.; 30 lbs. 10 oz. 11 pwt. 6 grs.; 85 lbs. 11 oz. 13 pwt. 4 grs.; 91 lbs. 7 oz. 7 pwt. 23 grs.

AVOIRDUPOIS WEIGHT.

Ton	*cwt.*	*qr.*	*lbs.*	*oz.*	*dr.*
40	18	2	15	14	15
80	19	3	17	13	12
67	11	1	12	9	7
79	17	1	23	15	13
93	13	2	26	10	11

Ton	*cwt.*	*qr.*	*lbs.*	*oz.*
19	16	23	3	13
14	13	1	19	12
29	11	2	12	11
39	17	1	16	9
47	19	2	19	15

APOTHECARIES' WEIGHT.

℔	℥	ʒ	℈	*gr.*
29	10	7	2	17
25	11	6	1	13
37	8	4	1	9
71	5	3	2	11
89	6	5	1	10

℔	℥	ʒ	℈	*gr.*
99	11	3	1	19
102	9	7	2	5
81	4	6	1	13
120	7	3	2	18
341	6	1	3	16

CLOTH MEASURE.

Yd.	*qr.*	*na.*	*E. E.*	*qr.*	*na.*	*E. Fr.*	*qr.*	*na.*
125	3	2	176	4	3	69	3	2
300	1	3	57	3	1	76	5	3
159	2	2	102	1	2	57	4	1
260	1	1	69	2	2	89	2	1
191	3	3	267	4	1	97	1	2
357	4	2	179	2	3	88	3	3

WINE MEASURE.

Tun	*hhd.*	*gal.*	*qt.*	*pt.*	*Tun*	*hhd.*	*gal.*	*qt.*	*pt.*
66	2	57	2	1	86	3	39	3	2
79	3	60	3	0	121	2	51	1	2
88	1	49	1	1	67	1	19	0	3
91	2	38	2	1	76	1	29	1	2
72	3	20	1	0	167	2	38	2	1
61	1	39	1	1	120	0	31	1	2

ALE AND BEER MEASURE.

Hhd.	*gal.*	*qt.*	*pt.*	*Hhd.*	*gal.*	*qt.*	*pt.*
102	21	2	1	171	29	1	1
201	39	3	0	169	49	3	0
310	42	2	0	289	38	1	1
412	38	1	1	169	42	1	1
121	39	2	1	128	31	2	1

16. Add 49 bushels, 3 pecks, 4 quarts, 1 pint; 39 bu 1 pk. 5 qt. 1 pt.; 59 bu. 2 pk. 3 qt. 0 pt.; 40 bu. 7 pk. 2 qt. 1 pt.; 150 bu. 0 pk. 6 qt. 1 pt.; 69 bu. 1 pk. 2 qt. 0 pt.

17. Add 360 degrees, 15 miles, 6 furlongs, 16 poles, 13 feet, 6 inches; 240 deg. 19 m. 5 fur. 29 p. 11 ft. 5 in. 2 b.; 159 deg. 51 m. 7 fur. 32 p. 14 ft. 7 in. 2 b.; 201 deg. 63 m. 3 fur. 15 p. 12 ft. 9 in. 2 b.

18. Add 971 miles, 6 furlongs, 11 poles, 3 yards, 1 foot; 239 m. 5 fur. 9 p. 2 yd. 2 ft.; 269 m. 7 fur. 31 p. 1 yd. 2 ft.; 67 m. 6 fur. 9 p. 2 yd. 2 ft.; 691 m. 5 fur. 8 p. 2 yd. 2 ft.

19. Add 69 acres, 2 roods, 1 rod; 76 acr. 3 ro. 39 rd.; 88 acr. 1 ro. 32 rd.; 150 acr. 3 ro. 29 rd.

20. Add 150 years, 221 days, 13 hours, 31 minutes, 29 seconds; 230 yr. 300 d. 23 h. 49 m. 59 s.; 191 yr. 149 d. 21 h. 39 m. 23 s.; 359 yr. 75 d. 23 h. 59 m. 19 s.

COMPOUND SUBTRACTION.

Art. 137.—1. If a picture-book cost 4*d.* and a spelling-book 11*d.*, how much more does one cost than the other?

2. James bought a book for 9*d.* and sold it for 1*s.* How much did he gain by the bargain?

3. From 2*s.* 6*d.*, take 1*s.* 8*d.*

4. From 8*s.* 9*d.* 3*qrs.*, take 6*s.* 8*d.* 2*qrs.*

5. From 4 qts., take 3 pts.

6. If a bushel of rye be worth 7*s.* 6*d.*, and a bushel of corn 6*s.* 4*d.*, how much more is the rye worth than the corn?

7. How much more is wheat worth at 9*s.* 8*d.* per bushel, than corn at 7*s.* 6*d.* per bushel?

8. How much more is 2 bushels 2 pecks, than 1 bushel 3 pecks?

9. From £29 9*s.* 6*d.* 3*qrs.*, take £23 10*s.* 7*d.* 2*qrs.*

Operation.

£	*s.*	*d.*	*qrs.*
29	9	6	3
23	10	7	2
5	18	11	1

In this example, we write the difference between 2 and 3 farthings in the line of farthings, and proceed to the column of pence; we carry none, because we borrowed none—but 7*d.* from 6*d.* cannot be obtained; we therefore borrow as many pence as make a shilling, and say, 7 from 12—the remainder 5, we add to 6, in the upper line, and write 11 in the column of pence. We now carry 1 to the column of shillings, which is equal to the 12 pence we borrowed, and say, 11 from 9, which cannot be obtained; again we must borrow as many of the denomination we have to subtract as make one of the next higher, which is 20*s.*, and say, 11 from 20, and 9 remain, which added to 9 in the upper

QUESTIONS.—1. What does Compound Subtraction teach? 2. Rule? 3. If the number in the upper line be less than the one standing under it, how may you proceed? 4. Why do you carry 1 to the next left-hand column?

line, is 18, which must be written in the column of shillings. Lastly, the 20*s.* which we borrowed, we pay by carrying 1 to the line of pounds, which must be subtracted as in simple subtraction. Hence,

Art. 138.—COMPOUND SUBTRACTION teaches to find the difference between two compound sums, or quantities.

RULE.

I. *Write the less number under the greater, so that numbers of the same denomination may stand directly under each other.*

II. *Begin to subtract with the lowest denomination, and take the lower line from the one above it; proceed in this way with all the denominations.*

III. *Should the number in the upper line be less than the one standing under it, borrow as many units as make* 1 *in the next higher denomination.*

IV. *From the units borrowed, subtract the lower number, and to the difference add the upper number; write their sum under the figures subtracted, observing to carry* 1 *to the next left-hand column.*

Proof—The same as Simple Subtraction.

EXAMPLES.

TROY WEIGHT.

Lbs.	*oz.*	*pwt.*	*gr.*		*Lbs.*	*oz.*	*pwt.*	*gr.*
91	10	19	21		39	11	14	20
87	11	15	19		37	11	15	19

AVOIRDUPOIS WEIGHT.

Ton	*cwt.*	*qr.*	*lbs.*	*oz.*	*dr.*		*Ton*	*cwt.*	*qr.*	*lbs.*
122	11	3	22	13	12		39	11	14	20
110	13	2	23	14	13		37	9	15	19

APOTHECARIES' WEIGHT.

℔	℥	ʒ	℈	*gr.*		℔	℥	ʒ	℈	*gr.*
21	10	7	2	16		33	9	6	1	13
19	9	6	1	17		29	7	7	0	14

WINE MEASURE.

Gal.	qt.	pt.	gi.
77	2	29	3
59	3	49	2

Hhd.	gal.	qt.	pt.	gi.
600	3	59	2	0
459	3	47	3	1

ALE AND BEER MEASURE.

Hhd.	gal.	qt.	pt.
981	49	1	1
392	51	3	0

Hhd.	gal.	qt.	pt.
1000	37	3	0
999	49	2	1

11. From 31 tuns, 3 hhd. 15 gal., take 29 tuns, 2 hhd. 26 gal.
12. From 39 yds. 3 qr. 2 na., take 27 yds. 2 qr. 3 na.
13. From 127 E. E. 3 qr. 2 na., take 121 E. E. 4 qr. 3 na.
14. From 247 E. Fl. 0 qr. 2 na., take 159 E. Fl. 2 qr. 1 na.
15. From 671 E. Fl. 4 qr. 3 na., take 582 E. Fl. 5 qr. 2 na.
16. From 971 mi. 6 fur. 11 p. 3 yds. 1 ft., take 439 mi. 5 fur. 12 p. 4 yds. 2 ft.
17. From 69 acr. 2 ro. 31 rd., take 49 acr. 3 ro. 37 rd.
18. From 150 yrs. 221 d. 13 h. 31 m. 29 s., take 130 yrs. 129 d. 14 h. 39 m. 41 s.
19. From 260° 15 mi. 5 fur. 16 p. 13 ft. take 150° 17 m. 6 fur. 17 p. 12 ft.
20. From 240° 49′ 31″ take 159° 59′ 41″.
21. From 9s. 21° 31′ 42″ take 7s. 22° 36′ 37″.
22. A note dated Feb. 3d, 1826, was paid March 12th, 1837. How long was it from the first date until it was paid?

The time from one date to another may be found by subtracting the former date from the latter, observing to number the months in their order; thus, January, 1st month; February, 2d month, etc.

A. D., 1837	3d mo.	12th day.
A. D., 1826	2d mo.	3d day.
Ans. 11 ys.	1 mo.	9 days.

OBS.—The month, in casting interest, is reckoned 30 days.

23. What is the time from June 3d, 1835, to July 15th, 1837? *Ans.* 2 yrs. 1 m. 12 d.

24. The latitude of a certain place is 42° 50′ north; that of another place is 39° 37′; what is the difference of latitude?

Ans. 3° 13′.

25. What is the difference of longitude between 39° 40′, and 29° 49′ west? *Ans.* 9° 51′.

As every circle, whether greater or less, is divided into 360 equal parts, or degrees, consequently, the circle described by the revolution of the earth on its axis every 24 hours, contains 360 equal parts, or degrees; and as 360 degrees are described in 24 hours, it is plain that in 1 hour, $\frac{1}{24}$ of 360, or 15 degrees, would be described; and, also, if 15 degrees be described in 1 hour, or 60 minutes, it is equally plain that 1 degree would be described in $\frac{1}{15}$ of 60 minutes, or in 4 minutes, and 1 minute of a degree in 4 seconds. Hence,

Art. 139.—To reduce longitude to time, we have the following

RULE.

Multiply the longitude in degrees and minutes by 4, and we have the time in minutes and seconds.

EXAMPLE.

Reduce 14° 15′ to time.

```
14°  15′
      4
---------
57′   0″ Ans.
```

Art. 140.—To find the difference of time between any two places, having the time of one place given, and their difference of longitude.

RULE

Reduce the longitude to time, and add it to the given time, if the longitude of the place whose time is required be east of the place whose time is given; and subtract it, if the longitude be west.

Obs.—The reason of this is, because the farther we go east, the later is it in the day; and the farther west, the earlier in the day. That is, when it is 12 o'clock, at noon, in London, 15 degrees east of London it would be 1 o'clock, P. M.; and 15 degrees west of London it would be but 11 o'clock, A. M.

Questions.—5. How is the time from one date to another found? 6. How many degrees in a circle? 7. How many degrees does the earth describe in one hour, in its revolution round the sun? 8. In one minute? 9. In one second? 10. What is the rule for finding the difference of time between two places, the longitude being known?

26. When it is 12 o'clock in London, what is the hour in Boston, 70 degrees west longitude from London?
Ans. 7 o'clock 20 m.

27. When it is 12 o'clock in Boston, what is the time in London, lon. 70 deg. east? *Ans.* 4 o'clock 40 m.

COMPOUND ADDITION AND SUBTRACTION.

ADDITION.	SUBTRACTION.
Art. 141.—1. A man bought a horse for £32 10*s.* and a pair of oxen for £24 11*s.* 6½*d.* How much did both cost?	**Art. 142.**—2. If a pair of oxen and a horse cost £57 1*s.* 6*d.* 2*qrs.*, and the horse cost £32 10*s.*, what was the cost of the oxen?
3. If I purchase a farm for £1092 4*s.* 8*d.*, for how much must I sell it to gain £57 19*s.* 8*d.*?	4. If I sell a farm for £1150 4*s.* 4*d.*, and gain £57 19*s.* 8*d.* by the bargain, what did the farm cost?
5. A pipe of wine sprang a-leak, and 31 gal. 1 qt. 1 pt. were lost, and there remained 86 gal. 2 qts. 1 pt. How many gallons were there at first?	6. From a pipe of wine containing 118 gallons there leaked out 31 gal. 1 qt. 1 pt. How many remained?
7. There was a silver tankard which weighed 4 lbs. 3 oz., the lid weighed 6 oz. 4 pwt. 6 grs. How much did both weigh?	8. If the weight of a silver tankard and lid be 4 lbs. 9 oz. 4 pwt. 6 grs., and the lid alone weigh 6 oz. 4 pwt. 6 grs., what was the weight of the tankard?
9. A merchant bought a quantity of sugar; sold 9 cwt. 3 qrs. 25 lbs.; had 7 cwt. 2 qrs. 17 lbs. left. How much did he buy?	10. A merchant bought 17 cwt. 2 qrs. 14 lbs. of sugar; sold 9 cwt. 3 qrs. 25 lbs. How much had he left?
11. From a piece of cloth were sold 6 yds. 2 qrs., and there remained 32 yds. 2 qrs. 2 na. How much was there at first?	12. If from a piece of cloth containing 39 yds. 2 na., were sold 6 yds. 2 qrs., how many remained?
13. A farmer has two mowing fields; one contains [illegible]8 acres, 3 ro., the other 12 acres, 2 ro. 24 rds. How many acres in both?	14. A farmer has two mowing fields, containing 31 acres, 1 ro. 24 rds.; one contains 12 acres, 2 ro. 24 rds. How many acres does the other contain?
15. A note dated July 20, 1834, was paid in 9 mo. 46 d. At what time was it paid?	16. A note dated July 20, 1834, was paid June 6, 1835. How long was it on interest?

COMPOUND MULTIPLICATION AND DIVISION.

MULTIPLICATION.

Art. 143.—1. If a bushel of oats cost 3*s.* 6*d.*, how much will two bushels cost?

3. How much must be paid for 4 books, at 4*s.* 3*d.* each?

5. What will 5 yds. of cloth cost at 3*s.* 8*d.* per yard?

7. How much beer in 8 bottles, each containing 2 qts. 1 pt. 2 gi.?

9. If 1 gallon of molasses cost 2*s.* 8*d.* 3*qrs.*, what will 8 gallons cost?

	s.	*d.*	*qr.*
	2	8	3
			8
£1	1	10	0

If one gallon cost 2*s.* 8*d.* 3*qrs.*, it is evident that 8 gallons will cost 8 times as much. We begin to multiply with the lowest denomination, which is farthings. 8 times 3*qrs.* are 24*qrs.*=6*d.* 0*qr.* Place a cipher in the column of farthings, and proceed to multiply the column of pence, reserving the 6*d.* found in 24*qrs.* to be added; 8 times 8*d.* are 64*d.*, and 6*d.* added are 70*d.*=5*s.* 10*d.* Write the 10*d.* in the column of pence, and reserve the 5*s.* to be added to the column of shillings. Lastly, 8 times 2*s.* are 16*s.*, and 5*s.* added are 21*s.*=£1 1*s.* and 10*d.*, the answer.

DIVISION.

Art. 144.—2. If 2 bushels of oats cost 7*s.*, how much are they per bushel?

4. If 4 books cost 17*s.*, what will 1 book cost?

6. If 5 yds. of cloth cost 18*s.* 4*d.*, how much is it per yard?

8. If 8 bottles contain 22 qts., how much does 1 contain?

10. If 8 gallons of molasses cost £1 1*s.* 10*d.*, what cost 1 gallon?

	£.	*s.*	*d.*	*qr.*
8)	1	1	10	0
	0	2	8	3

If the price of 8 gallons be divided into 8 parts, it is evident that one of these parts would be the price of one gallon. Thus, 1 pound divided by 8, gives a cipher as a quotient figure, which must be written under the column of pounds, and 1 pound remains, which must be reduced to shillings: 1×20=20*s.*, and 1*s.* added=21*s.* Dividing 21*s.* by 8, we have 2 as a quotient figure, and 5*s.* remainder, which reduced to pence, 5×12=60, and 10*d.* added=70*d.*, which divided by 8=8*d.* and 6*d.* over; reduce 6*d.* to *qrs.*, 6×4=24*qrs.*, divided by 8=3*qrs.*; we have then, £0 2*s.* 8*d.* 3*qrs.*, the answer.

Art. 145.—Compound Multiplication is when the multiplicand consists of different denominations.

RULE.

Multiply the price by the quantity. When the quantity does not

Art. 146.—Compound Division is when the dividend consists of different denominations.

RULE.

Divide the price by the quantity. When the quantity does not exceed 12, *set down the price of one yard, one pound, or one gallon, etc., and the quantity under the lowest for a multiplier, observing to carry as in Compound Addition.*

EXAMPLES.

1. What will 9 yds. of cloth cost at 5*s*. 6*d*. per yard?

3. What will 8 cwt. of cheese cost, at £1 10*s*. 5*d*. per cwt.?

5. What will 24 yards of cloth cost at 15*s*. 3*d*. per yard?

£	s.	d.	
	15	3	
		6	
4	11	6	
		4	
18	6	0	*Ans.*

When the multiplier is greater than 12, *and is a composite number, multiply by its component parts, as in the last example,* $6 \times 4 = 24$.

7. What is the weight of 56 casks of raisins, each weighing 1 cwt. 2 qrs. 12 lbs.?

9. How much will 66 acres of land come to, at £7 9*s*. 6*d*. per acre?

11. What will 108 boxes of sugar weigh, each weighing 2 cwt. 1 qr. 14 lbs.?

13. What will 112 yds. of cloth cost, at £1 10*s*. 6*d*. per yard?

15. How much cloth will be required to make 121 coats, if, to make one, it requires 3 yds. 3 qrs. 3 na.?

17. What is the value of 336

exceed 12, *divide by the whole quantity at once. Divide the highest denomination by the divisor; then, multiply the remainder, if any, by that number which expresses how many of the next lower denomination make one of that, adding to the product the next lower denomination; divide this sum by the given divisor, and so proceed.*

EXAMPLES.

2. If 9 yards of cloth cost £2 9*s*. 6*d*., what will 1 yard cost?

4. If 8 cwt. of cheese cost £12 3*s*. 4*d*., what is it per cwt.?

6. If 24 yards of cloth cost £18 6*s*. what will one yard cost?

	£	s.	d.	
6)	18	6	0	
4)	3	1	0	
	0	15	3	*Ans.*

When the divisor is greater than 12, *and is a composite number, divide by its component parts.*

8. If 56 casks of raisins weigh 90 cwt., what is the weight of one cask?

10. If 66 acres of land cost £493 7*s*., what will 1 acre cost?

12. If 108 boxes of sugar weigh 256 cwt. 2 qrs., what is the weight of one box?

14. If 112 yards of cloth cost £170 16*s*., what is it per yard?

16. If it take 476 yds. 1 qr. 3 na. to make 121 coats, how much will it require to make one?

18. If 336 yards of cloth cost

QUESTIONS.—1. What is Compound Multiplication? 2. Compound Division? 3. Rule for Compound Multiplication? 4. Rule for Compound Division? 5. How do you proceed when the multiplier is a composite number? 6. When the divisor is a composite number? 7. How do you proceed when the multiplier is greater than 12, and not a composite number? 8. How, when the divisor is not a composite number?

yards of cloth, at 2*s.* 5*d* per yard ?

19. What will 153 barrels of sugar weigh, each barrel weighing 3 cwt. 1 qr. 14 lbs. ?

As 153 is not a composite number, we will first find the weight of 100, then of 50, then of 3 ; the several products added will be the answer. Thus :

```
cwt. qr.  lbs.        cwt. qr.  lbs
  3   1   14×3=  10    0   14
          10
 33   3    0×5=168     3    0
          10
337   2    0  weight of 100.
168   3    0  weight of  50.
 10   0   14  weight of   3.
516   1   14  weight of 153.
```

The above may be given in the form of a rule.

When the multiplier is not a composite number, and is hundreds, multiply by 10, *and this product by* 10, *which will give the product of* 100, *and this by the number of hundreds. For tens, multiply the product of ten by the number of tens ; for units, multiply the multiplicand. The several products added will be the answer sought.*

£40 12*s.*, what is the cost of one yard ?

20. If 153 barrels of sugar weigh 516 cwt. 1 qr. 14 lbs., what is the weight of one barrel ?

When the divisor is not a composite number, divide by the whole divisor at once, after the manner of Long Division.

Thus, taking the last question,

```
      cwt.  qr.  lbs.
153)516    1    14(3 cwt.
    459
     57
      4
153)229(1 qr.
    153
     76
     28
    622
   152
153)2142(14 lbs.
    153
     612
     612
```

The divisor, 153, is contained in 516 three times, and there is a remainder of 57. That is, if 153 barrels weigh 516 cwt., 1 barrel weighs 3 cwt., and 57 remainder, which are parts of a cwt. and must be reduced to quarters, the next lower denomination ; therefore, multiply 57 by 4, and to the product add the one quarter, and divide the amount, 229, by 153. We now have 1 qr. as the quotient, and a remainder of 76, which must be reduced to pounds by multiplying it by 28, and adding the 14 lbs. to the product. Again, dividing by 153, we have 14 lbs. as the quotient. The several quotients, 3 cwt. 1 qr. 14 lbs., are the answer.

21. How much will a man spend in a year, if he spend 4*d.* a day ?

22. If a man in one year spend £6 1*s.* 8*d.* how much will he spend in a day ?

23. What is the value of 1900 yards of linen at 5*s.* 8½*d.* per yard ?

24. If 1900 yards of linen cost £542 5*s.* 10*d.*, what will one yard cost ?

25. What will 68 hogsheads of lime cost, at £1 1*s.* 6¼*d.* per hhd.?

26. If 68 hhds. of lime cost £73 3*s.* 5*d.*, what is it per hhd.?

27. What is the value of 26 yards of silk, at 9*s.* 6½*d.* per yard?

28. If 26 yards of silk be worth £12 8*s.* 1*d.*, what will 1 yard be worth?

29. How many gallons of beer in 14 bottles, each containing 3 qts. 1 pt. 1 gill?

30. If 14 bottles of beer contain 12 gal. 2 qts. 1 pt. 2 gills, how much does 1 bottle contain?

31. What is the weight of 6 chests of tea, each weighing 3 cwt. 2 qrs. 9 lbs.?

32. If 6 chests of tea weigh 21 cwt. 1 qr. 26 lbs., what is the weight of 1 chest?

33. How many acres in 9 fields, each containing 12 acr. 2 ro. 25 rds.?

34. If in 9 fields there are 113 acr. 3 ro. 25 rds., how many in 1 field?

35. How many cords of wood in 37 piles, each containing 8 cords, 28 ft.?

36. If 37 piles of wood contain 304 cords and 12 ft., how much in 1 pile?

37. How much will 17 casks of nails weigh, each weighing 1 cwt. 2 qrs. 16 lbs. 3 oz.?

38. If 17 casks of nails weigh 27 cwt. 3 qrs. 23 lbs. 3 oz., what will 1 cask weigh?

39. How many bushels of apples can be put into 125 barrels, each containing 3 bu. 1 pk. 5 qts.?

40. If 125 barrels contain 425 bush. 3 pks. 1 qt., how much does 1 contain?

41. If a ship sail 2 deg. 30 m. 10 sec. in 1 day, how far will she sail in 30 days?

42. If a ship sail 75 deg. 5 m. in 30 days, how far will she sail in 1 day?

43. If 3 men build 14 rds. 8 feet of wall in one day, how much will they build in 26 days?

44. If 3 men build 376 rods, 10 feet, in 26 days, how much do they build in 1 day?

45. If 1 yard of cloth cost £2 2*s.* 6*d.*, what will 229 yards cost?

46. Bought 229 yards of cloth for £486 12*s.* 6*d.*; what did it cost per yard?

47. The moon passes through 1 sign of the zodiac in 2 days, 6 h. 38 m. 34 sec. In what time does it pass through 12 signs?

48. If the moon pass through 12 signs of the zodiac in 27 days, 7 h. 42 m. 48 sec., in what time does it pass through 1 sign?

49. If one gallon of molasses cost 4*s.* 2*d.* 2*qrs.*, what will 1000 gallons cost?

50. If 1000 gallons of molasses cost £210 8*s.* 4*d.*, what is it per gallon?

51. If 1 pound of tea cost 8*s.* 5*d.* 2*qrs.*, what will 108 lbs. cost?

52. If 108 pounds of tea cost £45 13*s.* 6*d.*, what will 1 pound cost?

53. If 1 quintal of fish cost 23*s.* 9*d.*, what will 345 quintals cost?

54. If 345 quintals of fish cost £409 13*s.* 9*d.*, what was it per quintal?

Art. 147.—A concise view of the application of the principle employed in the addition of simple numbers to compound numbers and fractions:—

Add 2 tens and 2 units.

Operation.

tens. units.
2 + 2
10
22 *Ans.*

Add £2 and 2 shillings.

Operation.

£ s.
2 + 2
20
42 *Ans.*

Add $\frac{1}{2}$ and $\frac{1}{4}$.

Operation.

$\frac{1}{2}\times 2=\frac{2}{4}$, and $\frac{2}{4}+\frac{1}{4}=\frac{3}{4}$ *Ans.*

In each case it appears that the numbers to be added must be reduced to the lowest denomination mentioned; and also, that they are reduced by multiplying the higher by that number which expresses how many of the lower make one of the higher.

SUPPLEMENT TO COMPOUND NUMBERS.

Art. 148.—1. What is the weight of two pieces of gold, one weighing 1 lb. 0 oz. 6 pwt. 4 grs.; the other, 2 lbs. 3 oz. 8 pwt. 16 grs.? *Ans.* 3 lbs. 3 oz. 14 pwt. 20 grs.

2. A man has one wedge of gold, weighing 25 lbs. 3 oz. 12 pwt., and another weighing 1 lb. 11 oz. 12 pwt. 7 grs. What is the weight of the two?

Ans. 27 lbs. 3 oz. 4 pwt. 7 grs.

3. A silversmith had a quantity of silver, weighing 21 lbs. 9 oz. After refining it by melting, it weighed 15 lbs. 10 oz. 11 pwt. 19 grs. What was lost by melting?

Ans. 5 lbs. 10 oz. 8 pwt. 5 grs.

4. What is the sum and difference of 3 lbs. 10 oz., and 2 lbs. 11 oz. 7 pwt. 4 grs.?

Ans. { Sum: 6 lbs. 9 oz. 7 pwt. 4 grs.
Difference: 10 oz. 12 pwt. 20 grs.

5. What will 13 lbs. of coffee cost, at 1*s.* 2*d.* 3*qrs.* per pound? *Ans.* 15*s.* 11*d.* 3*qrs.*

6. What will 47 yards of cloth cost, at 17*s.* 9*d.* per yard? *Ans.* £41 14*s.* 3*d.*

7. How much will 10 cwt. of lead cost, at 7*d.* per lb.? *Ans.* £32 13*s.* 4*d.*

8. What is the value of 7 cwt. of sugar, at 4¾*d.* per lb.? *Ans.* £15 10*s.* 4*d.*

9. What is the weight of 4 hogsheads of sugar, each weighing 7 cwt. 3 qrs. 19 lbs.? *Ans.* 31 cwt. 2 qrs. 20 lbs.

10. Bought 1½ doz. large silver spoons, each weighing 3 oz. 5 pwt.; two doz. teaspoons, each weighing 15 pwt. 14 grs.; three silver cups, each weighing 9 oz. 7 pwt.; two silver tankards, each 21 oz. 15 pwt.; 6 silver porringers, each 11 oz. 18 pwt. What is the weight of the whole? *Ans.* 18 lbs. 4 oz. 3 pwt.

11. If 6 ells cost £5 17*s.* 6*d.*, what will 1 ell cost? *Ans.* 19*s.* 7*d.*

12. What must a man spend per month, to spend £17 14*s.* 6*d.* in a year? *Ans.* £1 9*s.* 6½*d.*

13. If 8 cwt. of cocoa cost £15 17*s.* 4*d.*, what is it per pound? *Ans.* 4*d.* 1*qr.*

14. If 132 bushels of oats cost £20 12*s.* 6*d.*, what is the cost of one bushel? *Ans.* 3*s.* 1*d.* 2*qrs.*

15. If 147 bushels of corn cost £47 12*s.* 5*d.*, what does it cost per bushel? *Ans.* 6*s.* 5*d.* 3*qrs.*

16. If 1 acre produce 152½ bushels of oats, how much will a square rod produce? *Ans.* 3 pks. 6 qts. 1 pt.

17. How much wood in 11 piles, each containing 120 cords, 7 cord-feet, 11 solid feet? *Ans.* 1330 cords, 4 cord-feet, 9 solid ft.

18. Multiply £86 12*s.* 6*d.* by 9; divide the product by 6; multiply the quotient by 4; divide the product by 12, and give the result?

19. If it take a printer 297 h. 59 m. 24 sec. to set 108 pages, how long will it take him to set 1 page? *Ans.* 2 h. 45 m. 33 sec.

20. A person wishes to draw a pipe of wine into bottles, containing a quart, 2 quarts, 1½ pint, ½ pint, of each an equal number. How many must he have?

Art. 149.—When it is required to find how many times several quantities, each an equal number, may be had in a given quantity—

RULE.

Reduce the given quantity to the lowest denomination mentioned, for a dividend, and each of the other quantities to the same denomination, and add them together for a divisor. The quotient will be the answer.

1 quart = 8 gills.
2 quarts = 16 gills.
$1\frac{1}{2}$ pints = 6 gills.
$\frac{1}{2}$ pint = 2 gills.
32 gills.

The 1 pipe reduced to gills equals 4032 gills, and $4032 \div 32 = 126$ bottles, the *Answer.*

21. How many bushel, half bushel, and peck baskets, of each an equal number, will it take to contain 175 bushels?

22. There are four fields, one containing 10 acres, 2 roods; another 9 acres; another 11 acres, 3 roods; another 6 acres, 3 roods, 30 rods. How many shares, of 65 rods each?

Ans. 94.

23. A man left $1043.28 to be divided as follows: His wife is to have two thirds; of the other third, his sister is to have one-half, and the remainder is to be divided between two nephews and nine distant relatives. To one nephew he gives 3 shares, to the other 2, to each of the relatives 1 share each. What is the share of each respectively?

Ans.
Wife,	$695.52
Sister,	173.88
Nephew,	37.26
do.	24.84
Relatives,	12.42

24. What will 156 acres of land cost, at £5 6*s.* 9*d.* 2*qrs.* per acre? *Ans.* £832 19*s.* 6*d.*

25. A. values a piece of land at $120, B. at $100, C. at $110. What is the average judgment?

A. 1 $120
B. 1 100
C. 1 110
3) 330
$110 *Ans.*

The average is found by dividing the sum of the several judgments by the number of judges.

26. Two gentlemen wished to exchange vehicles. One was a gig, the other was a wagon; but not being able to agree as

QUESTION.—9. What is the rule for the 20th example?

to the conditions, referred the matter to A., B. and C., who decided as follows: A. said the owner of the gig should pay the owner of the wagon \$20, and B. said he should pay \$15; but C. said the owner of the wagon should pay the owner of the gig \$10. What is the average judgment?

Ans. The owner of the gig must pay $\$8\frac{1}{3}$.

In cases where the judgment of the referees is part on one side of the question, and part on the other, subtract one side from the other, and divide the remainder by the number of referees, and the quotient will be the answer.

27. A. and B. wish to exchange watches, but cannot agree upon the difference. They refer the matter to C., D. and E., and agree to abide by their decision. C. gives his opinion that A. should give B. \$3. D. thinks the difference in B.'s favor is \$4; but E. takes the other side of justice, and says B. should pay A. at least \$1. What is the average judgment?

Ans. A. must pay \$2.

SUPPLEMENT TO FRACTIONS.

Art. 150.—*A factor may be transferred from the numerator of a fraction to its denominator, and from its denominator to its numerator, without altering the value of the fraction.* Thus, $\frac{4}{8}=\frac{2\times2}{8}=\frac{2}{\frac{8}{2}}=\frac{2}{4}=\frac{1}{2}$. Let it be required to separate the terms of $\frac{4}{15}$ into their prime factors, and transfer the factor 2 from numerator to denominator, and the factor 5 from denominator to numerator. Thus, $\frac{4}{15}=\frac{2\times2}{5\times3}=\frac{\frac{2}{5\times3}}{2}=\frac{\frac{2}{5}}{\frac{3}{2}}=\frac{2}{5}\div\frac{3}{2}=\frac{2}{5}\times\frac{2}{3}=\frac{4}{15}$.

On what principle is the factor 2 transferred from numerator to denominator, and the factor 5 from denominator to numerator?

On what page is the principle first illustrated?

Repeat the language.

Separate the terms of the following fractions into their prime factors, transfer as above, and illustrate.

$\frac{6}{12}$, $\frac{14}{28}$, $\frac{9}{27}$, $\frac{10}{48}$, $\frac{12}{25}$.

EXAMPLES.

1. What is the sum of $\frac{3}{4}$ and $\frac{7}{8}$? *Ans.* $\frac{13}{8}=1\frac{5}{8}$.

13

2. If a man receive $\frac{2}{3}$ of a dollar for 1 day's work, $\frac{3}{4}$ for another, and $\frac{5}{6}$ for another, how much does he receive for the 3 days' work? *Ans.* \$$2\frac{1}{4}$.

3. If I receive $\frac{5}{7}$ of a cord of bark from one man, and $\frac{6}{9}$ from another, what part of a cord do I receive from both? *Ans.* $1\frac{8}{21}$.

4. Add $5\frac{1}{2}$, $6\frac{1}{4}$, $2\frac{1}{8}$, $1\frac{3}{4}$. *Ans.* $15\frac{5}{8}$.

5. From $\frac{7}{8}$ take $\frac{1}{2}$. *Ans.* $\frac{3}{8}$.

6. From $12\frac{1}{2}$ take $10\frac{7}{8}$. *Ans.* $1\frac{5}{8}$.

7. Bought a piece of land containing $49\frac{14}{15}$ acres. Sold $21\frac{3}{5}$ acres. How many were there left? *Ans.* $28\frac{1}{3}$.

8. From $\frac{1}{4}$ of a day take $\frac{9}{40}$ of a minute. *Ans.* 5 h. $59\frac{31}{40}$ m.

9. From $\frac{7}{8}$ of $\frac{5}{6}$ of $\frac{24}{35}$ take $\frac{1}{6}$ of $\frac{3}{4}$ of $\frac{4}{5}$. *Ans.* $\frac{2}{5}$.

10. A man had 3 bags of money, containing in all 450 lbs.; in the first bag he had $230\frac{17}{27}$ lbs., in the second $100\frac{11}{81}$. How many were there in the third? *Ans.* $119\frac{19}{81}$.

11. From $\frac{11}{12}$ of a pound take $\frac{1}{3}$ of $\frac{3}{4}$ of $\frac{3}{7}$ of $\frac{7}{8}$ of $\frac{8}{8}$ of 8 shillings. *Ans.* £$\frac{211}{240}$.

12. If a bushel of corn cost $\frac{5}{6}$ of a dollar, what will $\frac{1}{2}$ of a bushel cost? *Ans.* \$$\frac{5}{12}$.

13. If $\frac{1}{2}$ of a bushel of corn cost $\frac{5}{12}$ of a dollar, how much must be paid for 1 bushel? *Ans.* \$$\frac{5}{6}$.

14. If $\frac{1}{19}$ of a hogshead cost £$\frac{35}{38}$, what will be the cost of $\frac{69}{70}$ of a hhd.? *Ans.* £$17\frac{1}{4}$.

15. Multiply 30 by $4\frac{1}{2}$ of $\frac{2}{3}$ of $\frac{3}{4}$ of $\frac{4}{5}$, and divide the product by $\frac{1}{2}$ of 8 of $\frac{7}{8}$ of $2\frac{1}{4}$. *Ans.* $6\frac{6}{7}$.

16. Divide $\frac{1}{8}$ of $\frac{2}{3}$ by $\frac{2}{3}$ of $\frac{3}{4}$. *Ans.* $\frac{1}{6}$.

17. Reduce $\frac{8}{9}$ of a pound, avoirdupois, to the fraction of a cwt. *Ans.* $\frac{1}{126}$.

18. Multiply $\frac{7}{8}$ of a day, reduced to minutes, by the fraction $\frac{1}{4}$. *Ans.* 315 m.

19. Reduce 3*s.* 6*d.* to the fraction of a pound. *Ans.* £$\frac{7}{40}$.

20. Reduce $\frac{1}{126}$ cwt. to the fraction of a pound, avoirdupois. *Ans.* $\frac{8}{9}$.

21. What is the value of $\frac{15}{16}$ of a dollar? *Ans.* 5*s.* $7\frac{1}{2}$*d.*

22. What is the value of $\frac{12}{17}$ of a Julian year? *Ans.* 257 d. 19 h. 45 m. $52\frac{16}{17}$ sec.

23. What is the value of $\frac{9}{14}$ of a guinea? *Ans.* 18*s.*

24. Reduce 4 cwt. 2 qrs. 12 lbs. 14 oz. $12\frac{4}{13}$ drs. to the fraction of a ton. *Ans.* $\frac{3}{13}$.

25. Reduce 16 h. 36 m. $55\frac{5}{13}$ s. to the fraction of a day. *Ans.* $\frac{9}{13}$.

26. Reduce 2 qrs. 9 lbs. 10 oz. $7\frac{2}{29}$ drs. to the fraction of a cwt. *Ans.* $\frac{17}{29}$.

27. If 100 oranges cost $10\frac{1}{3}s.$, how many hundred may be bought for $105\frac{1}{12}s.$? *Ans.* $10\frac{1}{4}$.

28. How much will $\frac{1}{4}$ cwt. cost, at $15\frac{3}{4}s.$ per cwt.? *Ans.* 3*s.* $11\frac{1}{4}d.$

29. If $\frac{3}{4}$ of a yard cost 18*d.*, what will 1 yard cost? *Ans.* 2*s*

30. If $\frac{3}{7}$ of $\frac{4}{5}$ of $\frac{7}{8}$ of a ship be worth $\frac{7}{9}$ of $\frac{6}{7}$ of $\frac{9}{16}$ of the cargo, valued at $36,000, what is the value of the ship? *Ans.* $45,000.

CIRCULATING DECIMALS.

Art. 151.—CIRCULATING, or RECURRING DECIMALS, are those that consist of a repetition of a number of digits, as .646464, etc., .4127127127, etc.; in fact, every decimal that is not finite is a circulating decimal, or is such, that if continued far enough, the same figures will again recur; but it is only those of which the periods of circulation consist of a few figures, that generally receive the definition of Circulating Decimals.

When the circulation consists of the same digit repeated, it is called a Simple Circulate, and is distinguished by a point placed over it; thus, .111, etc. $= .\dot{1}$; .333 $= .\dot{3}$, etc. When the period of circulation consists of more than one digit, it is called a Compound Circulate, and is distinguished by a point over the first and last repeating figure; thus, .234234234, etc. $= .\dot{2}3\dot{4}$. A Mixed Circulate is that which has other figures in it that are not repeated, as .7848484, etc., and these are represented thus, $.7\dot{8}\dot{4}$.

As all operations, as multiplication, division, etc., of these numbers may be performed by the same rules which are given for common decimals, and as but few cases occur in which those rules are not to be preferred, some rules only will be given for the reduction of circulating decimals to vulgar fractions, leaving the student to apply the rules.

If, however, he should wish to pursue the subject farther, he can find his curiosity amply gratified by consulting the following authors: Brown, Cunn, Malcolm, Emerson, Donn, and particularly Henry Clarke; also, Dr. Wallis, all of whom have treated at some length the theory of Circulating Decimals.

REDUCTION OF CIRCULATING DECIMALS.

Art. 152.—To reduce a simple, or compound circulate, to its equivalent fraction.

RULE.

Take the given decimal, considered as a whole number, for the numerator; and as many 9's as there are places in the circulate, for the denominator. When there are any integral figures in the circulate, as many ciphers must be annexed to the numerator as the highest place in the repetend is distant from the decimal point.

EXAMPLES.

1. The circulate—$.\dot{6}=\frac{6}{9}=\frac{2}{3}$.
2. - - $.\dot{3}\dot{6}=\frac{36}{99}=\frac{4}{11}$.
3. - - $.\dot{0}\dot{9}=\frac{9}{99}=\frac{1}{11}$.
4. - - $2.\dot{0}6\dot{3}=2\frac{63}{999}=2\frac{7}{11}$.
5. - - $\dot{1}.6\dot{2}=1\frac{23}{37}$.

Art. 153.—To reduce a mixed circulate to its equivalent fraction.

RULE

Subtract the finite part of the expression, considered as a whole number, from the whole mixed repetend, taken in the same manner for the numerator; and to as many 9's as there are repeating places in the circulate, annex as many ciphers as there are finite decimal places for a denominator; thus—

1. The circulate—$.13\dot{8}=\frac{138-13}{900}=\frac{125}{900}=\frac{5}{36}$.
2. Again—$2.4\dot{1}\dot{8}=2\frac{418-24}{990}=2\frac{414}{990}=2\frac{23}{55}$.

Obs.—This rule, as it is not of great practical utility, may be passed over until the review.

RATIO AND PROPORTION.

Art. 154.—We arrive at a knowledge of particular quantities by comparing them with other quantities, which are either equal to, or greater or less than those which are the objects of inquiry. We may inquire, how much greater one quantity is than another; or how many times the one contains the other. The answer to either of these inquiries is termed a ratio of the two quantities. One is called *arithmetical,* and the other *geometrical* ratio.

Art. 155.—*Arithmetical ratio is the difference between two quantities.* Thus, the arithmetical ratio of 6 to 3 is 3. It is sometimes expressed by two points placed horizontally between the two quantities; thus, $6 \cdot\cdot 3=3$, which is the same as $6-3=3$.

Art. 156.—*Geometrical ratio is the quotient arising from dividing one quantity by another.* Thus, the ratio of 6 to 3 is $\frac{6}{3}$, or 2. Geometrical ratio is expressed by two points placed one over the other, between the two quantities compared; thus $6 : 3=2$. If the ratio is not specified, it is always understood to be geometrical.

The two quantities taken together, are called a couplet.

The number which is compared, being placed first, is called the antecedent, and that with which it is compared, the consequent.

Of these three, the antecedent, the consequent, and the ratio, any two being given, the other may be found.

EXAMPLES.

1. If the antecedent be 16, and the consequent 4, what is the ratio? *Ans.* 4.

2. If the antecedent be 18, and the ratio 3, what is the consequent? *Ans.* 6.

Art. 157.—*Inverse, or reciprocal ratio, is the ratio of the reciprocals of two quantities.*

OBS.—The reciprocal of any quantity is a unit divided by that quantity. Thus, the reciprocal of 4 is $\frac{1}{4}$, the reciprocal of 3 is $\frac{1}{3}$.

The reciprocal ratio of 6 to 3 is $\frac{1}{6}$ to $\frac{1}{3}$; that is, $\frac{1}{6}\div\frac{1}{3}$, which

QUESTIONS.—1. What is Ratio? 2. What is arithmetical ratio? 3. What is geometrical? 4. What is compound ratio?

is equal to $\frac{3}{6}$. Hence, a reciprocal ratio is expressed by inverting the terms of the couplet. The reciprocal ratio of antecedent to consequent, is the direct ratio of consequent to antecedent. The direct ratio of 6 to 3 is $\frac{6}{3}=2$. The reciprocal ratio of 6 to 3 is $\frac{3}{6}=\frac{1}{2}$.

Art. 158.—*Compound ratio is the ratio of the products of the corresponding terms of two or more simple ratios.*

Thus, the ratio of 9 : 3 is 3.
And the ratio of 6 : 2 is 3.
54 : 6 = 9.

OBS. 1.—A compound ratio is not different in its nature from a simple ratio. The term *compound* is used merely to denote the origin of the ratio.

Art. 159.—*In a series of ratios, if the consequent of each preceding couplet is the antecedent of the following one, the ratio of the first antecedent to the last consequent is equal to that which is compounded of all the intervening ratios.* Thus,

12 : 6
6 : 18
18 : 3
3 : 4

Art. 160.—*If we multiply all the antecedents together, and all the consequents together, it will be found that the ratio of the products of the antecedents to the product of the consequents, is equal to the ratio of* 12, *the first antecedent, to* 4, *the last consequent, which is* $\frac{12}{4}=3$.

OBS. 2.—Rejecting all the antecedents but the first, and all the consequents but the last, is cancelling equal factors from dividends and divisors. (See Art. 42.)

Art. 161.—*If, in the several couplets, the ratios are equal, the sum of all the antecedents has the same ratio to the sum of all the consequents, which any one of the antecedents has to its consequent.* Thus,

12 : 6 = 2
10 : 5 = 2
8 : 4 = 2
6 : 3 = 2
36 : 18 = 2

OBS. 3.—It will be observed, in this example, that the terms of the ratio are not used as factors. The ratio is, therefore, not a compound ratio.

It has already been shown (Art. 44) that to multiply the dividend with a given divisor, is the same as to multiply the quotient, and to multiply the divisor with a given dividend, is the same as to divide the quotient. In Fractions the same principle was recognised, with this difference only in the mode of expression; we substituted *numerator* for dividend, and *denominator* for divisor. We shall now substitute *antecedent* for numerator or dividend, *consequent* for denominator or divisor, and *ratio* for value of the fraction.

Art. 162.—*To multiply the antecedent, or to divide the consequent, is the same as to multiply the ratio.* (See Art. 44.)

Thus the ratio of	12 : 6 is 2
Multiply the antecedent by 2, the ratio of	24 : 6 is 4
Divide the consequent by 2, the ratio of	12 : 3 is 4

Art. 163.—*To divide the antecedent, or to multiply the consequent, is the same as to divide the ratio.*

Thus, the ratio of	8 : 4 is 2
Divide the antecedent by 2, the ratio of	4 : 4 is 1
Multiply the consequent by 2, the ratio of	8 : 8 is 1

Art. 164.—*To multiply both antecedent and consequent by the same quantity, does not affect the ratio.*

Thus, the ratio of	6 : 3 = 2	the same ratio.
Multiply both terms by 3,	18 : 9 = 2	
Divide both terms by 3,	2 : 1 = 2	

The ratio of two *fractions*, which have a common denominator, is the ratio of their numerators. (See Art. 76.) Thus, $\frac{2}{3} : \frac{1}{3} = 2$.

The direct ratio of two fractions, which have a common numerator, is the reciprocal ratio of their denominators. (See Art. 77.) Thus, $\frac{3}{7} : \frac{3}{4} = \frac{4}{7}$.

Art. 165.—A factor may be transferred from antecedent to consequent, and from consequent to antecedent, without altering the ratio; observing, that when a factor is transferred, it becomes a divisor, and when a divisor is transferred, it becomes a factor. (See Art. 150.)

Thus, the ratio of	$16 : 2 \times 4 = 2$	the same ratio.
Transferring the factor 2,	$\frac{16}{2} : 4 \quad = 2$	

Art. 166.—It may be observed, in regard to ratio, that it exists only between quantities of the same nature, or, the things compared must be so far alike that one may be said to be larger or smaller than the other. For example, a rod can-

not be said to be longer than an hour, nor can there be a comparison between them in any respect, for there is no common property. But a rod can be said to be longer than a foot, for it is made up of feet. There may be, however, a relation between the *numbers* which stand for quantities of a dissimilar nature. Thus, the ratio of 16 to 8 is 2. Now, 16 may stand for rods, and 8 for hours, which things bear no relation to each other.

The subject of ratio is of incalculable importance, since it lies at the foundation of all arithmetical investigation. The practical nature of ratio will be seen by the following example.

1. If 6 yards of cloth cost 30 dollars, what will 12 yards cost? The ratio of 12 : 6 is 2, which shows that 12 is twice as large as 6. It is, therefore, plain that the cost of 12 yards will be as much greater than the cost of 6 yards, as 12 is greater than 6. Therefore, $30\times2=60$, the cost of 12 yards. Again, if we know the price of 1 yard, we can repeat this price 12 times, and thus obtain the price of 12 yards. If 6 yards cost 30 dollars, it is evident that one-sixth of 30 will be the cost of 1 yard. Although, strictly speaking, there is no relation between the cost and the number of yards, yet the ratio of 30 to 6, considered as *numbers* merely, is a number which will represent the cost of 1 yard. Therefore, $30 : 6=5$, the cost of 1 yard, and $5\times12=60$, the cost of 12 yards, as before.

If we now compare the cost of the second with the cost of the first piece, we shall find that the ratio is equal to the ratio of the length of the second piece, to the length of the first piece. Thus, $12 : 6=2$, and $60 : 30=2$.

When two or more couplets of numbers have equal ratios, these numbers are said to be proportionals. Hence, (Art. 167,) *Proportion is an equality of ratios.*

Arithmetical Proportion is an equality of arithmetical ratios, and Geometrical Proportion is an equality of geometrical ratios. Proportion may be expressed, either by the common sign of equality, or by four points placed between the couplets. Thus—

$8 \cdot\cdot 6=4 \cdot\cdot 2$, or $8 \cdot\cdot 6 :: 4 \cdot\cdot 2$, arithmetical proportion.

$12 : 6=8 : 4$, or $12 : 6 :: 8 : 4$, geometrical proportion.

The latter is read,—the ratio of 12 to 6 equals the ratio of 8 to 4, or 12 is to 6 as 8 is to 4.

The first and last terms are called the *extremes*, and the others the *means.*

Art. 167.—The number of terms must be at least four, for the equality is between the ratios of the couplets; and each couplet must have an antecedent and consequent. There may be, however, a proportion between three quantities; for one of the quantities may be repeated, so as to form the two terms. Thus, 6 : 12 : : 12 : 24.

Art. 168.—If four numbers are in *geometrical* proportion, the product of the extremes is equal to the product of the means. Thus, 12 : 8 : : 15 : 10, for $12 \times 10 = 8 \times 15$.

Art. 169.—By multiplying the extremes and means together, a proportion is reduced to an equation. When the product of any two numbers is equal to the product of any other two, the numbers may be formed into a proportion by taking the factors on one side of the equation for the extremes, and those on the other for the means. Thus, $4 \times 3 = 6 \times 2$. Making 4 and 3 constitute the extremes, and 6 and 2 the means, we have the following proportion; 4 : 2 : : 6 : 3. Form proportions of the following equations:

$$6 \times 8 = 4 \times 12$$
$$3 \times 12 = 4 \times 9$$
$$4 \times 7 = 14 \times 2$$
$$8 \times 9 = 12 \times 6$$

Art. 170.—In compounding proportion, equal factors may be rejected from antecedents and consequents. Thus:

$$12 : \not{4} : : 9 : \not{3}$$
$$\not{4} : \not{8} : : \not{3} : \not{6}$$
$$\underline{\not{8} : 20 : : \not{6} : 15}$$
$$12 : 20 : : 9 : 15$$

Art. 171.—If the corresponding terms of two or more ranks of proportional quantities be multiplied together, the products will be proportional. Thus:

$$12 : 4 : : 6 : 2$$
$$\underline{10 : 5 : : 8 : 4}$$
$$120 : 20 : : 48 : 8$$

Art. 172.—If the terms in one rank of proportionals be divided by the corresponding terms in another rank, the quotients will be proportional.

Thus, 12 : 6 : : 18 : 9
6 : 2 : : 9 : 3

Then, $\frac{12}{6} : \frac{6}{2} : : \frac{18}{9} : \frac{9}{3}$

Art. 173.—If to or from the terms of any proportion, there be added or subtracted the corresponding terms of any other proportion, having the same ratio, their sums or remainders will be proportional. Thus,

$$\begin{array}{r} 14:7::16:\ 8 \\ 4:2::\ 6:\ 3 \\ \hline 18:9::22:11 \\ 10:5::10:\ 5 \end{array}$$

Art. 174.—A factor may be transferred from one mean to the other, or from one extreme to the other, without altering the ratio, 16 : 8 : : 12 : 6.

The scholar may be exercised upon the foregoing proposition, in the following manner :

Teacher. What are the factors of the antecedent of the first couplet?
Scholar. 4 and 4.
T. Transfer one of these to the consequent, and illustrate.
S. $4 : \frac{8}{4} :: 12 : 6$. To divide antecedent and consequent by the same quantity, does not affect the ratio.
T. How does it appear that the antecedent has been divided?
S. We have removed from it the factor, 4. Removing a factor from any quantity, divides by that factor.
T. What are the factors of the antecedent of the second couplet?
S. 4 and 3.
T. Transfer the 4 to the consequent, and illustrate.
S. $4 : \frac{8}{4} :: 3 : \frac{6}{4}$.
T. Remove the denominators from the consequents.
S. 4 : 8 : : 3 : 6.
T. What effect upon the consequents has removing the denominators?
T. Is there a proportion between the four following numbers? 4 : 2 : : 6 : 3. Illustrate.
S. $4\times3=2\times6$.
T. Remove the factor, 4, from the left of the equation, and illustrate.
S. $3=\frac{2\times6}{4}$. To divide both members of the equation by the same quantity, does not affect the equation.
T. Do the four following numbers, 16 : 8 : : 12 : 6, constitute a proportion?
S. They do.
T. How do you know?
S. The ratios between the couplets are equal.
T. Divide the consequents by 2, and will there then be a proportion?
S. There will.
T. Are not the ratios affected?
S. They are.
T. Why then is not the proportion destroyed?
S. The ratios are still equal.
T. In what, therefore, does proportion consist?

S. In equality of ratios.

T. How do you ascertain when the ratios are equal?

S. By dividing the antecedents by the consequents, or by dividing consequents by antecedents, or by multiplying the extremes together and the means together.

T. How do you reduce a proportion to an equation? How do you form a proportion from an equation? Reduce to an equation the following proportion; 5 : 10 : : 4 : 8. Add 2 to each member of the equation. Is the equation affected? Why not? Add 2 to one member, and 3 to the other. Is the equation now affected? Repeat the axiom.

Let the teacher multiply exercises of this kind.

Art. 175.—Inverse, or reciprocal proportion, is an equality between a direct and reciprocal ratio. Thus, $4 : 2 : : \frac{1}{3} : \frac{1}{6}$. That is, 4 is to 2 as 3 is to 6 reciprocally. Sometimes the order of the terms is inverted, without writing them in the form of a fraction. Thus, 4 : 2 : : 3 : 6, inversely. In this case the first is to the second as the fourth is to the third.

We have seen that a factor may be removed from antecedent to consequent, and the reverse, and the proportion still be preserved.

Art. 176.—The terms of the proportion may also be changed, provided that the equality of the ratios be not affected. Thus,

12 : 8 : : 15 : 10
12 : 15 : : 8 : 10
8 : 10 : : 12 : 15
8 : 12 : : 10 : 15
10 : 8 : : 15 : 12
10 : 15 : : 8 : 12
15 : 10 : : 12 : 8
15 : 12 : : 10 : 8

In all these changes the product of the extremes will be found equal to the product of the means. If, therefore, we have the product of the extremes, and one of the means, it is easy to find the other. We can, therefore, find any one term of the proportion when we know the other three, for the term sought must be one of the extremes, or one of the means. The operation by which, three terms being given, a fourth proportional is found, is called the "Rule of Three," or "Rule of Proportion." There must always be three terms or numbers given, two of which are of the same kind, and the other of the kind of the answer required.

SIMPLE PROPORTION, OR RULE OF THREE.

Art. 177.—Proportion is of two kinds, Direct and Inverse. Proportion is direct, when the ratios are in the order in which the question is proposed; Inverse, when one of the ratios is inverted. A question is known to belong to Direct Proportion when *more* requires *more*, or *less* requires *less*. More requires more, when the second term is greater than the first, and requires that the fourth be greater than the third. Less requires less, when the second term is less than the first, and requires that the fourth be less than the third.

1. If 3 men build 12 rods of wall in a given time, how many rods will 6 men build in the same time?

In this question the ratios are in the order in which the question is proposed, 3 : 12 : : 6 : the answer. More requires more: for, evidently, 6 men will perform more labor in the same time than 3 men.

We may employ the same numbers in the proposal of a different question, and the ratios will be inverted.

2. If 3 men perform a certain amount of labor in 12 days, in how many days would 6 men perform the same?

In this question more requires less: for 6 men would require less time to perform the same amount of labor than 3 men: 3 is to 6 reciprocally as 12 is to the answer, $\frac{1}{3} : \frac{1}{6} : : 12 :$ the answer.

Of the three terms given in Proportion, two are called the terms of *condition*, and one the term of *demand*. Thus, 3 men, 12 days, are the terms of *condition*, and 6 men the term of *demand*.

It may be observed that, in *Proportional* questions, the term of demand is the only term which presents any difficulty. The two other terms simply state a fact, or the condition upon which the conclusion rests, and are to be employed as the means of solving the difficulty. The answer to the question proposed, in *Direct Proportion*, depends upon the ratio of the term of *demand* to that term of the *condition*, which is of the

---

QUESTIONS.—5. What is Proportion? 6. What are the first and last terms called? 7 Having the extremes and one of the means given, how may the other mean be found? 8. What is the Rule of Three? 9. How are the ratios? 10. What is meant by the order in which the question is proposed? 11. How is a question known to belong to the Rule of Three Direct? Illustrate. 12. Rule for stating the question?

same name or kind. In *Inverse Proportion*, the answer depends not upon the ratio of *demand* to *condition*, but of *condition* to *demand*. In either case, this ratio multiplied into that term, which is of the same name or kind as the answer required, gives the desired result.

Solution of Question 1st.

$\frac{6}{3}\times 12=24$ *Ans.*

The ratio of 6 men to 3 men expresses how much more labor 6 men can perform in a given time, than 3 men, $\frac{6}{3}=2$. They would perform twice as much. The first step in solving a question is to find the ratio; the second, to multiply. If 3 men build 12 rods, then 6 men will build $\frac{6}{3}$ of $12=24$ rods.

Solution of Question 2d.

$\frac{3}{6}\times 12=6$ *Ans.*

The ratio of 3 men to 6 men, expresses how much less time 6 men would require than 3 men, to perform the same amount of labor.

If 3 men will perform a certain amount of labor in 12 days, 6 men will perform the same in $\frac{3}{6}$ of $12=6$ days. Also, if it takes 3 men 12 days to perform a certain amount of labor, it will take 1 man 3 times as long, $3\times 12=36$, and 6 men one-sixth as long as 1 man, $36\div 6=6$ days.

3. If 5 tons of hay cost 10 dollars, how many dollars will 20 tons cost?

Does this question belong to Direct or Inverse Proportion?

How do you know?

What is meant by more requiring more? Illustrate by example 3d.

Which is the term of demand?

Which are the terms of condition? State the question according to the example given.

What is the first step in the solution? What is the second?

Which term presents the difficulty?

How are the other terms to be employed?

4. If 7 men reap a field of grain in 14 days, how many men can reap the same in 21 days?

Does this question belong to Direct or Inverse Proportion?

How do you know?

What is meant by more requiring less?

How do you know when more requires less?

Are the ratios direct or reciprocal?

What do you mean by a reciprocal ratio?

Is the ratio that of the demand to the condition, or of the condition to the demand? State the question, and illustrate.

Art. 178.—It has been stated that, in Proportion, we have either the two extremes and one of the means, or the two means and one of the extremes given, to find the other.

1. The extremes in a proportion are 12 and 4, one of the means 6. What is the other mean?

2. The extremes in a proportion are 12 and 4, and one of the means 8. What is the other mean?

3. The means in a proportion are 8 and 6, and one of the extremes is 4. What is the other extreme?

4. The means in a proportion are 8 and 6, and one of the extremes is 12. What is the other extreme?

5. If a man travel in 3 days 60 miles, how many miles can he travel in 9 days?

Operation.

$\frac{9}{3}\times 60=180$ *Ans.*

If he travel 60 miles in 3 days, the ratio of 9 to 3 shows how many more miles he could travel in 9 days than in 3 days.

6. The extremes in a proportion are 12 and 9, and one of the means is 4. What is the other mean?

7. The extremes in a proportion are 12 and 9, and one of the means is 27. What is the other mean?

From the preceding illustrations of ratio and proportion, we derive the following

RULE.

I. *Write that term of the condition which is of the same name or kind as the fourth term, or answer required, for the third term. If the Proportion be Direct, write that number which expresses the demand for the second, and the remaining term of the condition for the first. If the Proportion be Inverse, write the number which expresses the demand, for the first, and the remaining term of the condition for the second term.*

II. *Multiply the second and third terms together, and divide the product by the first.*

III. *If the first or second terms consist of different denominations, reduce both to the lowest mentioned. If the third term consist of different denominations, reduce it to the lowest, or the lower to a fraction of the highest.*

IV. *Since the second and third terms are factors of a dividend, and the first term is the divisor, any factor common to the first and second or first and third terms, may be rejected.*

Obs.—For the statement of a question proportionally, the foregoing is the rule, but for practice, the following has many advantages: 1st, It is more convenient for cancelling; 2d, We avoid all fractions in the operation; 3d, We can often avoid the labor of the reduction of the terms.

RULE FOR CANCELLING.

Draw a perpendicular line, and place the sign of the answer on the left, (if the answer is to be in dollars, this $ is called the sign; if pounds, this £,) and the term which expresses the demand on the right of the line, and that term of the condition, which is of the same name or kind, on the left, and the remaining term of the condition, on the right. Reduce, cancel, multiply, and divide, as before directed.

If we do not wish to consider whether the proportion be direct or inverse, the following may be adopted:—

RULE.

Write that number which is of the same name or kind as the answer to the question, for the third term of the proportion. If the answer is greater than the third term, write the greater of the remaining terms for the second, and the less for the first, but if the answer is less than the third term, write the less of the remaining terms for the second term, and the remaining for the first. Reduce the terms, cancel, multiply, and divide, as before directed.

EXAMPLE.

Art. 179.—If 3 yards of cloth cost 18 dollars, what will 9 yards cost?

Operation 1st.	*Operation 2d.*
3	6
$\not{3} : \not{9} :: 18$	$\not{3} : \not{18} :: 9$
3	6
54 *Ans.*	54 *Ans.*

Having stated the question and solved it, the student will give the following illustration at the black-board. This question belongs to Direct Proportion; more requires more—the third term is greater than the first, and requires that the fourth be greater than the second. Having placed the name of the answer for the third term, we write 9, the number expressing the demand, for the second term, and 3, the remaining term of the condition, for the first. The answer depends upon the ratio of demand to condition. The ratio of 9 to 3 shows how much more 9 yards will cost than 3 yards; but 3 : 18 : : 9 : the answer. The ratio of 18 : 3 expresses the cost of one yard. Hence, by analysis, if 3 yards cost 18 dollars, one yard will cost $\frac{18}{3} = 6$, and 9 yards will cost $6 \times 9 = 54$.

Teacher. In what does Proportion consist?

T. How many ratios are there?

Scholar. Two.

T. How many circumstances in the given question affect the answer

S. One.

T. What is it?

S. The difference between the number of men in the demand and condition.

Operation 3d.

$	9 3
3	18
	54 *Ans.*

This mode of stating the question will admit of the same genera illustration as the other. But a variety of illustrations may be employed. The following may be adopted. We place the *sign* of the answer on the left of the line, and first dispose of the term of demand, because it is the term which presents the difficulty. We place it on the right of the line, because it is to be a dividend. It is also properly the antecedent in the last couplet, whose consequent is the term sought. We therefore place that term which is of the same name, on the left, for a divisor, and close the statement by placing the name of the answer on the right. The first step in the solution is to find the ratio of demand to condition, which we find to be 3. The second, to multiply the ratio into the name of the answer. We thus obtain 54 dollars, the answer.

T. Do we necessarily, in the solution, first find the ratio of demand to condition?

S. We do not.

T. Why not?

S. We may reject from either factor of the dividend, a factor equal to the divisor: for $18 \times 3 = 9 \times 6 = 54$.

T. If we reject the factor 3 from the divisor, and from 18, one of the factors of the dividend, how will the question read? If we reject the factor 3 from the divisor, and from 9, one of the factors of the dividend, how will the question then read?

T. Prove your answer to be right, and also show the connection between antecedent and consequent.

S. We may substitute in the place of the sign of the answer, the answer itself; we shall then have the means on the right, and the extremes on the left of the line. We commence with the answer, or last consequent, and draw a line to its antecedent, or the demand, thence to the left, connecting it with the condition, or first antecedent, thence to its consequent on the right, then back to the point of commencement. The statement then reads, consequent, antecedent, antecedent, consequent.

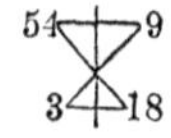

Thus antecedents are connected with antecedents, and consequents with consequents.

EXAMPLES.

1. If 12 yards of cloth cost \$48, what will 4 yards cost? *Ans.* \$16.

2. If 4 bushels of wheat cost \$8, how much will 16 bushels cost? *Ans.* \$32.

3. If a man earn \$24 in 12 days, how much does he earn in 6 days? *Ans.* \$12.

4. If 8 yards of cloth cost \$12, how much will 10 yards cost? *Ans.* \$15.

5. If 10 yards of cloth cost \$15, how much will 8 yards cost? *Ans.* \$12.

6. If 6 acres of land are bought for \$180, for how much may 15 acres be bought? *Ans.* \$450.

7. If 15 acres of land cost \$450, what will 6 acres cost? *Ans.* \$180.

8. If 18 yards of cloth cost \$36, what will 20 yards cost? *Ans.* \$40.

9. If 7 men be paid \$8$\frac{2}{5}$ for a certain amount of labor, what ought 25 men to receive at the same rate? *Ans.* \$30.

OBS. 1.—Mixed numbers must be reduced to improper fractions, and the numerators placed on that side of the line where the whole numbers would be placed. Let it be remembered, that the numerator of a fraction always occupies the same side of the line which a whole number would occupy, standing in the place of the fraction. (See question 14, below.)

10. If 2 horses plough 5$\frac{1}{9}$ acres in a day, how many acres would 18 horses plough in the same time? *Ans.* 46 acres.

11. If 3$\frac{1}{3}$ dollars will buy 6$\frac{3}{4}$ yards of cloth, how many yards will \$40 buy? *Ans.* 81 yards.

12. If \$12$\frac{3}{7}$ buy 4$\frac{1}{14}$ yards of cloth, how many yards will \$174 buy? *Ans.* 57 yards.

13. If 1$\frac{4}{11}$ of a bushel of wheat cost \$2$\frac{5}{8}$, how much will 60 bushels cost? *Ans.* \$115$\frac{1}{2}$.

14. If $\frac{1}{4}$ of a yard of cloth cost $\frac{2}{3}$ of a dollar, how much will $\frac{7}{8}$ of a yard cost? *Ans.* \$2$\frac{1}{3}$.

Read this question thus: What will 7 yards cost, if 1 yard cost 2 dollars?

QUESTIONS.—13. When either of the terms is a compound quantity, what is the rule? 14. Analyze question 2d. 15. Rule for mixed numbers? 16. On which side of the line is the numerator of a fraction to be placed?

Statement.

$	7
1	2

Now write each denominator on the side of the line opposite its numerator, thus:

Operation.

$	7
~~8~~	
1	~~4~~
3	~~2~~
3	$7 = 2\frac{1}{3}$ *Ans.*

15. If 3 horses consume $4\frac{1}{2}$ tons of hay in 4 months, how many tons will 22 horses consume in the same time?

Ans. 33 tons.

16. If 1 yard of ribbon cost 8 pence, how many dollars will 72 yards cost?

Ans. $8.

Operation.

How many	$	~~72~~	yd.
yd.	1	8	*d.*
d.	~~12~~	1	*s.*
s.	~~6~~	1	$
		$8	*Ans.*

OBS. 2.—When the answer is required in a different denomination from that given in the supposition, follow the tables from the denomination given to the denomination required. In the last example, the price of 1 yard is 8 pence. The answer is required in dollars; therefore, continue the statement by saying, 12 pence make 1 shilling, and 6 shillings make 1 dollar, the denomination required. Then 6 times 12 on the left cancels 72 on the right; 8 being the only number left on the right of the line, and there being no number on the left greater than 1, 8 is the answer in dollars.

17. If 1 pint cost 10*d.*, what will 3 hhds. cost in pounds?

Operation.

How many	£	~~3~~	hhds.	
hhd.	1	63	gal.	Reducing hogsheads to pints, see Red. Descend.
gal.	1	~~4~~	qts.	
qt.	1	~~2~~	pts.	
pt.	1	~~10~~	*d.*	Reducing pence to £ see Red. Ascend.
d.	~~12~~	1	*s.*	
s.	~~20~~	1	£	
		£63	*Ans.*	

In this example the first and third terms are of different denominations, and the second term is different from the answer sought; therefore, follow the tables until you find the name of the answer required, observing to commence each successive step on the left with the

QUESTIONS.—17. Repeat the Obs. under question 16. 18. What is the ratio of each term in the demand to each corresponding term in the supposition, in question 16?

denomination last placed on the right, thus: How many £ will 3 hhds. cost?

If	1	hhd.	is	63	galls.
and	1	gal.	is	4	qts.
and	1	qt.	is	2	pts.
and	1	pt.	is	10	pence.
and	12	*d.*	is	1	*s.*
and	20	*s.*	is	1	£

The facility of passing from one denomination to another will be readily seen in the statement of this question; and also that the process of Reduction Descending and Ascending employed in the ordinary mode, is rendered unnecessary by cancelling; thus, 3 times 4, on the right, are equal to 12 on the left, and 2 times 10 on the right are equal to 20 on the left; 63 being the only number left greater than 1, and standing on the right, it is the answer in pounds.

18. If 3 hhds. cost £63, what will 1 pint cost in pence? *Ans.* 10*d.*

Statement.

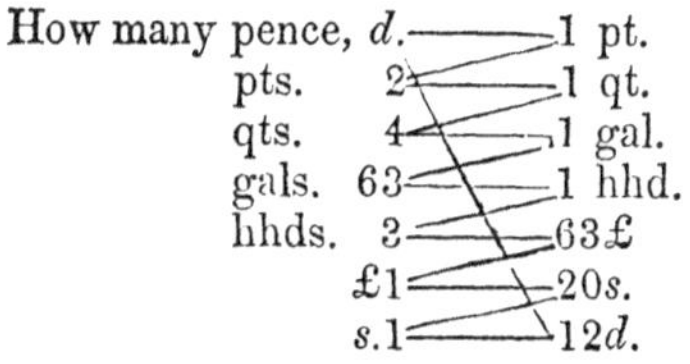

How many pence,	*d.*	1 pt.
pts.	2	1 qt.
qts.	4	1 gal.
gals.	63	1 hhd.
hhds.	3	63£
	£1	20*s.*
	*s.*1	12*d.*

Having made the statement, the connection between the numbers may be shown by connecting, with a continued line, first, the sign of the answer on the left with the demand on the right, and the demand with a number of the same name on the left, and this again with its equal, or its value, (as 3 hhds. with £63, its price,) on the right; and thus on from left to right alternately, until you come to the name of the answer, on the right; then returning with the line to the sign of the answer where you commenced.

19. If 1 yard of cloth cost $13\frac{1}{3}$ shillings, how much will 12 ells English cost in pounds? *Ans.* £10.

Operation.

$13\frac{1}{3}=\frac{40}{3}$

How many	£	12 Ell E.
Ell E.,	1	5 qrs.
Qrs.,	4	1 yd.
Yd.,	1	40*s.* 10
	3	
Shil.,	20	£1
		£10 *Ans.*

20. If 12 ells English cost £10, what will 1 yard cost? *Ans.* $13\frac{1}{3}$*s.*

21. If 1 pint of rye cost 2 pence, what will 7 chaldrons cost in guineas? *Ans.* $85\frac{1}{3}$ guineas.

22. If 1 pint cost 10*d.*, what will 3 hhds. cost in dollars? *Ans.* \$210.

23. If 1 nail cost 3 farthings, how many pounds will 40 yards cost? *Ans.* £2.

24. If 7 chaldrons cost $85\frac{1}{3}$ guineas, what will 1 pint cost in pence? *Ans.* 2*d.*

25. How many pounds will 3 tons of lead cost, at 2 farthings per ounce? How many guineas? How many dollars? *Ans.* £224; 160 guineas; $\$746\frac{2}{3}$.

26. What will 20 dozen pairs of gloves cost, at 4*s.* 6*d.* per pair? *Ans.* £54.

OBS. 3.–4 shillings 6 pence may be reduced to pence=54 pence; or thus, $4\frac{1}{2}$ shillings=$\frac{9}{2}$s.

27. How many yards of cloth may be bought for £32 10*s.*, at 12*s.* 6*d.* per ell English? *Ans.* 65 yds.

28. If $\frac{1}{4}$ of a yard cost $\frac{5}{6}$ of a shilling, how many guineas will 42 yards cost? *Ans.* 5 guineas.

29. If 96 lbs. of red lead cost £3 12*s.*, what is 1 lb. worth? *Ans.* 9*d.*

30. What is the value of $1\frac{1}{4}$ cord of bark, if $4\frac{1}{2}$ cords be worth \$20.25? *Ans.* $\$5.62\frac{1}{2}$.

31. If $6\frac{1}{2}$ yards cost \$3, what will $9\frac{1}{4}$ yards cost? *Ans.* \$4.269+.

32. How many miles will a man travel in $\frac{5}{8}$ of a day, if in $\frac{1}{2}$ of a day he travel 5120 rods? *Ans.* 20 miles.

33. If $\frac{1}{19}$ of a barrel of flour serve 3 men 1 day, how much will be sufficient to serve 402 men the same length of time? *Ans.* $7\frac{1}{19}$ barrels.

34. A man owning $\frac{3}{5}$ of a coal mine, sold $\frac{3}{4}$ of his share for \$36000. What was the value of the mine? *Ans.* \$80000.

35. What will 8 bales of cloth cost, each bale containing 12 pieces, and each piece 27 yards, at \$54 per piece, and what will be the cost of 1 yard? *Ans.* The whole \$5184, and 1 yd. \$2.

36. If \$100 gain \$6 in one year, how much will \$450 gain in the same time? *Ans.* \$27.

37. If 17 tons 12 cwt. of iron cost £165, what will be the cost of 2 cwt.? *Ans.* 18*s.* 9*d.*

38. If 112 lbs. of beef cost 18*s.* 8*d.*, what is 1 lb. worth? *Ans.* 2*d.*

39. If a man travel 200 miles in 15 days, what is the average distance for every 3 days? *Ans.* 40 m.

40. If $\frac{1}{6}$ of a yard cost $\frac{1}{5}$ of a pound, what will 60 yards cost in dollars? *Ans.* \$240.

41. If 30 horses consume 70 bushels of oats in 4 weeks, how many bushels will 9 horses consume in the same time?
Ans. 21 bushels.

42. A merchant bought a number of bales of velvet, each containing $129\frac{17}{27}$ yds., at the rate of \$7 for 5 yds. Sold the same at the rate of \$11 for 7 yds., and gained \$200 by the bargain. How many bales were there? *Ans.* 9 bales.

Operation.

FIRST STATEMENT.

Yds. in a bale.
$129\frac{17}{27}=\frac{3500}{27}$

How many \$	5 yds.
yds. 7	11 \$
7	$55=\$7\frac{6}{7}$.

Sold 5 yds. for	$\$7\frac{6}{7}$.
Bought 5 yds. for	\$7.
Gained on 5 yds.	$\$\frac{6}{7}$.

SECOND STATEMENT.

Bales	200\$
\$ \$ 6	7
	5 yds.
yds. 3500	27 9
	1 bale.
	9 bales *Ans.*

OBS. 4.—In the above question it is necessary to make two statements. First, to find the gain on 5 yards. We then say, on how many bales is \$200 gained, if $\frac{6}{7}$ of a dollar is gained on 5 yards, and $\frac{3500}{27}$ yards make 1 bale?

43. If 500 men consume $102\frac{54}{73}$ barrels of flour in 9 months, how many barrels will 365 men consume in the same time?

Operation 1*st.*

$$102\frac{54}{73}=\frac{7500}{73}$$

$$500:365::\frac{7500}{73}$$

Then, $5\times73=365$

The first and second terms are now equal, and may be rejected. The answer is therefore 75. A divisor transferred from one antecedent to the other, becomes a multiplier.

Operation 2*d.*

$$102\frac{54}{73}=\frac{7500}{73}$$

How many bbls.	365 men 5
Men 500	
	7500 bbls.
73	
	75 bbls. *Ans.*

44. A merchant bartered $5\frac{8}{9}$ cwt. of sugar, at $6\frac{3}{4}d.$ per lb., for tea at $8\frac{5}{8}s.$ per lb. How much tea did he receive ?

Operation 1st.

$5\frac{8}{9}=\frac{53}{9}$ cwt. $\frac{53}{9}$ of $\frac{112}{1}=\frac{5936}{9}$

$6\frac{3}{4}=\frac{27}{4}d.$ $\frac{69}{8}$ of $\frac{12}{1}d.=\frac{828}{8}$

$8\frac{5}{8}=\frac{69}{8}s.$

$\frac{27}{4}\times\frac{5936}{9}=\frac{160272}{36}$

Invert the divisor.

$\frac{8}{828} : \frac{160272}{36} : : 1$

```
  8      :   160272  : : 1
 ---         ------
 828           36
  36
 ----
 4968   160272
 2484        8
 29808)1282176(43 1/69 Ans.
```

Operation 2d.

$5\frac{8}{9}=\frac{53}{9}$ cwt.

$6\frac{3}{4}=\frac{27}{4}d.$

$8\frac{5}{8}=\frac{69}{8}s.$

Tea.	53 sugar.
9	
cwt. 1	112 lbs. $7\times8\times53=2968$
lb. 1	27d. 3
4	
d. 12	1s.
s. 69	8
	1 lb. tea.
69	$2968=43\frac{1}{69}$ *Ans.*

45. If I buy $3\frac{1}{2}$ lbs. of sugar for 25 cents, what part of a ton can I buy for \$6 ?

Operation 1st.

lbs. ton.

$3\frac{1}{2}=\frac{7}{2}$ of $\frac{1}{28}$ of $\frac{1}{4}$ of $\frac{1}{20}=\frac{7}{4480}$

```
  2
 ---
  56
   4
 ---
 224
  20
 ----
 4480
25 : 600 : :  7/4480
      7
25)4200(168
   25
   170
   150
    200
    200
```

$8\Big|\frac{168}{4480}\Big|\frac{21}{560}\Big|\frac{3}{80}$ *Ans.* (7)

Operation 2d.

lbs.

$3\frac{1}{2}=\frac{7}{2}$ \$

Ton.	6.00
cts. 25	7 lbs. 12 3
2	
lbs. 28	1 qr.
qrs. 4	1 cwt.
cwt. 20	1 ton.
80	$3=\frac{3}{80}$ *Ans.*

46. If 2 lbs. of sugar cost $\frac{1}{4}$ of a dollar, what will 100 lbs. of coffee cost, if 8 lbs. of sugar are worth 5 lbs. of coffee ?

9*

FIRST STATEMENT.

Operation 1*st*.

cof. sug. cof.
$\cancel{5} : 8 :: \cancel{100}\ 20$
8
160 lbs. sugar.

Operation 2*d*.

What will \$	$\cancel{100}$ coffee 20
coffee $\cancel{5}$	$\cancel{5}$ sugar
sugar $\cancel{2}$	1\$
$\cancel{4}$	
	\$20 *Ans.*

SECOND STATEMENT.

sug. sug. cts.
$\cancel{2} : \cancel{160} :: 25$
80 80
\$20.00 *Ans.*

47. If $\frac{1}{3}$ lb. less by $\frac{1}{6}$ cost $13\frac{1}{5}d.$, what will 14 lbs. less by $\frac{1}{5}$ of 2 lbs. cost? *Ans.* £4 9*s.* $9\frac{3}{25}d.$

48. A merchant failing in trade owes \$6000. His property amounts to \$2400. What does his creditor receive to whom he owes \$500, and what does he pay on the dollar?

Ans. { Creditor, \$200. On dollar, 40 cts.

49. Bought $\frac{1}{2}$ yd. of cloth for \$$\frac{3}{8}$; sold $\frac{1}{3}$ of $\frac{1}{7}$ of $\frac{1}{8}$ for $\frac{5}{8}s.$ Did I make or lose? *Ans.* Made 10 cents +.

50. If 40 yards of cloth cost \$32, what will 1 ell English cost? *Ans.* \$1.

51. A mercer bought $3\frac{1}{2}$ pieces of silk, each piece containing $24\frac{1}{3}$ yds., at 6*s.* 6*d.* per yd. What did the whole cost him? *Ans.* £27 13*s.* 7*d.*

52. If $\frac{5}{8}$ of a gallon cost £$\frac{5}{8}$, what will $\frac{5}{9}$ of a tun cost? *Ans.* £140.

53. If $2\frac{1}{2}$ yds. of cloth cost 60 cents, what will $125\frac{3}{4}$ yds. cost? *Ans.* \$30.18.

54. If $\frac{8}{9}$ of a cord of wood is worth \$6, what is 40 cords worth? *Ans.* \$270.

INVERSE PROPORTION.

(For the rule, see Art. 178.)

EXAMPLES.

Art. 180.—1. If 4 men build a wall in 20 days, in how many days could 8 men build the same wall?

Operation.

In how many days.	
	~~4~~ m.
~~2~~ m. ~~8~~	~~20~~ d. 10
	12 days *Ans.*

If 4 men can build a wall in 20 days, 8 men would build the same in $\frac{4}{8}=\frac{1}{2}$ of the time. ½ of 20, or 20 multiplied by the ratio of 4 to 8=10 days, the answer.

2. If 6 men mow a field in 21 days, in how many days would 9 men mow the same? *Ans.* 14 days.

3. If it take 9 men 14 days to mow a field, how long would it take 6 men to mow the same field? *Ans.* 21 days.

4. If it take 6 men 21 days to mow a field, how many men would mow the same in 14 days? *Ans.* 9 men.

5. If a man perform a journey in 6 days, when the days are 16 hours long, in how many days can he perform the same when the days are 12 hours long? *Ans.* 8 days.

6. If 1 cwt. be transported 150 miles for 1 guinea, how far can 6 cwt. be carried for the same money? *Ans.* 25 miles.

7. How many yards of carpeting, ½ yard in width, will cover a room 30 feet long and 20 feet wide? *Ans.* 133⅓ yds.

8. What must be the length of a garden, 16 rods in breadth, to contain 2 acres? *Ans.* 20 rods.

9. How many yards of lining, ¾ yard wide, will it take to line a cloak 4½ yds. long and 1¼ yd. wide? *Ans.* 7½ yds.

10. If I lend a friend $200 six months, how long ought he to lend me $1000 to repay the kindness, allowing the month to be 30 days? *Ans.* 36 days.

11. Suppose 800 men were placed in a garrison, with provision sufficient to last them 2 months, how many must depart that the provision may last them 5 months? *Ans.* 480 men.

12. A ship's company of 15 persons is supposed to have bread to last their voyage, allowing each person 8 ounces per day. They pick up a crew of 5 persons in distress, whom they permit to share their daily allowance with them. What will be the allowance of each person? *Ans.* 6 ounces.

13. When wheat is sold at 93 cts. per bushel, the penny loaf weighs 12 ounces. What must it weigh when wheat is $1.24 per bushel? *Ans.* 9 ounces.

14. How many yards of cloth, 1⅓ yd. in width, are equal in measure to 30 yds. 1 ell English in width? *Ans.* 25 yds.

15. How long must a board, 4½ inches in breadth, be, to contain a square foot? *Ans.* 32 inches.

16. A certain building was raised in 8 months by 120 workmen. How many workmen could have done the same amount of labor in 2 months? *Ans.* 480 men.

17. How much in length that is 16 rods in width will it take to make an acre? *Ans.* 10 rods.

18. There is a cistern having a pipe which will empty it in 6 hours. How many pipes of the same capacity will empty it in 20 minutes? *Ans.* 18 pipes.

19. If 30 men can perform a piece of work in 11 days, how many men will accomplish another piece of work, 4 times as large, in a fifth part of the time? *Ans.* 600 men.

COMPOUND PROPORTION.

Art. 181.—When a proportion is formed by the combination of two or more simple proportions, it is called Compound Proportion, or Double Rule of Three.

1. If 8 men consume 24 bushels of wheat in 5 months, how many bushels will 4 men consume in 15 months?

In this question, the number of bushels consumed depends on two circumstances—the number of men, and the time. We may consider the circumstances separately, and solve the question by two statements in the Single Rule of Three. First, the number of men. If 8 men consume 24 bushels in 5 months, how many bushels will 4 men consume in the same time?

Operation 1st.

$$\begin{array}{l} \quad\quad\quad 2) \\ 2\,\not{8} : \not{4} : : 24(12 \quad \textit{Ans.} \end{array}$$

Secondly, the time. If 4 men consume 12 bushels in 5 months, how many bushels will the same number of men consume in 15 months?

Operation 2d.

$$\begin{array}{r} 3 \quad\quad\quad \\ \not{5} : \not{1}\not{5} : : 12 \\ 3 \\ \hline 36 \; \textit{Ans.} \end{array}$$

QUESTION.—1. How is Compound Proportion formed?

The first operation is in Simple Proportion, because we employed but one simple ratio as a multiplier upon 24 bushels, the name of the answer, viz., the ratio of 4 men to 8 men.

The second operation is also in Simple Proportion, for the same reason. The ratio employed is the ratio of 15 months to 5 months.

We may now unite these two statements in one, applying the rule already given in Simple Proportion. Thus,

2 8 : 4 :: }
3 } $24 \times 3 \div 2 = 36$ *Ans.*
5 : 15 :: }

Or thus:

b.	4
2 8	15 3
5	24 12
	36 *Ans.*

Here we have two terms of demand, viz., 4 men and 15 months; and two terms of condition, 8 men and 5 months.

The ratio of 4 to 8 is $\frac{1}{2}$, and the ratio of 15 to 5 is 3. If we multiply these two simple ratios together, we have a compound ratio, which, multiplied into 24 bushels, gives the answer. $\frac{1}{2} \times 3 = \frac{3}{2}$, and $24 \times \frac{3}{2} = 36$, the answer. It is the use of a compound ratio which constitutes Compound Proportion.

All questions in Compound Proportion may be solved by two or more statements in Simple Proportion, or they may be analyzed thus: If 8 men consume 24 bushels in 5 months, 1 man would consume $\frac{1}{8}$ of $24 = 3$ bushels, and in 1 month $\frac{1}{5}$ of $3 = \frac{3}{5}$ of 1 bushel. Then 4 men would consume 4 times $\frac{3}{5} = \frac{12}{5}$ in 1 month, and in 15 months 15 times $\frac{12}{5} = 36$ bushels, the answer.

Art. 182.—*Compound Proportion teaches to solve by one statement questions which would require two or more by Simple Proportion.*

OBS. 1.–The student should be required, first, to solve the question by analysis, then by proportion.

2. If a man build 27 rods of wall in 3 days, when the days are 12 hours long, how many rods can he build in 9 days, when the days are 16 hours long?

If a man in 3 days build 27 rods, in one day he would build $\frac{1}{3}$ of $27 = 9$ rods. If in one day, 12 hours long, he build 9 rods, in one hour he would build $\frac{9}{12}$ of a rod, and in 16 hours, $\frac{9}{12} \times 16 = \frac{144}{12} = 12$ rods. If in 1 day, 16 hours long, he build

QUESTIONS.—1. What does Compound Proportion teach? 2. What constitutes Compound Proportion?

12 rods, in 9 days he would build $12 \times 9 = 108$ rods, the answer.

In this example, 3 days, 12 hours long, are equal to $12 \times 3 = 36$ hours; and 9 days, 16 hours long, are equal to $16 \times 9 = 144$ hours. We have, then, this proportion; 36 h. : 144 h. :: 27 rds. : 108 rods, for $\frac{144}{36} = 4$, and $\frac{108}{27} = 4$. The ratio of the time in the demand, to the time in the supposition, is the same as the ratio of the term sought to the rods in the conditions of the question. That is, the ratio of 144 hours to 36 hours, expresses how many more rods can be built in 9 days, 16 hours long, than in 3 days, 12 hours long.

It will be perceived, that the ratio of the time in the demand, to the time in the supposition, is the product of two simple ratios. It is a ratio of the ratio of days to days, and hours to hours, (a ratio produced by the multiplication of simple ratios is called a compound ratio.) Thus, if a man in 3 days, 12 hours long, build 27 rods of wall, the amount of wall built in 9 days, (the days being of equal length,) is expressed by the ratio of 9 to 3, $\frac{9}{3} = 3$; that is, he could build 3 times the number of rods, $27 \times 3 = 81$ rods; but the days are 16 hours long; this circumstance, again, affects the result, and is expressed by the ratio of 16 to 12, $\frac{16}{12} = 1\frac{1}{3}$; that is, the amount of labor performed in 16 hours is greater by $\frac{1}{3}$ than the labor performed in 12 hours; $\frac{1}{3}$ of 81 rods $= 27$, and $27 + 81 = 108$ rods. If we multiply these simple ratios, $3 \times 1\frac{1}{3} = 4$, we have a compound ratio, the same as above, and $27 \times 4 = 108$ rods, the same answer. The ratios of the days to days, and hours to hours, may be expressed thus. If, in 3 days, 12 hours long, 27 rods of wall are built, how many rods can be built in $\overset{\text{days.}}{\frac{9}{3}}$ of $\overset{\text{hours.}}{\frac{16}{12}}$? $\frac{9}{3}$ of $\frac{16}{12} = \frac{144}{36} = 4$, and $27 \times 4 = 108$, the answer.

Obs. 2.–The teacher will now call upon some member of the class, to select the terms, and form first a simple proportion, and then a compound, in the following manner, and illustrate as he proceeds; 3 : 9 :: 27 : This statement involves Simple Proportion. The days are considered of equal length. There is but one circumstance that affects the answer, viz.: the difference in the number of the days. This affects it in a threefold ratio. The ratio of 9 : 3=3, which shows how many more rods could be built in 9 days, than in 3 days. But the days are not of equal length. This circumstance must also be considered. We will therefore introduce into the statement another simple ratio. The ratio of 16 to 12, which shows how many more rods could be built in 16 hours, than in 12.

$$\left.\begin{array}{rrcl} & & \cancel{3} & \\ & \cancel{3} : & \cancel{9} & :: \\ & & 4 & \\ \text{Thus,} & \cancel{3}\ \cancel{12} : & \cancel{16} & :: \end{array}\right\} \begin{array}{r} 27 \\ 4 \\ \hline 108\ \textit{Ans.} \end{array}$$

Multiplying antecedents and consequents by antecedents and consequents, the ratios are compounded, and thus the question becomes Compound Proportion.

T. How does it appear that antecedents and consequents have been multiplied?

S. Introducing a factor, multiplies by that factor, and cancelling equal factors from antecedents and consequents, does not affect the ratios. In this example, the factors are all cancelled but 4 and 1. 4 is therefore the compound ratio.

T. Does this question involve Direct or Inverse Proportion? How do you know? How many simple ratios are there? Upon how many circumstances does the answer depend? What are they? In what ratio does the first circumstance affect the answer? The second? In what both combined? What is this ratio called? What is a compound ratio? Does it differ, in itself considered, from a simple ratio?

Question 2.—If a man build 27 rods of wall in 3 days, when the days are 12 hours long, in how many days can he build 108 rods, when the days are 16 hours long?

$$\left.\begin{array}{rcl} \cancel{27} : & \cancel{108} & :: \\ & 3 & \\ \cancel{4}\ \cancel{16} : & \cancel{12} & :: \end{array}\right\} 3\times 3=9\ \textit{Ans.}$$

Let the student state and illustrate this question, in the following manner:

This question involves Inverse Proportion—more requires less. It would require a less number of days to perform the same amount of labor, when they are 16 hours long, than when they are 12. The number of the days will be inversely as their length. We therefore place the 16 hours in the demand, for the first term, and the 12 hours in the condition for the second. We reject the factor 4, which is common to 16, the antecedent, and to 12, the consequent. Then $4\times 27=108$, which is an antecedent and consequent, and therefore may be rejected. The ratios are now compounded, and the proportion reads, 1 : 3 :: 3 : 9 the answer.

T. By what rule are the foregoing questions stated?

S. By the rule given for Simple Proportion.

3. If 3 men can build 360 rods of wall in 24 days, how many rods can 8 men build in 27 days?

4. How many men will it take to build a wall, 75 rods long, 8 feet high, 3 feet thick, in 6 days, working 9 hours each day, if 20 men can build a wall 100 rods long, 6 feet high, 4 feet thick, in 12 days, working 12 hours each day?

Statement.

100 ⎫		75 ⎫		
6 ⎪	:	8 ⎪	: :	
4 ⎬		3 ⎬	20 men.	
6 ⎪	:	12 ⎪	: :	
9 ⎭		12 ⎭		

Operation.

How many men?	75	5 × 8 = 40 men *Ans.*
5 100	8	
3 6	3	
4	12 3	
6	12 2	
9	20 men.	
	40 men *Ans.*	

5. If $\frac{3}{4}$ of a yard of cloth, $\frac{7}{8}$ yd. wide, cost £$\frac{3}{5}$, what is the value of $\frac{5}{8}$ yard, $1\frac{3}{4}$ yard wide, of the same quality?

6. If a man travel 240 miles in 12 days, when the days are 12 hours long, how far can he travel in 27 days, when the days are 16 hours long?

Art. 183.—In Proportion, both Simple and Compound, the terms in the supposition and demand may be distinguished by *cause* and *effect*, or *producing* and *produced terms.* That which causes any thing, or produces an effect, as men, time, length, breadth, depth, etc., may be denominated a *producing term.* Thus, in the foregoing question, among the terms of supposition, one man, 12 days, 12 hours long, are the joint cause, or *producing terms*, and miles the effect, or *produced term.* Among the terms of demand, 27 days, 16 hours long, are the joint cause, or the *producing terms*, and the rods required are the effect, or the *produced term.* In all questions in Proportion the answer required will be either *cause* or *effect*, (a *producing* or *produced term.*) Hence,

Art. 184.—When the term required is a produced term,

Draw a perpendicular line, and place all the terms of demand on the right of the line, and all the corresponding terms of the condition on the left, closing the statement by placing the name of the answer on the right.

QUESTIONS.—1. What is meant by *producing* and *produced* terms? 2. Rule, when the term required is a produced term?

Obs. 1.—All questions under the foregoing head will be found to be in *Direct Proportion.*

7. If 4 students spend £19 in 3 months, how many pounds will 8 students spend in 9 months ? *Ans.* £114.

Operation.

	How many £ ?	8 students. } Producing terms
Producing terms of {	Students 4	9 months. } of the demand.
the supposition. {	Months 3	£19. { Produced term of the supposition.
		£114 *Ans.*

8. If 7 men can reap 84 acres in 12 days, 12 hours long, how many acres can 20 men reap in 5 days, 14 hours long ? *Ans.* $116\frac{2}{3}$ acres.

9. If 8 reapers receive £3 4*s.* for 4 days' work, how much ought 20 reapers to receive for 15 days' work ? *Ans.* £30.

10. If 3 men receive £8.9 for 19.5 days' labor, how much ought 20 men to receive for 100.25 days' labor ? *Ans.* £305 0*s.* 8*d.* 1*qr.*

11. If 20 cwt. may be carried 80 miles for $35, how much will it cost to transport 40 cwt. 100 miles ? *Ans.* $$87\frac{1}{2}$.

12. If the freight of 9 hhds. of sugar, each weighing 12 cwt., 20 leagues, cost £16, what must be paid for the freight of 50 tierces, each weighing $2\frac{1}{2}$ cwt., 100 leagues ?

Obs. 2.—Hundred weight and distance are the producing terms, and the money received, the produced term.

13. If $2\frac{1}{4}$ yards of cloth, $1\frac{3}{5}$ yards wide, cost £$3\frac{3}{5}$, how much will $36\frac{1}{2}$ yards, $1\frac{1}{2}$ yards wide, cost ? *Ans.* £$54\frac{3}{4}$.

14. If 24 bushels of wheat be consumed by 8 persons in 5 months, how many persons will consume 36 bushels in 15 months ?

Operation.

How many persons ?	36 bushels 4.
3 bushels 24	5 months.
3 men 15	8 persons.
	4 persons *Ans.*

Art. 185.—When the term required is a producing term,

Draw a perpendicular line, and place the produced term of the demand and the producing terms of the supposition on the right, and the remaining terms on the left of the line, and proceed as before.

OBS.—This head will be found to correspond with *Inverse Proportion*, and may be applied to both Simple and Compound.

15. If 6 men build a wall, 20 feet long, 6 feet high, and 4 feet thick, in 16 days, working 12 hours in a day, in how many days will 24 men build a wall 200 feet long, 8 feet high, and 6 feet thick, working 10 hours in a day? *Ans.* 96 days.

Operation.

	Left	Right	
How many days?		200	Produced terms of demand.
Produced terms of supposition.	20	8	
	6	6	
Producing terms of demand.	4	6	Producing terms of supposition.
	4 24	12	
	10	16	
		96 days *Ans.*	

16. If 3 men, in 24 days, 9 hours long, can dig 328 rods of trench, 6 feet wide and 4 feet deep, how many men will it take to dig a trench 984 rods long, 9 feet wide, and 8 feet deep, in 27 days, when the days are 12 hours long?

Ans. 18 men.

17. If a man can travel 240 miles in 16 days, when the days are 14 hours long, how many days will it take him to travel 720 miles, when the days are 12 hours long?

Ans. 56 days.

18. If 98 lbs. of bread be sufficient to serve 7 men 14 days, how many days will 63 lbs. serve 21 men?

Ans. 3 days.

19. If 40 men in 15 days, 12 hours long, build a wall 200 feet long, 12 feet high, and 5 feet thick, how many hours long must the day be, that 20 men, in 12 days, may build a wall 100 feet long, 10 feet high, and 6 feet thick?

Ans. 15 hours.

20. If 20 men in 12 days, 15 hours long, can build a wall 100 feet long, 10 feet high, and 6 feet thick, in how many days, of 12 hours long, can 40 men build a wall 200 feet long, 12 feet high, and 5 feet thick? *Ans.* 15 days.

21. If 16 compositors set 150 pages of types, each page

QUESTION.—3. Rule, when the term required is a *producing* term?

consisting of 48 lines, and each line of 50 letters, in 8 days, 10 hours long, how many compositors will be required to set 500 pages of 72 lines each, and 45 letters in a line, in 6 days, 8 hours long? *Ans.* 120 compositors.

SUPPLEMENT TO THE RULES OF PROPORTION.

EXAMPLES.

Art. 186.—1. If I can hire 30 horses pastured 7 weeks for £6, how many weeks may 5 horses be pastured for £4 5*s.* $8\frac{4}{7}$*d.*? *Ans.* 30 weeks.

2. If 8 men build 48 rods of fence in 1 day, how many rods will 24 men build in the same time? *Ans.* 144 rods.

3. If 24 men build 144 rods of fence in 1 day, how many rods will 8 men build in the same time? *Ans.* 48 rods.

4. How many men will it require to build 144 rods of fence in 1 day, if 8 men build 48 rods in the same time? *Ans.* 24 men.

5. If 20 men build a mill in 160 days, in how many days could 25 men build the same mill? *Ans.* 128 days.

6. If 25 men build a mill in 128 days, how many men will build the same mill in 160 days? *Ans.* 20 men.

7. How many cords of wood may be bought for £40, if $2\frac{2}{3}$ cords cost $6? *Ans.* $53\frac{1}{3}$ cords.

8. How many bushels of wheat may be bought for £40, if 1 bushel of wheat be worth 2 bushels of rye, and 4 bushels of rye be worth 5 bushels of corn, and 8 bushels of corn be worth 16 bushels of oats, and 1 bushel of oats be worth 2 shillings? *Ans.* 80 bushels.

9. If 18 cords of oak wood be worth $26\frac{2}{3}$ cords of hemlock, and a cord of hemlock $14\frac{2}{5}$ shillings, how much will $4\frac{1}{2}$ cords of oak cost in cents, and how many guineas?

Ans. { 1600 cents. / $3\frac{3}{7}$ guineas. }

10. How many stoves may be bought for 672 shillings, if 8 fire-frames are worth $3\frac{1}{3}$ stoves, and 16 fire-frames are worth 32 guineas? *Ans.* 5 stoves.

11. How much will $450 gain in a year, if $100 in the same time gain $8? *Ans.* $36.

12. A merchant owning $\frac{2}{5}$ of a vessel, sold $\frac{5}{8}$ of his share for $934. What was the value of the ship? *Ans.* $3736.

13. How many guineas will $8\frac{2}{5}$ yards of cloth cost, if $\frac{7}{8}$ of a yard cost $\frac{5}{8}$ of a dollar? *Ans.* $1\frac{2}{7}$ guineas.

14. If $\frac{11}{13}$ of a pound of sugar cost $\frac{7}{15}$ of a shilling, what will $\frac{32}{43}$ of a pound cost? *Ans.* 4*d.* $3\frac{1657}{2365}$ *qrs.*

15. If $2\frac{1}{2}$ lbs. of tobacco cost 4*s.* 6*d.*, how much will 180 lbs. cost in dollars? *Ans.* $54.

16. If when wheat is 4*s.* 6*d.* per bushel, the penny loaf weigh 12 oz., what ought it to weigh when wheat is $.50 per bushel? *Ans.* 18 oz.

17. How many yards of cloth, $\frac{3}{4}$ yd. wide, are equal in measure to 30 yds. $1\frac{1}{2}$ yds. in width? *Ans.* 60 yards.

18. If 40 bushels of grain will pay a debt when the price is 60 cents per bushel, how many bushels will it take when the price is $1.20? *Ans.* 20 bushels.

19. How far may 30 cwt. be transported for $8, if $2\frac{1}{2}$ cwt. be carried 180 miles for the same money? *Ans.* 15 miles.

20. How many yards, of $\frac{3}{4}$ yd. wide, will it take to line 850 suits of clothes, each suit to contain $3\frac{1}{2}$ yards of cloth, $1\frac{3}{4}$ yds. in width? *Ans.* 6941 yds. 2 qrs. $2\frac{2}{3}$ nails.

21. If 30 horses consume 600 bushels of oats in 8 months, how much will each horse consume per day? *Ans.* $2\frac{2}{3}$ quarts.

22. A man owns $\frac{1}{2}$ of a ship, which ship is valued at $\frac{1}{4}$ of the ship and cargo—the latter worth $96000. What is the value of $\frac{1}{4}$ of his share? *Ans.* $4000.

23. If 12 men consume $\frac{1}{2}$ of $\frac{2}{3}$ of $\frac{3}{4}$ of $\frac{4}{5}$ of 30 bushels of wheat, in $\frac{1}{2}$ of $\frac{14}{39}$ of $\frac{6}{7}$ of $\frac{8}{15}$ of 20 months, how much will 4 men consume in $\frac{3}{8}$ of $\frac{6}{7}$ of $\frac{16}{19}$ of $\frac{38}{39}$ of $\frac{14}{15}$ of 40 months?
Ans. 12 bushels.

24. If 4 men spend $\frac{3}{4}$ of $\frac{8}{9}$ of $\frac{6}{7}$ of $\frac{14}{15}$ of £30 in $\frac{7}{11}$ of $\frac{9}{13}$ of $\frac{26}{27}$ of $\frac{11}{7}$ of 9 days, how many dollars will 21 men spend in $\frac{3}{4}$ of $\frac{14}{15}$ of $\frac{6}{7}$ of $\frac{4}{12}$ of 45 days? *Ans.* $420.

25. If a man travel 336 miles in 14 days, when the days are 18 hours long, in how many days will he travel 672 miles, the days being 12 hours long? *Ans.* 42 days.

26. If a man travel 240 miles in 12 days, when the days are 12 hours long, in how many days will he travel 720 miles, when the days are 16 hours long? *Ans.* 27 days.

27. If a family of 9 persons spend $450 in 7 months, how much would be sufficient to maintain them 8 months, if 5 persons more were added to the family? *Ans.* $800.

28. What is the value of 1 grain of gold, if $17\frac{2}{3}$ lbs. be worth £10224? *Ans.* $2\frac{1}{106}$*s.*

EXCHANGE.

Art. 187.—Exchange is the act of paying or receiving the money of one country for its equivalent in the money of another country, by means of *Bills of Exchange.* It comprehends both the reduction of moneys and the negotiation of bills. It determines the comparative value of the currencies of different nations, and shows how foreign debts may be discharged, and remittances made from one country to another, without the risk, trouble, or expense of transporting specie or bullion.

When the United States were British colonies, the sterling value of the pound was the same in all the colonies; but the legislatures of the different colonies emitted bills of credit, which afterwards depreciated in their value—in some states more, and in others less.

Art. 188.—The following table exhibits the number of shillings in a dollar in each of the states.

TABLE I.

To exchange from / *to*	New Eng. States and Virginia.	Penn'a, N. Jersey, Delaware, and Maryland.	New York and North Carolina.	South Carolina and Georgia.
New England States and Va.	Dollar 6*s*. 0*d*.	Add $\frac{1}{4}$.	Add $\frac{1}{3}$.	$\times 7$ and $\div 9$
Pennsylvania, New Jersey, Delaware and Maryland	Subtract $\frac{1}{5}$.	Dollar 7*s*. 6*d*.	Add $\frac{1}{15}$.	$\times 3\frac{1}{9}$ and $\div 5$
New York and North Carolina.	Subtract $\frac{1}{4}$.	Subtract $\frac{1}{16}$.	Dollar 8*s*. 0*d*.	$\times 7$ and $\div 12$
South Carolina and Georgia.	Add $\frac{2}{7}$.	$\times 5$ and $\div 3\frac{1}{9}$.	$\times$ by 12 $\div$ by 7.	Dollar 4*s*. 8*d*.

Obs.—The value of a dollar in any state is found either opposite to that state, or under it in the table.

As the *number* of shillings in a dollar is different in different states, the value of the dollar being the same, it follows, that the *value* of the shilling is different; and as the number of shillings in a pound is the same, the value of the pound must differ in the ratio of the shillings. That is, if 6 shillings in New England, and 8 shillings in New York, make a dollar, then a pound in New York is to a pound in New England, as 8 is to 6; $\frac{8}{6}=\frac{4}{3}$; or £4 in New York are equal to £3 in New England.

Art. 189.—The relative value of the pound in different states may be seen by the following Table.

TABLE II.

	£ £	
New Jersey, Pennsylvania, Delaware, and Maryland,	15=16	New York and North Carolina.
	5= 4	New England and Virginia.
	45=28	South Carolina and Georgia.
New England States and Virginia,	4= 5	New Jersey, Pennsylvania, &c.
	3= 4	New York and North Carolina.
	9= 7	South Carolina and Georgia.
New York and N. Carolina,	16=15	New Jersey, Pennsylvania, &c.
	4= 3	New England and Virginia.
	12= 7	South Carolina and Georgia.
S. Carolina and Georgia,	28=45	New Jersey, Pennsylvania, &c.
	7=12	New York and North Carolina.
	7= 9	New England and Virginia.

Art. 190.—The following table shows the value of pounds, shillings, and pence in each of the United States, according to their respective currencies.

TABLE III.

New Jersey, Pennsylvania, Delaware, and Maryland.	New York and North Carolina.	N. England States and Virginia.	S. Carolina and Georgia.
£ 3=$8.	£ 2=$5.	£ 3=$10.	£ 7=$30.
s. 3=40 cts.	*s.* 2=25 cts.	*s.* 3=50 cts.	*s.* 7=1.50 cts.
d. 9=10 cts.	*d.* 24=25 cts.	*d.* 18=25 cts.	*d.* 14=25 cts.

DOMESTIC EXCHANGE.

Art. 191.—To reduce the currency of one state to that of another.

RULE.

Draw a perpendicular line, and place the demand of the question on the right, and the supposition on the left, as in the Rule of Three Direct.

EXAMPLES.

1. What sum in Georgia, is equal to £1800 New England currency?

Operation.

How many Georgia £	£1800 N. E.	
If N. E.	£9	7 G. 200
		£1400 *Ans.*

In stating the question, say how many pounds Georgia currency are equal to £1800 New England, if £9 New England are equal to £7 in Georgia. (See Table II.)

2. How many pounds in New York currency will £240 New Jersey currency make? *Ans.* £256.

3. What sum in Virginia is equal to £375 16*s.* 9*d.* New York currency? *Ans.* £281 17*s.* 6*d.* 3*qrs.*

4. What is the value in New Jersey currency of a bill of exchange for £375 10*s.*, on a cotton dealer in Georgia? *Ans.* £603 9*s.* 7*d.* 3*qrs.*

5. A manufacturer in Massachusetts sends to Georgia a lot of shoes, which amount to £420 7*s.* What is the value in New England currency? *Ans.* £540 9*s.*

6. A manufacturer in New Jersey consigns to his agent in Charleston a quantity of *ready-made* clothing, which, when sold, and the charges deducted, amounted to £532 11*s.* What is the value in N. Jersey currency? *Ans.* £855 17*s.* 8*d*+.

7.

EXCHANGE FOR £320 10*s.* 6*d.* *Boston, July 26th,* 1837.

Twelve days after sight, please pay to PETER FINCH, or order, three hundred twenty pounds, ten shillings and sixpence, value received, and place the same to the account of your

Ob't Servant,

To PETER FINCH. ISAAC WATERFORD, JR.

What sum in New York currency will discharge this bill? *Ans.* £427 7*s.* 4*d.*

QUESTIONS.—1. What does Exchange teach? 2. Why is the value of the pound different in different states? 3. Is the value of a dollar everywhere the same? 4. Is the value of a shilling the same? 5. What is Domestic Exchange? 6. Rule of statement?

8. What sum in South Carolina currency is equal to £429 7*s.* 3*d.* in New England? *Ans.* £333 18*s.* 11*d.* 2*qrs.*

9. What sum in Pennsylvania is equal to £259 15*s.* 9*d.* Georgia currency? *Ans.* £417 10*s.* 3*d.* 2*qrs.*

10. *Philadelphia, June 1st,* 1837.

EXCHANGE FOR £240 10*s.* PENNSYLVANIA CURRENCY.

Sixteen days after sight, pay to GEORGE SIMPSON, or order, two hundred and forty pounds, ten shillings, Pennsylvania currency, as per advice from

Yours, etc.,

To THOMAS SMART, Merchant, N. Y. JOSIAH LITTLE.

What sum in New York currency will discharge the above bill? *Ans.* £256 10*s.* 8*d.*

FOREIGN EXCHANGE.*

Art. 192.—All foreign coins, by a late act of Congress of the United States, are prohibited being a lawful tender. The gold coins of Great Britain and Portugal, of their present standard, are valued at the rate of 100 cents for every 27 grains, or $88\frac{8}{9}$ cents per dwt. The gold coins of France, of their present standard, are to be valued at the rate of 100 cts. for $27\frac{1}{2}$ grains, or $87\frac{1}{4}$ cts. per dwt. The gold coins of Spain, of their present standard, are to be valued at the rate of 100 cts. for $28\frac{1}{2}$ grains, or 84 cts. per dwt.

In England, Ireland, and the English West India islands, accounts are kept in pounds, shillings, pence, and farthings; though the intrinsic value, in each place, is not the same.

Exchange is said to be at par between two countries, or states, when the money given in one is equivalent in value to that received for it in another.

The course of exchange is fluctuating, being above or below par, according to the occurrences of trade, or the demand for money.

A Bill of Exchange is a written order for the payment of a certain sum of money, at an appointed time. It is a mercantile contract, in which four persons are mostly concerned, viz.:

---

* This rule may be omitted until the review.

First. The drawer, who receives the value, and is also called the *maker* and *seller* of the bill.

Second. The debtor in a distant place is one on whom the bill is drawn, and who is called the *drawee.* He is also called the *acceptor*, after he accepts the bill, which is an engagement to pay it when due.

Third. The person who gives the value for the bill, and is called the *buyer, taker*, and *remitter*.

Fourth. The person to whom the bill is ordered to be paid, who is called the *payee*, and who may, by endorsement, pass it to any other person.

Art. 193.—The following tables show the par value of foreign money in the United States.

TABLE IV.

Coins current in the United States, with their Sterling and Federal value.

NAMES OF COIN.	Standard weight		Sterling money of Great Britain.			N. E. States.			N. York and N. Carolina.			N. Jers'y Penn., Del., and Maryl'd.			S. Carolina and Georgia.			Federal value.		
Gold.	*pwt.*	*g.*	£	*s.*	*d.*	£	*s.*	*d.*	£.	*s.*	*d.*	£.	*s.*	*d.*	£	*s.*	*d.*	$	*c.*	*m.*
A Johannes,	18	0	3	11	0	4	16	0	6	8	0	6	0	0	4	0	0	16	00	0
A half Johann.	9	0	1	16	0	2	8	0	3	4	0	3	0	0	2	0	0	8	00	0
A Doubloon,	16	21	3	6	0	4	8	0	5	16	0	5	12	6	3	10	0	14	93	3
A Moidore,	6	18	1	7	0	1	16	0	2	8	0	2	5	0	1	8	0	6	00	0
An Eng. Guin.	5	6	1	1	0	1	8	0	1	17	0	1	15	0	1	1	9	4	66	7
A French do.	5	5	1	1	0	1	7	6	1	16	0	1	14	6	1	1	5	4	60	0
A Span. Pistole	4	6	0	16	6	1	2	0	1	9	0	1	8	0	0	18	0	3	77	3
A French do.	4	4	0	16	0	1	2	0	1	8	0	1	7	6	0	17	6	3	66	7
Silver.																				
English or Fr. Crown,	18	0	0	5	0	0	6	8	0	8	9	0	8	3	0	5	0	1	10	0
Doll. of Spain Sweden, or Denmark,	17	6	0	4	6	0	6	0	0	8	0	0	7	6	0	4	8	1	00	0
Eng. Shilling,	3	18	0	1	0	0	1	4	0	1	9	0	1	8	0	1	1	0	22	2
A Pistareen,	3	11	0	0	10$\frac{3}{4}$	0	1	2	0	1	7	0	1	6	0	0	11	0	20	0

All other gold coins, of equal fineness, at 89 cents per dwt., and silver at $1.11 cents per oz.

QUESTIONS.—1. What is foreign exchange? 2. When is exchange said to be at par? 3. What is the meaning of *par?* Answer: it is a *Latin* word, which signifies *equal.*

TABLE V.

Art. 194.—*Value of Foreign Coins in Federal Money, as established by a late Act of Congress.*

	d.	c.	m.
Pound Sterling*	4	44	4
Pound, of Ireland	4	10	0
Pagoda, of India	1	94	0
Tale, of China	1	48	0
Mill-ree, of Portugal	1	24	0
Ruble, of Russia	0	66	0
Rupee, of Bengal	0	55	5
Guilder, of the United Netherlands	0	39	0
Mark Banco, of Hamburgh	0	33	5
Livre Tournois, of France	0	18	5
Real Plate, of Spain	0	10	0

* £1 sterling before 1832 was $4.444; since that time $4.80.

TABLE VI.

Art. 195.—*Moneys of different countries.*

FRANCE.

12 derniers = 1 sol.
20 sols = 1 livre = 18½ cts.
3 livres = 1 crown.

Obs.—The above is according to the old system; the present method of keeping accounts in France is in francs and centimes, or hundredth parts, thus:

10 centimes = 1 decime.
10 decimes = 1 franc = $.1873125.

The 5-franc piece weighs 25 grammes, or 386.1 grains Troy, and is equal in value to $.9365625.

SPAIN.

The money of Spain is of two kinds; one is called vellon, the other plate money. Accounts are most generally kept in rials and marvadies vellon.

4 marvadies vellon, or 2⅛ marvadies of plate, } = 1 quarta.
8½ quartas, or 34 marvadies vellon,[1] } = 1 rial vellon.
15 rials vellon, - - = 1 peso, or current dollar.
16 quartas, or 34 marvadies of plate, } = 1 rial of plate = 10 cts.

8	rials of plate, - -	=1	piastre=80 cts.
10	rials of plate, - -	=1	dollar=$1.00.
5	piastres, - - -	=1	Spanish pistole=$4.

ITALY.

12	derniers	=1	sol.
20	sols	=1	livre.
5	livres	=1	piece of eight at Genoa.
6	livres	=1	piece of eight at Leghorn.
6	solidi	=1	gross.
24	grosses	=1	ducat.

PORTUGAL.

400 reas=1 crusado.
1000 reas=1 millrea=$1.24.

The reas and millreas are imaginary pieces of money; the real moneys of Portugal are as follows:

Silver.

1	crusado	=400 reas	=50 cts.
12	vintin piece	=280 reas	=30 cts.
5	do.	=100 reas	=12½ cts.
2½	do.	= 50 reas	= 6¼ cts.

Gold.

1	double johannes	=25	millreas,	600	reas=$32.
1	single do.	=12	do.	800	reas=$16.
	half do.	= 6	do.	400	reas=$8.
	quarter do.	= 3	do.	200	reas=$4.
	eighth do.	= 1	do.	600	reas=$2.
	festoon, or $\frac{1}{16}$	=		800	reas=$1.
1	moidore	= 4	do.	800	reas=$6.

HOLLAND.

8	pennings	=1	groat - - -	=01 ct.
2	groats	=1	stiver - - -	=2*d*., or 2 cts.
6	stivers	=1	shilling - -	=12 cts.
20	stivers	=1	florin, or guilder	=40 cts.
2½	florins	=1	rix dollar - -	=$1.00.
6	florins	=1	pound Flemish	=$2.40.
5	guilders	=1	ducat - - -	=$2.00.

DENMARK.

16 schillings = 1 mark = \$0.33⅓.
3 marks = 1 rix dollar = \$1.00.
6¼ marks = 1 ducat = \$2.08⅓.

RUSSIA.

3 copecs = 1 altima.
10 do. = 1 grivena.
50 do. = 1 politin.
2 politin = 1 ruble = 75 cts.
2 rubles = 1 ducat.

CHINA.

10 caxa = 1 candareen = \$.0148.
10 candareens = 1 mace = \$.148.
10 mace = 1 tale = \$1.48.

BARBARY.

10 aspers = 1 rial = \$0.12½.
2 rials = 1 double = .25 cts.
4 doubles = 1 dollar = \$1.00.
24 medins = 1 chequin = .75 cts.
32 do. = 1 dollar = \$1.00.
180 aspers = 1 zequin = \$2.25.
15 doubles = 1 pistole = \$3.75.

TURKEY.

3 aspers = 1 para = \$0.00652925.
40 paras = 1 piastre = \$0.26117647.

BILLS OF EXCHANGE.

Art. 196.—To find the value of bills of exchange above par.

EXAMPLES.

1. What is the value of a bill of exchange for \$860, at 5 per cent. above par?

Operation.

What value \$	860 \$
If \$100	105 \$
\$	903 *Ans.*

16*

2. A. of Boston, is indebted to C. of London, £1000. How much sterling must be remitted, exchange being 50 per cent.?
Ans. £1500.

3. B. of New York, is indebted to D. of Liverpool, £650 sterling, to discharge which he purchases a bill at 3 per cent. above par. How many dollars does he give for it?
Ans. $2975.258.

Art. 197.—To find the value of bills of exchange below par.

1. What sum sterling money is equal to £340 6*s.* 4*d.* Massachusetts currency, exchange 40 per cent.?
Ans. £204 3*s.* 9*d.* 2*qrs.*

Reduce the shillings and pence to the decimal of a pound by inspection.

Operation.

```
    |340.317
100 |60
----|--------
    |204.1902=£204 3s. 9d. 2qrs. Ans.
```

2. D. in Philadelphia, owes E. in London, £600 sterling, to discharge which he purchases a bill at 3 per cent. below par. How many dollars must he give? *Ans.* $2586.666+.

(3.)

EXCHANGE FOR £540 8*s.* 9*d.* STERLING. *Boston,* ——.

At thirty days' sight, pay to TIMOTHY DICKS, or order, five hundred and forty pounds, eight shillings, and ninepence, value received, and place the same to the account of

To JOHN JOHNSON, Merchant, Liverpool. JAMES STRIKER

What is the value of this bill in Pennsylvania currency, exchange at 56 per cent.?

4. A. of Cork, draws upon B. of London, for £870 12*s.* 4*d.* Irish, exchange at 8 per cent. How much sterling will discharge this bill?

(5.)

EXCHANGE FOR 2446 LIVRES, 6 SOLS, 4 DERNIERS.

Thirty days after sight of this my second of exchange, first of same tenor and date not paid, pay to TITUS TRUE, or order, two thousand four hundred forty-six livres, six sols, four derniers, value received, and place the same to my account.

PETER J. TUTTLE.

To TRUSTRAM CROCKER, Merchant, Paris.

How much sterling is the above bill, and how much in Pennsylvania currency?

6. A merchant in Toulon is indebted to a merchant in Boston 8462 francs, 20 centimes. What is the amount in federal money?

7. A. of Albany, buys a draft on C. of Paris, of 8846 francs, 34 centimes, for $1600. What is the rate of exchange?

8. Q. of Barcelona, is indebted to P. of New York, 925 piastres, 3 rials, 24 marvadies plate. How much, in federal money, is Q. charged in P.'s book? *Ans.* $740.37.

9. C. of Ireland, remits to D. of London, £345 10*s.* Irish. With how much sterling must C. be credited, exchange being 8 per cent.? *Ans.* £319 18*s.* 1*d.* $3\frac{4}{9}$*qrs.*

10. A bill for 3625 pesos, 4 rials, 31 marvadies, being remitted to Cadiz, what sum New Jersey money is equal to it, at 7*s.* 6*d.* per peso? *Ans.* £1359 9*s.* 11*d.* 1*qr.*

11. A Virginia merchant shipped tobacco to Norway worth £1673 18*s.*, Virginia currency. How many rix dollars, at 6*s.* each, must he receive? *Ans.* $5579.666.

12. A. in Philadelphia, owes B. of Amsterdam, $750. How many guilders is it, at 40 cts. per guilder? *Ans.* 1875.

13. What sum must be paid in Savannah for an invoice of goods charged at 490 florins, 15 stivers, allowing the exchange at 40 cents per florin, and freight and duties 30 per cent.? *Ans.* 255.19.

14. A merchant in Philadelphia receives of a merchant in Amsterdam an invoice of goods, amounting to 12340 florins, 19 stivers, 12 pennings. How much must be remitted, in Pennsylvania currency, to discharge the bill, at $35\frac{1}{4}$*d.* per florin, and what sum in sterling, exchange at 38*s.* 6*d.* Flemish per pound sterling? *Ans.* £941 12*s.* 0*d.* 0*qrs.* sterling.

15. In 16745 marks, how many dollars, allowing $33\frac{1}{3}$ cents per mark? *Ans.* $5581.666.

16. In 2045 piastres, 9 rials plate, how many dollars? *Ans.* $1636.90.

17. What will 8400 arsheens of ravens duck cost, at 15 rubles for 45 arsheens, in rubles, and also in federal money? *Ans.* 2800 rubles; $2100.

18. A. of Bordeaux, draws on B. of Liverpool, for 1400 crowns, at 55*d.* sterling per crown; for the value of which B. draws again on A. at 56*d.* sterling per crown; besides com-

mission of $\frac{1}{2}$ per cent. What did A. gain or lose by this transaction? *Ans.* Gained 18 crowns.

19. In $1820, how many pagodas of India?

20. In $605, how many rupees of Bengal?

21. How many dollars in 4678 tales, 8 mace, 7 candareens?

22. In 8000 aspers, how many dollars? *Ans.* $100.

23. A merchant in Philadelphia imported from England 700 ells of cloth, at 5 shillings sterling per ell. The cost of transportation and duty, on the whole amount, was 35 per cent., the exchange at par. For how many cents must 1 yard be sold in Philadelphia, to gain $12\frac{1}{2}$ per cent.?

Ans. 135 cents.

REDUCTION OF CURRENCIES.

Art. 198.—Reduction of Currencies teaches to reduce pounds, shillings, pence, &c., to federal money, and the reverse.

RULE.

Reduce the dollar to the fraction of a pound; and, if there be shillings, pence, and farthings, in the given sum, reduce them to the decimal of a pound, by inspection, and proceed as in the Rule of Three.

1. In £63, New England and Virginia currency, how many dollars?

Operation.

How many $	6̸3̸ £ 21
If £3̸	10
	1$
	$210 *Ans.*

If $1 is £$\frac{3}{10}$, then it is evident that the quotient of £1 divided by $\frac{3}{10}$ would be the number of dollars in 1 pound, and so of any number of pounds.

2. In $210, how many pounds, New England and Virginia currency?

Operation.

How many £	$21̸0̸
If $1	3£
1̸0̸	
	£63 *Ans.*

If £$\frac{3}{10}$ is $1, then the product of £$\frac{3}{10}$ multiplied by any number of dollars, will be the number of pounds required.

Questions.—1. What does Reduction of Currencies teach? 2. Rule for reducing pounds, shillings, and pence, to Federal Money? 3. Rule for reducing Federal Money to pounds, shillings, pence, etc.?

3. In £240 10*s.* Virginia, &c., currency, how many dollars, cents, and mills ?

4. In $801.666+ how many pounds and shillings, New England, &c., currency ?

5. Reduce £210 15*s.* Virginia, &c., currency, to federal money.

6. Reduce $702.50 to Virginia, &c., currency.

7. What sum in federal money is equal to £300 10*s.* 6*d.*, New England and Virginia currency ?

8. What sum in New England and Virginia currency is equal to $1001.75 ?

9. Reduce £380 9*s.* North Carolina currency, to federal money.

10. Reduce $951.125 to North Carolina currency.

11. Reduce £67 5*s.* 3*d.*, New Jersey currency, to federal money.

12. Reduce $179.366+ to New Jersey currency.

13. What sum in federal money is equal to £67 5*s.* 3*d.*, South Carolina and Georgia currency ?

14. What sum in South Carolina and Georgia currency is equal to $288.265 ?

15. Reduce £102 17*s.* 4*d.*, South Carolina, &c., currency, to federal money.

16. Reduce $440.858 to South Carolina, &c., currency.

17. Reduce £630 6*s.* 4½*d.*, Pennsylvania currency, to federal money.

18. Reduce $1680.85 to Pennsylvania, &c., currency.

19. Reduce £600 10*s.* 6*d.* sterling, to federal money, the dollar being 4*s.* 6*d.*

20. Reduce $2669 to sterling money.

21. What sum in federal money is equal to £126 14*s.* Canada and Nova Scotia currency, the dollar being 5*s.* ?

22. What sum in Canada and Nova Scotia is equal to $506.80 ?

23. Reduce £346 16*s.*, New York, &c., currency, to federal money.

24. Reduce $867 to New York and North Carolina currency.

25. Reduce £125 7*s.* 9*d.*, Maryland, &c., currency, to federal money.

26. Reduce $334.365+ to Maryland, &c., currency.

27. Reduce £501 3*s.* 9*d.*, Massachusetts currency, to federal money.

28. Reduce $1670.623 to Massachusetts currency.

29. Reduce 450*d.*, New Jersey, &c., currency, to cents. (See Table III.)

Operation.

How many cents?	450 pence.
d. 9	10 cents.
	500 cts. *Ans.*

30. Reduce 500 cents to New Jersey currency.

Operation.

How many pence?	50~~0~~
cents ~~10~~	9*d.*
	450 pence.

31. Reduce 540 pence, New England currency, to federal money.

32. Reduce 750 cents to pence, New England currency.

PERCENTAGE.

Art. 199.—The consideration of profit and loss adds the chief interest to all business operations. It is necessary, therefore, that there should be some standard by which all should agree to make their estimates: 100 has been adopted, and hence gain and loss are said to be so much per centum; that is, so much by the hundred. The gain or loss *per centum* is called *percentage.* The individual who makes 20 per cent. profit on his goods, makes a high *percentage;* and he who makes but 4 per cent. makes a low *percentage.* Since, therefore, 100 denominates or is the denominator of the gain, it is plain the gain itself will be the numerator. If the gain be 5 per cent., it would be expressed thus: $\frac{5}{100}=.05$.

1 per cent. equals $\frac{1}{100} = .01$
2 per cent. equals $\frac{2}{100} = .02$
3 per cent. equals $\frac{3}{100} = .03$
4 per cent. equals $\frac{4}{100} = .04$
5 per cent. equals $\frac{5}{100} = .05$
6 per cent. equals $\frac{6}{100} = .06$

Whatever, therefore, be the amount of capital invested, the gain or loss will be so many hundredths of the capital, to be added to, or subtracted from it. The gain or loss on any sum is to be calculated by the rule for the multiplication of decimals.

1. What is 1 per cent. of 20 dollars? *Ans.* 20 cents.
2. What is 2 per cent. of 40 dollars? *Ans.* 80 cents.
3. What is 3 per cent. of 50 dollars? *Ans.* $1.50.
4. What is 4 per cent. of 75 dollars? *Ans.* $2.00.
5. What is 5 per cent. of 90 dollars? *Ans.* $4.50.
6. What is 6 per cent. of 100 dollars? *Ans.* $6.00.
7. What is 7 per cent. of 250 dollars? *Ans.* $17.50.
8. What is 8 per cent. of 375 dollars? *Ans.* $30.00.

Under the general head *Percentage,* may be reckoned *Interest, Discount, Insurance, Commission, Loss* and *Gain.*

INTEREST.

Art. 200.—INTEREST is a premium paid, or an allowance made by the borrower to the lender, for the use of a certain sum of money.

The money lent, upon which interest is to be received, is called the *principal.*

The premium paid for the use of the *principal,* is called the *interest.*

The sum paid on $100, or 100 cents, or £100 per annum, is called the *rate* per cent., or per centum. (*Per centum* signifies by the hundred; *per annum,* by the year.)

The principal and interest added together, is called the *amount.*

OBS.—The rate of interest established by law in the New England states, is 6 per cent. In New York the legal interest is 7 per cent. In England it is 5 per cent. When the rate is not mentioned in this work, 6 per cent. is understood.

Interest is either *simple* or *compound.*

SIMPLE INTEREST.

Art. 201.—SIMPLE INTEREST is that which arises from the principal only.

What is the interest of $12 for 1 year, at 6 per cent?

Operation.	
$12	If the interest of $1 for 1 year be six cents, or $\frac{6}{100}$ of a dollar, then the interest of $12 would be 12 times .06, or $.06\times12=72$ cents.
.06	
.72 *Ans.*	

QUESTIONS.—1. What is interest? 2. What do you understand by the principal? 3. What do you understand by the rate per cent.? 4. What does *per cent.* signify? What, *per annum?* 5. What is the *amount?* 6. What is the legal rate of interest in New England? 7. What in New York? 8. What is *simple* interest? 9. Rule to obtain the interest for one year? 10. Why is the rate per cent. written so many hundredths of a dollar?

OBS. 1.–The rate per cent. is written as so many hundredths of a dollar: thus 6 per cent. is written .06; 7 per cent., .07; 5 per cent., .05; 2 per cent., .02. It is evident that the rate per cent. must be written so many hundredths, because, being so many cents on every hundred cents, it is so many 100ths of a dollar.

EXAMPLES.

1. What is the interest of \$30, for 2 years, at 5 per cent.?

Operation.	
\$30 .05 1.50 2 \$3.00	If the interest of \$1, for 1 year, at .05 per cent. be 5 cents, then the interest of \$30 will be $.05 \times 30 = 1.50$, and for 2 years, $\$1.50 \times 2 = \$3.00$. Hence, to compute the interest for 1 or more years, we have the following

RULE.

Multiply the principal by the rate, expressed as the decimal of a dollar, and the product will be the interest for one year. When the time is more than one year, multiply the interest for one year by the number of years.

2. What is the interest of \$45, for 1 year, at 6 per cent.?

Ans. \$2.70.

3. What is the interest of \$22.25 for 1 year, at $5\frac{1}{2}$ per cent.?

Reduced to a decimal.

$.05\frac{1}{2} = \frac{.11}{2} = .055$

Operation.	*Operation.*
\$22.25	Or thus: $\frac{1}{2}$)\$22.25
.055	$.05\frac{1}{2}$
11125	11125
11125	1112
\$1.22375	\$1.2237

OBS. 2.–The decimals below mills are not regarded in the answer in this, or the following questions. For pointing the product, see *Multiplication of Decimals.*

4. What is the interest of \$62.75, for 2 years, at 3 per cent.?

Ans. \$3.765.

5. What is the interest of \$535.42, for 4 years, at 2 per cent.?

Ans. \$42.833.

6. What is the interest of \$115.675, for 1 year, at $7\frac{1}{2}$ per cent.? at $6\frac{1}{2}$ per cent.? at $8\frac{3}{4}$? at $9\frac{1}{2}$? at $12\frac{2}{3}$ per cent.?

7. What is the interest of $450.50, for 3 years, at 6 per cent.? *Ans.* $81.09.

In the preceding examples, the interest has been computed for 1 or more years; but it is often necessary to calculate the interest for months and days. Now, as the interest on $1, at .06 per cent., for 1 year, or 12 months, is 6 cents, it is evident that it amounts to half a cent a month, or 12 half cents a year on a dollar. If, therefore, we multiply any number of dollars by half the number of months, we shall have the interest for the time in cents. Again: as 1 month is 30 days, and the interest for 1 month is $\frac{1}{2}$ cent, or 5 mills, for 1 day it would be $\frac{5}{30}=\frac{1}{6}$ of a mill. If, therefore, we multiply by $\frac{1}{6}$ of the days, we have the interest in mills; or, we may reduce the days to the fraction of a month, and multiply by half the fraction.

8. What is the interest of $40, for 1 year, 6 months, and 5 days?

Operation.

$\frac{5}{6}=\frac{1}{3}$ & $\frac{1}{2}$)$40
.090$\frac{5}{6}$
360
20
13
$3.633 *Ans.*

If the interest of $1 for 12 months be .06 cents, the interest for 6 months will be .03 cents, and for 5 days, $\frac{5}{6}$ of a mill; therefore, the interest of $1 for 12 mos., 6 mos. and 5 days, will be .06+.03+.000$\frac{5}{6}$=.090$\frac{5}{6}$=the same as one half the months, and one sixth of the days. Hence the

RULE.

Art. 202.—When there are months and days in the given time—*Multiply by half the number of months in the whole time, and one sixth of the days. If there be an odd month, call it* 30 *days, to which add the odd days, if any; and, dividing them by* 6, *write the quotient in the place of mills, in the multiplier.*

OBS. 1.—If the interest is required for a number of years, multiply the interest for 1 year by the number of years, and compute the interest for the months and days, as above directed.

EXAMPLES.

9. What is the interest of $275, for 2 years, 5 months, and 6 days? *Ans.* $40.15.

QUESTIONS.—11. What is the rule for pointing off the product? 12. What is the rule for computing interest for months and days? 13. Why do we multiply by one half the months, and one sixth of the days?

10. What is the interest of $749.605, for 3 years, 7 months, and 15 days?

11. What is the interest of $342, for 1 month, 15 days?

Operation.

```
 2)342
   .007½
 ------
  2.394
    171
 ------
$2.565 Ans.
```

OBS. 2.—As there is no even number of months, we supply the two first decimal places with ciphers, as a guide in pointing off the product.

12. What is the interest of $678.59, for 1 year, 3 months, and 11 days?

13. What is the amount of $678.59, on interest, for 1 year, 3 months, and 11 days? *Ans.* $730.728.

OBS. 3.—The *amount* is the principal and interest added together.

14. What is the interest of $600, for 27 days?

15. What is the amount of $750.60, on interest, for 18 mos. and 18 days?

16. What is the amount of $1000, on interest, for 4 years and 6 months? *Ans.* $1270.

17. A note for $450, on interest, was dated January 1st, 1835. What was due, principal and interest, March 16th, 1837? *Ans.* $509.625.

yrs.	mo.	d.	
1837	3	16	
1835	1	1	
2	2	15	time.

18. A note for $60.50, on interest, was dated Dec. 20, 1834. What was there due, principal and interest, Jan. 28, 1837? *Ans.* $68.143.

19. What is the amount of $879.30, on interest, 2 years, 5 months, and 19 days? *Ans.* $1009.582.

20. What is the interest of $375 for 7 days? *Ans.* $.437.

21. What is the interest of $89.285, for 1 year, 7 months, and 29 days? *Ans.* $8.913.

22. What is the interest of $336 for 5 months and 16 days? *Ans.* $9.296.

23. What is the amount of $1844.48, on interest 2 months and 21 days? *Ans.* $1869.38.

24. What is the amount of $2731.50, on interest 3 years, 9 months, and 26 days? *Ans.* $3357.924.

25. What is the amount of $1764, on interest from June 14, 1829, to July 14, 1837? *Ans.* $2619.54.

26. What is the interest of £240 8*s.* 6¾*d.*, for 1 year?

Operation.	
£240.428	Reduce the shillings, pence, and farthings, to the decimal of a pound by inspection, (see Art. 134;) then proceed as in Federal Money. The interest will be in pounds and decimal parts, which must be reduced to shillings.
.06	
14.42568=	
£14 8*s.* 6*d.* *Ans.*	

27. What is the interest of £379 15*s.*, for 1 year and 6 months? *Ans.* £34 3*s.* 6½*d.*

28. What is the interest of £416 12*s.* 6*d.*, for 10 months? *Ans.* £20 16*s.* 7½*d.*

29. What is the interest of £427 13*s.* 9*d.* 2*qrs.*, for 1 year and 8 months? *Ans.* £42 15*s.* 4½*d.*

30. What is the interest of £129 7*s.* 3*d.* 3*qrs.*, for 3 years, 7 months, and 5 days? *Ans.* £27 18*s.* 6½*d.*

31. What is the amount of £320 10*s.* 6*d.*, on interest for 2 years, 6 months, and 15 days? *Ans.* £369 8*s.* 1¼*d.*

32. What is the interest of £430 7*s.* 8*d.* 3*qrs.*, for 4 years, 3 months, and 20 days? *Ans.* £111 3*s.* 7¾*d.*

Art. 203.—When the rate of interest is any other than six per cent., and the time consists of years, months, and days,

RULE.

Find the interest first for 6 *per cent., and then for* 1 *per cent., and multiply the interest at* 1 *per cent. by the given rate, and the product will be the answer.*

EXAMPLES.

33. What is the interest of $680, for 1 year and 6 months, at 7 per cent.? *Ans.* $71.40.

Operation.

```
   $680
    .09
 6)61.20 interest at 6 per cent.
   10.20 interest at 1 per cent.
       7
  $71.40 interest at 7 per cent.
```

QUESTIONS.—14. What is the rule for computing interest on pounds, shillings, pence, etc.? 15. Rule, when the rate of interest is any other than 6 per cent.?

34. What is the interest of \$336.40, for 2 years, 8 months, and 3 days, at 3 per cent.? *Ans.* \$26.996

35. What is the interest of \$556.36, for 3 years, at 1 per cent.? *Ans.* \$16.69.

Obs.—The interest of any sum at 1 per cent., for 1 year, is the principal itself, with the separatrix moved two figures towards the left; therefore, to obtain the interest at 1 per cent., for any number of years, we have only to multiply by the number of years.

36. What is the interest of \$0.56 cents, for 5 years, 5 months, and 10 days, at 9 per cent.? *Ans.* \$.274.

37. What is the amount of \$1000, on interest for 5 years and 7 months, at $7\frac{1}{2}$ per cent.? *Ans.* \$1418.75.

38. What is the interest of \$1569.20, for 1 year, at 1 per cent.? *Ans.* \$15.692.

INTEREST BY CANCELLING.

RULE.

State the question, as in Direct Proportion, by placing the terms of demand on the right, and the terms of supposition on the left.

EXAMPLES.

Art. 204.—1. What is the interest of \$500, for 3 years, at 6 per cent.?

Obs. 1.—The terms of supposition in Interest are not expressed, being always 100 and 1 year. The foregoing question may be expressed thus:

What is the interest of \$500, for 3 years, if the interest of \$100 for 1 year be \$6?

Operation.

What interest ?	5~~00~~ \$
If \$1~~00~~	3 years.
Year 1	6 \$
	\$90 *Ans.*

2. What is the interest of \$720, for 1 year and 6 months, at 6 per cent.? *Ans.* \$64.80.

Question.—16. What is the interest of any sum for 1 year at 1 per cent.?

OBS. 2.—When the given time is months, weeks, or days, either less or greater than a year, reduce it to the lowest denomination, and 1 year, the time in the supposition, to the same denomination.

3. What is the interest of $642.255, for 2 years and 6 months? *Ans.* $96.338.

4: What is the interest of $1000.68, for 2 months and 15 days? *Ans.* $12.508.

5. What is the interest of $440, for 4 years, at 4 per cent.? *Ans.* $70.40.

6. What is the interest of $60.10, for 5 years, at 5 per cent.? *Ans.* $15.025.

7. What is the interest of $160, for 36 days, at 7 per cent.? *Ans.* $1.12.

8. What is the amount of $780, for 3 years and 4 months, at 3 per cent.? *Ans.* $858.

OBS. 3.—If we multiply the amount of $1 for the given time, by the given principal, the result will be the same as adding the principal to the interest. Thus, the amount of $1 for 3 years and 4 months, at 3 per cent., is $1.10, which, multiplied by $780, gives $858, the answer.

Art. 205.—When time, rate, and amount are given, to find the principal.

1. What principal will amount to $858, in 3 years and 4 months, at 3 per cent.?

The student will perceive, that this question is the reverse of question 8th, preceding, and also that 858 is there a product, of which 1.10, the amount of $1 for the given time, is a factor; therefore, if we divide 858 by 1.10, we shall obtain the other factor, or the principal required; $858 \div 1.10 = \$780$, the answer. Hence the

RULE.

Divide the given amount by the amount of $1 *for the given time, and the quotient will be the answer.*

EXAMPLES.

2. What principal will amount to $778.10, in 4 years and 3 months, at 6 per cent.? *Ans.* $620.

3. What principal will amount to $650, in 6 years, at 5 per cent.?

Operation by cancelling.

What principal.	$650 amount.
Amount, $1.30	$1 principal.
	$500 *Ans.*

4. What principal will amount to $738.40, in 7 years?
Ans. $520.

Art. 206.—When time, rate, and interest are given, to find the principal.

1. What principal will gain $27.52, in 1 year, 5 months, and 6 days?

We have seen, that the interest of a given principal for a given time, is the product of the interest of $1 for the same length of time, and the principal; therefore, if we divide $27.52 by .086, the interest of $1 for the given time, we shall obtain the principal required, as before. Hence the

RULE.

Divide the given interest by the interest of $1 *for the given time, and the quotient will be the answer.*

2. What principal will gain $19 in 4 months, at 6 per cent.?
Ans. $950.

3. What principal will gain $1500 in 5 years, at 6 per cent.?
Ans. $5000.

Art. 207.—When principal, interest, and time are given, to find the rate per cent.

1. If $50 in 6 months gain $1.50, what is the rate per cent.?

If the interest of $50, at 1 per cent., be 25 cents, then the quotient of $1.50, the whole interest, divided by 25 cents, will be the rate per cent. required. $1.50 \div 25 = 6$ per cent., the answer. Hence the

RULE.

Divide the given interest by the interest on the given principal, at 1 *per cent. for the given time, and the quotient will be the answer.*

2. If $300 gain $12 in 8 months, what is the rate per cent.?

Operation by cancelling.

What interest.	100 $
If $300	12 m.
m. 8	12 $ int.
Ans.	6 per cent.

3. If $740 gain $27.75 in 9 months, what is the rate per cent.? *Ans.* 5 per cent.

4. If $1000 gain $75 in 6 months, what is the rate per cent. ? *Ans.* 15 per cent.

Art. 208.—When principal, rate, and interest are given, to find the time.

1. In what time will $300 gain $12, at 6 per cent. ?

Operation.

In what time.	12$ Int.
Int. $18	1 year.
3	2=8 months.

Having found the interest of $300 for 1 year, the question may be expressed thus: In what time will $12 interest be gained, if $18 be gained in 1 year? It is evident, that the ratio of the interest for 1 year, is to the given interest, as 1 year is to the time required. Hence the

RULE.

Divide the given interest by the interest of the given principal for 1 *year, and the quotient will be the answer.*

2. In what time will $240 gain $4.80, at 6 per cent. ? *Ans.* 4 months.

3. In what time will $600 amount to $645, at 5 per cent. ? *Ans.* 1 year and 6 months.

4. In what time will $375 gain $28.12½, at 6 per cent. ? *Ans.* 1 year and 3 months.

5. The interest on a note of $225, at 4 per cent., was $11.40. What was the time ? *Ans.* 1 year, 3 months, 6 days.

PARTIAL PAYMENTS.

Art. 209.—When notes are paid within one year from the time they become due, it has been the usual custom to find the amount of the principal from the time it became due, until the time of settlement, and then to find the amount of each endorsement, from the time it was paid, until settlement, and to subtract their sum from the amount of the principal.

EXAMPLES.

Boston, January, 1, 1841.

For value received, I promise to pay SAMUEL FULTON, or order, two hundred and fifty dollars and forty cents, in three months, with interest afterwards. ELIHU JONES.

On the back of this note were the following endorsements: March 15, 1841, received one hundred and fifty dollars. June 10, 1841, received forty-five dollars. The balance on the note was paid January 1st, 1842. How much was the balance?

First payment,	$150	2d payment,	$45	Principal,	$250.40
Int. 9 m. 16 d.	7.15	Int. 6 m. 21 d.	1.507	Int. 9 m.	11.268
	$157.15		$46.507		$261.668
			157.15		203.657
		Amount of payments,	$203.657	Balance,	$58.011

Concord, Sept. 1, 1840.

For value received, I promise to pay JOHN FOSTER & Co., or order, one thousand dollars, on demand, with interest.

STEPHEN PAYWELL.

On this note are the following endorsements: March 1, 1841, received two hundred dollars. April 6, 1841, received one hundred and fifty dollars. July 5, 1841, received two hundred and forty dollars. What was there due at the time of settlement, which was August 15, 1841? *Ans.* $457.042.

If settlement is not made till more than a year has elapsed after the commencement of interest, the preceding mode of computing interest, when partial payments have been made, is not in strict conformity with law.

The methods of computing interest on notes and bonds differ in different places.

The United States Court, and the courts of several of the states, have established a general rule for the computation of interest, when partial payments have been made. The following is, in substance, the

RULE.

Compute the interest up to the time of the first payment; and if the payment exceed the interest, deduct the excess from the principal, and cast the interest on the remainder up to the second payment, and so on. If the payment be less than the interest, cast the interest up to the time when the sum of the payments shall exceed the interest; then deduct the excess from the principal, and proceed as before.

When a note is given specifying interest annually, simple interest is cast on the note to the time of final settlement; and also simple interest on the several sums of interest from the time they became due to the time of final settlement.

$3784.25.

(1.)

For value received, I promise to pay JAMES LARNED, or order, three thousand seven hundred eighty-four dollars and twenty-five cents, with interest.

July 10, 1826.

JOHN FISHER.

On this note were the following endorsements:

Jan. 16, 1827,	received	$148.21
Aug. 11, 1827,	"	50.00
Dec. 24, 1828,	"	2789.25
Feb. 12, 1830,	"	1000.00

What was due Dec. 14, 1830? *Ans.* $464.867.

The first principal, on interest from July 10, 1826,		$3784.25
Interest to Jan. 16, 1827, time of the first payment, (6 months, 6 days,)		117.311
		$3901.561
Payment exceeding the interest, Jan. 16		148.21
Remainder for a new principal		$3753.351
Interest from Jan. 16, 1827, to Dec. 24, 1828, (1 year, 11 months, 8 days,)		436.639
		4189.990
Payment, Aug. 11, less than the interest,	$50.00	
Payment, Dec. 24, exceeds the interest,	2789.25	
Sum of the payments,		2839.250
Remainder for a new principal		$1350.740
Interest from Dec. 24, 1828, to Feb. 12, 1830, (1 year, 1 month, 18 days,)		91.850
		1442.590
Payment, Feb. 12, exceeds the interest,		1000.000
Remainder for a new principal		$442.590
Interest from Feb. 12, 1830, to Dec. 14, 1830, (10 months, 2 days,)		22.277
Balance due Dec. 14, 1830		$464.867

What would have been due on the foregoing note at the time of final settlement, had annual interest been specified?

(2.)

$6420.50. For value received, I promise to pay THOMAS TERRIL, or order, six thousand four hundred twenty dollars and fifty cents, with interest.

SAMUEL ENGLISH.

May 4, 1830.

On this note were the following endorsements:

March	4, 1831,	received	$40.00
Dec.	1, 1831,	"	200.00
Feb.	10, 1832,	"	5000.00
June	28, 1833,	"	1534.25

What was the sum due March 1, 1834? *Ans.* $500.784.

3. A.'s note of $374.62 was given Jan. 1, 1834, on interest after 90 days. June 4, 1836, he paid $320. What was due August 15, 1837? *Ans.* $110.942.

4. B.'s note of $654.32 was given Dec. 12, 1831, on which was endorsed the interest for 18 months and 4 days. What was due on settlement, Nov. 20, 1833? *Ans.* $671.114.

COMMISSION, BROKERAGE, AND INSURANCE.

Art. 210.—COMMISSION and BROKERAGE are compensations of so much per cent. to factors and brokers, for their respective services in buying and selling goods, etc.

INSURANCE is an exemption from hazard, obtained by the payment of a certain sum, which is generally so much per cent. on the estimated value of the property insured.

Premium is the sum paid by the owner of the property, for the insurance.

Policy is the name given to the instrument, or writing, by which the contract of indemnity is effected between the insurer and insured.

The Policy should always cover a sum equal to the estimated value of the property insured, together with the premium: that is, a policy to secure the payment of $100, at 3 per cent., must be made out for $103.

QUESTIONS.—1. What are Commission and Brokerage? 2. What is Insurance? 3. What is a Premium? 4. What is a Policy? 5. What sum should the Policy cover? 6. Give the example.

RULE.

Method of operation the same as in Simple Interest.

EXAMPLES.

1. If a factor purchase goods to the amount of $1800, and I allow him $\frac{3}{4}$ per cent. for his services, what must I pay him?

Operation.

```
100|1800  9
2 4|3
---|---
  2|27
   |$13½  Ans.
```

2. What commission must a factor receive for selling goods to the amount of $864.78, at $4\frac{1}{2}$ per cent? *Ans.* $38.915.

3. What is the commission on $3784.22, at $12\frac{1}{2}$ per cent? *Ans.* $473.027.

4. A factor buys goods to the amount of $1200. What will be his commission, at $1\frac{1}{2}$ per cent.? *Ans.* $18.

5. What is the brokerage on 9798.67\frac{1}{2}$, at $5\frac{3}{4}$ per cent.? *Ans.* $563.423.

6. The value of a certain ship and cargo is $50000. What is the insurance, at 15 per cent.? *Ans.* $7500.

7. What is the duty on 4 boxes of tea, each weighing 1 cwt. 2 qrs. 14 lbs., at $1\frac{1}{2}$ cent per lb.? *Ans.* $10.92.

8. What may a broker demand on $1000 at 3 per cent.? *Ans.* $30.

9. What will be the premium for insuring a ship and cargo, valued at $57840, at $3\frac{1}{2}$ per cent.? *Ans.* $2024.40.

10. What may a broker demand on £320 10*s.* 6*d.*, at 4*s.* 3*d.* per cent.? *Ans.* £68 2*s.* 2*d.* 3*qrs.*

OBS.—The above example is not reduced to decimals by Inspection.

11. What will be the premium for insurance on property to the amount of $9248.28, at $\frac{1}{2}$ per cent.? at $\frac{3}{4}$ per cent.? at $\frac{1}{3}$ per cent.? at $\frac{2}{3}$ per cent.? at $\frac{7}{8}$ per cent.? at $\frac{4}{5}$ per cent.?

COMPOUND INTEREST.

Art. 211.—COMPOUND INTEREST is interest upon interest, or that which arises from making the interest a part of the principal, whenever it becomes due.

RULE.

Find the amount of the given principal for the first year, or the first stated time for the interest to become due, by simple interest, and make the amount the principal, for the next year, or stated period; and so on to the last. From the last amount, subtract the given principal, and the remainder will be the compound interest required.

EXAMPLES.

1. What is the compound interest of $200, for 3 years, at 6 per cent.?

Operation.

```
     $200, first principal.
      .06
    -----
    12.00 interest.  \ to be added.
    200   principal. /
    -----
    212, amount, or principal, for 2d year.
     .06
    -----
    12.72, compound interest, 2d year. \ to be
   212     principal,           "      / added.
   ------
   224.72, amount, or principal, for 3d year.
     .06
   ------
   13.4832, compound interest, 3d year. \ to be
  224.72    principal,           "      / added.
  --------
  238.2032, amount.
  200       first principal, subtracted.
  --------
  $38.2032, the compound interest required.
```

2. What is the compound interest on a note of $325, on interest 5 years? *Ans.* $109.92.

3. What is the compound interest of $680, for 4 years? *Ans.* $178.479.

4. What is the compound interest of $500, for 4 years, at 7 per cent. per annum?

5. What is the compound interest of $470, for 5 years, at 5 per cent. per annum?

6. To what sum will $478 amount, in 3 years, at 6 per cent., compound interest?

QUESTIONS.—1. What is Compound Interest? 2 What is the Rule?

TABLE,

Showing the amount of $1, or £1, for any number of years, not exceeding 30 years, at the rates of 5 and 6 per cent. compound interest.

Years.	5 per cent.	6 per cent.	Years.	5 per cent.	6 per cent.
1	1.05	1.06	16	2.18287+	2.54035+
2	1.1025	1.1236	17	2.29201+	2.69277+
3	1.15762+	1.19101+	18	2.40661+	2.85433+
4	1.21550+	1.26247+	19	2.52695+	3.02559+
5	1.27628+	1.33822+	20	2.65329+	3.20713+
6	1.34009+	1.41851+	21	2.78596+	3.39956+
7	1.40710+	1.50363+	22	2.92526+	3.60353+
8	1.47745+	1.59384+	23	3.07152+	3.81974+
9	1.55132+	1.68947+	24	3.22509+	4.04893+
10	1.62889+	1.79084+	25	3.36355+	4.29187+
11	1.71033+	1.89829+	26	3.55562+	4.54938+
12	1.79585+	2.01219+	27	3.73345+	4.82234+
13	1.88564+	2.13292+	28	3.92012+	5.11168+
14	1.97993+	2.26090+	29	4.11613+	5.41838+
15	2.07892+	2.39655+	30	4.32194+	5.74349+

OBS. 1.—Although the decimals, in the preceding numbers, are carried to five places, yet four are generally sufficient for most business operations.

7. What is the compound interest of $650, for 6 years, at 6 per cent.? *Ans.* $272.031.

By the foregoing table we find the amount of $1 for 6 years to be $1.41851; which, multiplied by $650, gives $922.031, the amount of $650 for 6 years, and $922.031 – 650 = $272.031, the interest required.

8. What is the compound interest of $350 for 2 years and 6 months? *Ans.* $55.057.

OBS. 2.—When there are months and days, first find the amount for the *years*, and on this amount cast the interest for the months and days; this, added to the amount, will give the answer.

9. What is the compound interest of $135, for 3 years, 6 months, and 6 days? *Ans.* $30.77.

10. What is the compound interest of $678.25, for 12 years and 6 months, at 5 per cent.? *Ans.* $570.236.

11. What is the compound interest of $579.75, for 20 years?—for 30 years?

QUESTION.—3. What is the Rule, when there are months and days?

12. What is the amount of a note of $150, for 4 years, at 6 per cent., compound interest? *Ans.* $189.37.

13. The amount of a certain note, at compound interest for 4 years, was $189.37+. What was the principal?

This question, it will be perceived, is the reverse of the last. If the amount required is obtained by multiplying the amount of $1 for the given time by the given principal, then it follows, that if we divide the given amount by the amount of $1 for the given time, we shall obtain the required principal.

14. What is the amount of $597.75, for 20 years, at 6 per cent., compound interest? *Ans.* $1917.061.

15. What is the amount of $1350, for 3 years, at 5 per cent., compound interest? *Ans.* $1562.793.

16. What is the amount of a note for $150, for 2 years, compound interest, the interest becoming due at the end of every 3 months? *Ans.* $168.967.

17. What is the compound interest of £240 10*s.* 6*d.*, for 2 years, at 6 per cent.? *Ans.* £29 14*s.* 6*d.* 3*qrs.*

18. What is the amount of £450, for 3 years, at 5 per cent., compound interest? *Ans.* £520 18*s.* 7*d.*

19. What is the amount of £256 10*s.* for 7 years, at 6 per cent., compound interest? *Ans.* £385 13*s.* 7½*d.*

DISCOUNT.

Art. 212.—Discount is an allowance made for the payment of money before it becomes due.

The *present worth* of a debt due at any future period, is so much money as, being put on interest, at a given rate per cent., will amount to the debt, when it becomes due.

1. A. holds B.'s note for $106, due in 1 year: What is the present worth of the note, discounting at 6 per cent.?

It is evident, that if B. pays A. $106 now, at the end of the year, when the note becomes due, A. will have the interest of $106 more than is his due; therefore, B. ought to pay him such a sum, as, being put on interest, would amount to $106

Questions.—1. What is *discount?* 2. What is *present worth?*

at the end of the year. If we divide 106 by the amount of $1 for 1 year, we shall have the principal, or that sum which being put on interest at the usual rate per cent., will amount to the debt when it becomes due. (See Art. 205.) $106÷$1.06=$100, the present worth of $106 due a year hence. From the above we derive the following

RULE.

Art. 213.—To find the present worth—*Divide the given sum by the amount of $1 for the given time, and the quotient will be the* PRESENT WORTH.

The *present worth,* subtracted from the debt, will leave the *discount.*

2. What is the present worth of $246.21, payable in 2 years and 8 months, discounting at 6 per cent.?
Ans. $212.25.

3. How much ready money will purchase a note of $1719.04, due 6 years hence, discounting at 6 per cent.? *Ans.* $1264.

4. Suppose I owe a note of $416, to be paid in 4 years and 2 months, and wish to pay it now, what must be discounted for present payment? *Ans.* $83.20.

5. How much ready money will purchase a note of $37.165, due 5 years, 1 month, and 18 days hence, discounting at 6 per cent.? *Ans.* $28.413+.

6. What is the present worth of a note of $840, payable one half in 10 months, the other half in 20 months, discount, 6 per cent. per annum? *Ans.* $781.818.

7. What is the present worth of $1500, due 40 years hence, discount, 12 per cent. per annum? *Ans.* $258.62+.

8. What is the discount of $420, due in 1 year and 6 months, at 6 per cent.? *Ans.* $34.679+.

9. What is the discount of $109.86, for 1 year, at 6 per cent.?
Ans. $6.219.

10. Bought goods to the amount of $1909.34, at four months' credit. How much ready money must I pay, discounting at $3\frac{1}{2}$ per cent.? *Ans.* $1887.322.

11. What is the present worth of £4000, payable in 9 months, at $4\frac{3}{4}$ per cent. discount?
Ans. £3862 8*s.* 0*d.* 2*qrs.* +

QUESTIONS.—3. Rule for finding the *present worth?* For finding the *discount?* 4. What is the difference between *interest* and *discount?*

The foregoing is the *correct* method of reckoning discount, yet the usual method in practice is to compute the interest for the time, and deduct it from the given sum. The *interest* thus found is called the *discount.*

The difference between interest and discount, on a small sum, for a short time, is inconsiderable; but the difference becomes very considerable when the sum is large and the time long for which the discount is to be made.

12. What is the difference between the interest and discount of $100 for 1 month, at 6 per cent.?

Ans. $2\frac{1}{2}$ mills, nearly.

13. What is the difference between the interest and discount of $649, for 3 years, at 6 per cent.? *Ans.* $17.82.

Art. 214.—*Bank discount* is the same as *simple interest* When a note is discounted at a bank, the interest is computed on the sum from the date of the note to the time when it becomes due, including three days of *grace*, and deducted as *discount.* Thus, if a note of $100 be discounted for 30 days, the interest is computed for 33 days. Custom has allowed to the borrower 3 days after the day on which the note becomes due, called *days of grace;* and as payment is generally withheld until the third day, it is justice that interest should be paid for these days.

If the payment of a note cannot conveniently be made at the proper time, the note may be taken up, if the bank allow the indulgence, by a new note, which must be presented on the day of discount immediately preceding the day on which the note would have become due, paying at the same time the discount, or interest, as before stated. Thus the borrower loses the discount on his note from the day on which he replaces it by another to the day on which it would have been to be paid.

The discount of any sum discounted for 30, 60, or 90 days, is found by multiplying by $\frac{1}{6}$ of the days. (See Art. 202.)

EXAMPLES.

14. What is the bank discount on a note of $714, for 30 days, at 6 per cent.?

15. What is the bank discount on a note of $1692, for 60 days, at 6 per cent.?

QUESTION.—5. What is the usual method in practice?

Operation.	*Operation.*
2)714	2)1692
.005½	.010½
3570	16920
357	846
$3.927 *Ans.*	$17.766 *Ans.*

16. What is the bank discount on a note of $784, for 90 days, at 6 per cent.?

Operation.

```
  2)784
    .015½
  -----
   3920
   784
     392
  -----
$12.152 Ans.
```

17. What is the bank discount on a note of $53, for 30 days? *Ans.* $.291+.

18. What is the bank discount on a note of $1092, for 30 days? *Ans.* $6.006.

19. What is the bank discount on a note of $2049, for 30 days? *Ans.* $11.269.

20. A.'s note of $561, for 60 days, is discounted at the bank, at 6 per cent. What ready money does he receive? *Ans.* $555.109.

21. B.'s draft for $150, drawn at 15 days' sight, is cashed at the bank, at 3 per cent. discount. How much money does he receive? *Ans.* $149.812+.

22. What is the bank discount on a note of $340, for 90 days, at 6 per cent.? *Ans.* $5.27.

23. What is the bank discount on a note of $632.75, for 90 days, at 6 per cent.? *Ans.* $9.807.

When a note is offered at the bank for discount, one or two endorsers are generally required; and the note is presented in one of the following forms:

QUESTIONS.—6. What is *bank discount?* 7. What is meant by *days of grace?* 8. How is discount found for 30, 60, and 90 days? 9. When a note is offered at a bank for discount, what is required?

$500. CONCORD, July 4th, 1849.

For value received, we, the subscribers, jointly and severally promise to pay the President, Directors, and Company of the New England Bank, or order, five hundred dollars, at said bank, on demand, with interest after sixty days.

When a note, called *business paper*, is offered for discount, it is generally made in the following form:

$350. BOSTON, August 6th, 1849.

Three months after date, I promise to pay to the order of Mr. JOHN SAVAGE, at the Commonwealth Bank, three hundred and fifty dollars, value received.

A. B.

In order to negotiate this note to an individual, or to procure a discount of it at a bank, the said Savage should endorse his name upon the back of the note, and such other names of endorsers should be procured as may be required; in which case, the promiser, or payer, A. B., is first liable for the note, and the note should be demanded of him, when it becomes due. If not paid, immediate notice should be given to the endorsers of the note; and on such demand and notice, the endorsers become liable for payment of the note; otherwise they are not holden.

The promiser, or payer of a note, is the individual who signs it. The promisee, or payee, is the person to whom the note is payable.

When a note is endorsed, the promisee, or payee, is always an endorser.

LOSS AND GAIN.

Art. 215.—LOSS AND GAIN teach to find what is gained or lost in the purchase and sale of goods; and also to regulate the price, so as to gain or lose, at a certain rate per cent.

1. If I purchase goods to the amount of $50, and sell the same for $60, what do I gain per cent.?

QUESTIONS.—10. What is the form of a note payable to the president, directors, &c., of a bank? 11. What is the form of a note called *business paper*? 12. Who is the promiser of a note? 13. Who the promisee?

It is evident that the gain on $1 would be $\frac{1}{50}$ as much as on $50. Since, then, the gain on $50 is $10, or the gain is $\frac{10}{50}$ of the cost, then $10÷50=20 cts. on a dollar, or 20 per cent., the *Answer*. Hence the

RULE.

When the prices at which goods are bought and sold are given, to find the gain or loss per cent.: Divide the gain *or* loss, *found by subtraction, by the* cost *of the article.*

EXAMPLES.

2. A merchant bought goods to the amount of $500, and sold the same for $700. What did he gain per cent.?

The question may be thus expressed, as in the Rule of Three: What is the gain on $100, if on $500 the gain be $200?

Operation.

What gain ?	~~100~~ $
If $ ~~500~~	~~200~~ 40
	40 per cent. *Ans.*

3. A merchant purchased goods to the amount of $342.25, and gains on the sale $41.07. What is the gain per cent.?

Ans. 12 per cent.

4. Bought flour to the amount of $840. Sold the same for $907.20. What do I gain per cent? *Ans.* 8 per cent.

5. Suppose a merchant purchase goods to the amount of $1000, and sell them for $910, what is the loss per cent.?

Ans. 9 per cent.

6. Bought fur caps for $7 apiece; sold them for $7.25. What was the whole gain in laying out $630, and what was the gain per cent.?

Ans. { Whole gain, $22.50.
Gain per cent., 3.57+.

7. What is the whole loss, and what is the loss per cent., in laying out $70 for hats, at $1.75 each, and selling them for 25 cents apiece less than cost?

Ans. { Whole loss, $10.
Loss per cent., $14\frac{2}{7}$.

QUESTIONS.—1. What is Loss and Gain? 2. How is the gain or loss per cent. found? 3. Having the gain or loss per cent., how is the price found at which an article is bought or sold?

8. Bought 100 yards of cloth, at $6.72 per yd., and sold the same for $8.40. What did I gain per cent?

Ans. 25 per cent.

Art. 216.—When the gain or loss per cent. is given, to find the price at which the goods are bought and sold.

RULE.

If the per cent. be gain, *add it to* 100; *if the per cent. be* loss, *subtract it from* 100, *and proceed as in the Rule of Three.*

EXAMPLES.

9. A merchant sold cloth, which cost $6.72 per yard, at 25 per cent. profit. For how much did he sell the cloth per yard? (See Interest, Obs. 3, Art. 204.)

Operation.

How many $	~~6.72~~ $ 168
~~4~~ $ ~~100~~	~~125~~ $ 5
	$8.40 *Ans.*

10. A merchant sold cloth at $8.40 per yard, and gained 25 per cent. What was the first cost?

Operation.

How many $	8.40 $
~~5~~ $ ~~125~~	~~100~~ $ 4
	$6.72 *Ans.*

11. If 1 tun of wine cost £40, for how much must it be sold to gain $6\frac{1}{4}$ per cent.? *Ans.* £42 10*s.*

12. Sold 10 yards of cloth for £4 16*s.*, and gained 10 per cent. What was the prime cost per yard? *Ans.* 8*s.* $8\frac{8}{11}$*d.*

13. Bought 7 tuns of wine, at $61.20 per hhd.; sold at 18 cents a pint. What was the whole gain, and how much per cent.?

Ans. { Whole gain, $826.56.
Gain per cent., $48.235.

14. Purchased 40 gallons of molasses, at 3*s.* per gallon. By accident, 6 gallons leaked out. At what rate must I sell the remainder per gallon to gain 10 per cent. upon the first cost, and give 8 months' credit? *Ans.* 4*s.* 0*d.* 1*qr.*+

15. If I sell a pound of silk for $12.72, and gain $1.20, how much should I gain in selling a bale which cost $1152?

Ans. $120.

16. Bought 300 lbs. of coffee, at 4*s.* 2*d.* per lb., ready money,

and sold the same for 5*s*. per lb., payable in 8 months. How much was gained upon the whole, and how much per cent.

Ans. { £9 12*s*. 3*d*. 2*qrs*.+ / $15\frac{3}{10}$ per cent.

17. Bought 50 yards of broadcloth, at $5 per yard, which I purpose to sell at 25 per cent. profit, ready money; but if I sell it on credit, I must have 5 per cent. extra. How must I sell it per yard, at 6 months, to make both these gains? *Ans.* $6.695.

18. If by selling tea at 57 cents per lb. I lose 3 cents, what is the loss per cent.? *Ans.* 5 per cent.

19. A merchant purchases 180 casks of raisins, at 16*s*. per cask; sells the same at 28*s*. per cwt., and gains 25 per cent. What is the weight of each cask? *Ans.* 80 lbs.

20. What will be the gain in selling $500 worth of flour, at 8 per cent. advance? *Ans.* $40.

21. Bought 1000 bushels of corn, for $1922.25. For how much must it be sold to gain 15 per cent.? *Ans.* $2210.587.

22. Bought 80 reams of paper, at $2.50 per ream. For how much must the whole be sold to lose 5 per cent.? *Ans.* $190.

23. A merchant bought 500 yards of broadcloth for $2125. For how much must he sell the whole to lose 10 per cent.? *Ans.* $1912.50.

24. If I buy 45 bushels of salt, at 95 cents per bushel, for how much must it be sold per bushel to gain 20 per cent.? *Ans.* $1.14.

25. Bought 64 bushels of wheat, at $1.75 per bushel. For how much per bushel must I sell it to lose 3 per cent.? *Ans.* $1.697.

STOCK.

Art. 217.—Stock is a general name for capital employed in trade, manufactures, insurance, banking, etc. Also, for money loaned to government, or property in a public debt.

The Capital Stock of a company, or corporation, is the whole amount originally invested by such company, or corpo-

---

Questions.—1. What is stock? 2. What is the capital stock of a company or corporation?

ration, which sum is divided into shares, and each holder receives a certificate of the number of shares to which he is entitled. If stock which cost $100 per share sells in the market for any thing more than that amount, it is said to be above *par*; that is, above the sum equal to the first cost—the term *par* signifying equality. If it sells for less than that amount, it is below par; and the amount above or below par is spoken of as so much per cent. If it sells for $6 on the $100 in advance, it is 6 per cent. above par. If it sells for so much less, it is so much below par.

EXAMPLES.

1. What is the value of $600 of stock, at 6 per cent. above par? *Ans.* $636.

2. What is the value of $2000 of railroad stock, at 87½ per cent.? *Ans.* $1750.

3. What is the value of $1500 of bank stock, at 108 per cent.? at 107 per cent.? at 115 per cent.? at 105 per cent.?

BARTER.

Art. 218.—Barter is the exchanging of one commodity for another, according to prices or values agreed upon by the parties.

RULE.

Divide the value of that article whose quantity is given, by the price of the article whose quantity is required; or, the question may be solved by the Rule of Three.

EXAMPLES.

1. How many pounds of coffee, at $13\frac{1}{3}$ cents per pound, must be given in barter for 1200 lbs. of sugar, at 8 cents per pound?

Operation.

$13\frac{1}{3}=\frac{40}{3} : 8 :: 1200$

1200

4|0)960|0

240

3

720 lbs. *Ans.*

By cancelling.

How many lbs. coffee?	120~~0~~ lbs. sug.
cts. ~~40~~	~~8~~ 2
$13\frac{1}{3}=\frac{40}{3}$	3
	1 lb. coffee.
	720 lbs. *Ans.*

2. How much tea, at 64 cents per pound, must be given in barter for 2 cwt. of chocolate, at 32 cents per pound?

Ans. 112 lbs.

3. How many pounds of lead, at 9 cents per pound, must be given for 783 lbs. of iron, at 6 cents per pound?

Ans. 522 lbs.

4. A. has broadcloth, at 16*s.* 6*d.* per yard. B. has linen, at 1*s.* 4*d.* per yard. How many yards of broadcloth must be given in exchange for 660 yards of linen? *Ans.* $53\frac{1}{3}$ yds.

5. A. bartered $53\frac{1}{3}$ yards of broadcloth, at 16*s.* 6*d.* per yard, for 660 yards of linen. What was the price of the linen?

Ans. 1*s.* 4*d.*

6. How much sugar, at 8 cents per pound, must be given in barter for $1\frac{3}{4}$ cwt. of cinnamon, at $54\frac{6}{7}$ cents per pound?

Ans. 12 cwt.

7. A. barters $1\frac{3}{4}$ cwt. of cinnamon, at $54\frac{6}{7}$ cents per lb., for 12 cwt. of sugar. What was the value of the sugar per pound? *Ans.* 8 cents.

8. A. has linen, worth 20*d.* per ell English, ready money, but in barter he will have 2*s.* B. has broadcloth, worth 14*s.* 6*d.* per yard, ready money. What ought to be the price of the broadcloth, in barter? *Ans.* 17*s.* $4\frac{4}{5}$*d.*

9. B. has coffee which he barters with C. at 10*d.* per lb. more than it cost him, for tea which cost 10*s.*; but in barter C. puts it at 12*s.* 6*d.* What was the first cost of the coffee?

Ans. 3*s.* 4*d.*

10. A. has 5 tons of butter, at $425 per ton, and $10\frac{1}{2}$ tons of tallow, £33 15*s.* per ton, which he barters with B. for 316 barrels of beef, at 21*s.* per barrel, and the remainder in cash. How much money does he receive? *Ans.* $2200.25.

11. C. and D. barter. C. has corn, at 75 cents, ready money, but in barter he will have $1. B. has rye, at 50 cents, ready money. What ought he to have for his rye, in barter?

Ans. $66\frac{2}{3}$ cents.

12. A. has rye at $1.44 per bushel, ready money, but in barter he will have $1.56 per bushel. D. has cotton, at 18 cents per pound, ready money. What price must the cotton be in barter, and how many pounds of cotton must be bartered for 100 bushels of rye?

Ans. { Cotton, $19\frac{1}{2}$ cents per pound.
800 lbs. for 100 bushels rye. }

13. B. gives C. 250 yards of drugget, at $18\frac{1}{2}d.$ per yard, for $308\frac{1}{3}$ lbs. of pepper. What does the pepper cost C. per lb.? *Ans.* 15*d.*

14. A. and B. barter. A. has 41 cwt. of hops, at $7.20 per cwt., for which B. gives him $96 in money, and the rest in prunes, at 10 cents per lb. What quantity of prunes does A. receive? *Ans.* 17 cwt. 3 qrs. 4 lbs.

15. How many acres of land, worth £40 10*s.* per acre, must be given for 600 acres, worth $8.50 per acre? *Ans.* $37\frac{7}{9}$.

16. A. has $7\frac{1}{2}$ cwt. of sugar, at 8*d.* per pound, for which B. gives him $12\frac{1}{2}$ cwt. of flour. How much per pound was the flour? *Ans.* $4\frac{4}{5}d.$

17. A. has corn, at $1.25, ready money, but in barter he values it at $1.50 per bushel. B. has cotton at 20 cents per pound, ready money. What should be the price of the cotton, in barter, and how many pounds must be given for 100 bushels of corn? *Answer to the last,* 625 lbs.

18. A. has cloth, valued at $4 per yard, ready money, but in barter he will have $4.50. B. has cloth at 2 pounds per yard, ready money; at what price ought B. to rate his cloth in barter, and how many yards must be given A. in exchange for 540 yards of cloth? *Answer to the last,* 324 yards.

19. D. has ribbon, at 2*s.* per yard, ready money, but in barter he will have 2*s.* 3*d.* E. has broadcloth, for which he will have in barter 36*s.* 6*d.* 3*qrs.* What ought to be the cash price of E.'s cloth, and how many yards of ribbon ought D. to give him for 488 yards of broadcloth?

Ans. { E.'s cloth, 32*s.* 6*d.*
7930 yards ribbon.

SUPPLEMENT

TO INTEREST, DISCOUNT, BARTER, AND LOSS AND GAIN.

Art. 219.—1. What is the interest of $365.25 for 1 year, 3 months, and 2 days? *Ans.* $27.515.

2. What will $1002.153 amount to in 4 years, 1 month, and 15 days, at simple interest? *Ans.* $1250.185.

3. What is the interest of $125000 for 1 day? *Ans.* $20.833.

4. How much will £300 amount to in $5\frac{3}{4}$ years, at $3\frac{1}{4}$ per cent.? *Ans.* £356 1*s.* 3*d.*

5. What is the amount of £10 15*s.* 6*d.*, for 16 years and 10 months? *Ans.* £21 13*s.* 1*d.* 3*qrs.*

6. How much will $185.26 amount to in 2 years, 3 months, and 11 days, at $7\frac{1}{2}$ per cent.? *Ans.* $216.944.

7. How much will $298.59 amount to, from May 19th, 1797, to Aug. 11, 1798, at 8 per cent.? *Ans.* $327.913.

8. What is the interest of $658, from Jan. 9th to the 9th of Oct. following, at $\frac{1}{2}$ per cent.? $2.467.

9. What principal will amount to $1319.90, in 5 years and 8 months? *Ans.* $985.

10. Took up a note, April 29, 1799, which amounted to $205.86, dated June 14, 1798, on interest at $5\frac{3}{4}$ per cent. What was the sum borrowed? *Ans.* $196.

11. A note of 6 years' standing amounted to £3810; the principal was £3000. What was the rate per cent.? *Ans.* $4\frac{1}{2}$.

12. At what rate per cent. will $420 amount to $520.80 in 8 years? *Ans.* 3 per cent.

13. At what rate per cent. will £413 12*s.* 6*d.* amount to £546 3*s.* 8*d.* in $4\frac{3}{4}$ years? *Ans.* $6\frac{3}{4}$.

14. In what time will $500 amount to $725, at 5 per cent.? *Ans.* 9 years.

15. In what time will a note of £420 amount to £520 16*s.*, at 3 per cent.? *Ans.* 8 years.

16. What will be the amount of $597.75, in 20 years, at 6 per cent., compound interest? *Ans.* $1917.077.

17. Gave a note for £450, payable in 3 years, at 5 per cent., compound interest. To what did it amount? *Ans.* £520 18*s.* $7\frac{1}{2}$*d.*

18. What is the amount of £217, for $2\frac{1}{4}$ years, at 5 per cent., interest payable quarterly? *Ans.* £242 13*s.* $4\frac{1}{2}$*d.*

19. Bought a quantity of goods, to the amount of £250, ready money, and sold them for £300, payable in 9 months. What was the gain in ready money, discounting at 6 per cent.? *Ans.* £37 1*s.* 7*d.* 1*qr.*

20. What is the present worth of $1000, payable one-half in 4 months, the other half in 8 months, discounting at the rate of 5 per cent.? *Ans.* $975.345.

21. How much tea, at 9*s.* 6*d.* per pound, must be given in barter for 156 gallons of wine, at 12*s.* $3\frac{1}{2}$*d.* per gallon? *Ans.* 201 lbs. $13\frac{27}{57}$ oz.

22. A. has 240 bushels of rye, at 90 cents per bushel, ready money, which he barters with B., at 95 cents, for wheat which cost 99 cents per bushel. How many bushels of wheat must he receive for his rye, and at what price? *Ans.* $218\frac{2}{11}$ bushels, at $\$1.04\frac{1}{2}$ per bushel.

23. A. and B. barter. A. has cloth which cost him 28*d.*, B.'s cost him 22*d.* B. puts his cloth at 25*d.*, in barter. How high must A. rate his cloth, to gain 10 per cent. in the trade? *Ans.* 35*d.*

24. Bought 100 yards of cloth, at \$2 per yard. How must I sell it per yard, to gain \$50? *Ans.* \$2.50.

25. Bought cloth at \$1.50 per yard, which, not proving so good as I expected, I am willing to lose $17\frac{1}{2}$ per cent. How must I sell it per yard? *Ans.* \$1.237+.

26. Bought 50 gallons of wine, at 4*s.* per gallon. By accident, 10 gallons leaked out. How must I sell the remainder per gallon, to gain 10 per cent. upon the whole cost? *Ans.* 5*s.* 6*d.*

27. A man sells a quantity of corn at \$1 per bushel, and gains 20 per cent. Some time after, he sold of the same to the amount of \$37.50, and gained 50 per cent. How many bushels were there in the last parcel, and at what rate did he sell it per bushel? *Ans.* 30 bushels, at \$1.25 per bushel.

EQUATION OF PAYMENTS.

Art. 220.—Equation of Payments is the method of finding the mean time for the payment of several debts due at different times.

1. If a man owes me \$10, to be paid in 4 months, and \$5, to be paid in 7 months, and he wishes to pay the whole at once, in what time should the whole be paid?

It is evident that the use of \$10 four months is the same as

---

Questions.—1. What is Equation of Payments? 2. Rule?

the use of \$1 forty months; and the use of \$5 seven months is the same as the use of \$1 for thirty-five months. Then, \$10+\$5=\$15, and 40+35=75 months. Thus it appears, that the use of \$10 for four, and \$5 for seven months, is the same as the use of \$1 for seventy-five months: \$15, therefore, may be used $\frac{1}{15}$ as long as \$1. That is, $\frac{1}{15}$ of seventy-five months, 75÷15=5 months, the answer. Hence the

RULE.

Multiply each payment by the time when it becomes due, and divide the sum of the products by the sum of the payments, and the quotient will be the time required.

2. A merchant has owing him \$420, to be paid as follows: \$100 in 8 months, \$100 in 2 months, and \$220 in 5 months. In what time ought the whole to be paid at once?

Ans. 5 months.

3. A. owes B. \$800, to be paid as follows: \$200 in 3 months, \$150 in 4 months, and the remainder in 8 months. What is the equated time for the payment of the whole?

Ans. 6 months.

4. A. owes B. \$380, to be paid as follows: \$100 in 6 months, \$120 in 7 months, and \$160 in 10 months. What is the equated time for the payment of the whole?

Ans. 8 months.

5. A merchant has owing him \$698, of which \$181 is to be paid at the present time, \$199 in 3 months, and \$318 in 8 months. What is the equated time for the payment of the whole? *Ans.* 4½ months.

6. A. owes B. \$500, of which ½ is to be paid in 3 months, ⅕ in 8 months, and the remainder in 2 months. What is the equated time for the payment of the whole?

Ans. 3 months, 21 days.

7. A. has owing him \$924, of which ⅓ is to be paid in 3 months, and ½ in 2 years. In what time ought the whole to be paid? *Ans.* 13 months.

8. I have three notes against a man: one of \$400, due in 5 months; one of \$500, due in 6 months; and the other of \$350, due in 9 months; and he wishes to pay the whole at once. In what time ought he to pay it? *Ans.* 6.52 months.

9. A. owes B. \$960, of which ½ is due in 3 months, ⅓ in 1½ months, ⅙ in 9 months. What is the equated time for the payment of the whole? *Ans.* 3 months, 15 days.

10. A merchant bought goods to the amount of \$3000, and agreed to pay \$500 ready money, \$600 in 4 months, and the remainder in 9 months; but they agree to make one payment of the whole. What is the equated time?

Ans. 6 months, 15 days.

FELLOWSHIP.

Art. 221.—FELLOWSHIP is a rule by which merchants and others, trading in company, may ascertain their respective gain or loss, in proportion to each man's share in the joint stock.

The money, or value of property vested in trade, is called the *Capital,* or *Stock.*

The gain or loss to be shared by the company is called the *Dividend.*

When the several stocks are employed without regard to time, it is called *Single Fellowship.*

1. Two men, A. and B., bought a horse for \$60, of which sum A. paid \$40, and B. paid \$20. They sold the horse for \$90. What was each man's share of the gain?

It is evident, that each man's share of the *gain* should bear the same ratio to the *whole gain,* that his *share* of the *stock* bears to the *whole stock.* Now, the whole stock was \$60, of which A. paid \$40; then A. paid $\frac{40}{60}=\frac{2}{3}$ of the whole stock, and B. paid $\$20=\frac{20}{60}=\frac{1}{3}$ of the whole stock. As the whole gain was \$30, A.'s share is $\frac{2}{3}$ of $30=\$20$, and B.'s share is $\frac{1}{3}$ of $30=\$10$. Hence the

RULE.

As the whole stock is to each man's stock, so is the whole gain or loss to each man's share of the gain or loss.

Or the question may be expressed thus: What gains each individual stock, if the whole stock give —— gain?

Whole stock \$60—A.'s stock \$40.

Whole gain \$30—B.'s stock \$20.

QUESTIONS.—1. What is Fellowship? 2. What is capital, or stock? 3. What is the dividend? 4. What is Single Fellowship? 5. What is the rule? 6. What is the method of proof?

Operation.

How much A.'s gain.	40$ A.'s stock.
Whole stock $60	30$ whole gain.
$	20 *Ans.* A.'s gain.

Operation.

B.'s gain.	20$ B.'s stock.
	30$
$60	
$	10 *Ans.* B.'s gain.

Proof—Add together the respective gains, and if the work be right, their sum will equal the whole gain.

2. A., B., and C. trade in company. A.'s stock is $240, B.'s $360, and C.'s $600. They gain $325. What is each man's share of the gain?

Ans. { A.'s gain, $65.00. B.'s gain, 97.50. C.'s gain, 162.50.

3. A. and B. bought a lot of land for $1280, of which B. paid $400, and A. the remainder. They sold it so as to gain $200. What was each man's share of the gain?

Ans. { A.'s gain, $137.50. B.'s gain, 62.50.

4. A. and B. owned a ship, valued at $72000—lost at sea; insurance $50000. What was each man's loss, supposing A. owned 3 times as much as B.?

Ans. { A.'s loss, $16,500. B.'s loss, 5,500.

5. A man dying, leaves property to the amount of $3000. A. has a note of $600 against the estate, B. has a note of $1800, and C. a note of $1600. How much must each lose?

Ans. { A.'s loss, $150. B.'s loss, 450. C.'s loss, 400.

6. Three partners, A., B., and C., shipped 216 horses for the south. A.'s share of the cost of the horses was $2880; B.'s, $5760; C.'s, $4320. During the voyage they were obliged to throw 90 overboard. How many horses did each partner lose?

Ans. { A. lost 20. B. lost 40. C. lost 30.

7. A. and B. trade in company. A.'s stock was 60 guineas, and his share of the gain was $\frac{5}{8}$. What was B.'s stock?

Ans. 36 guineas.

8. Three men gained in an adventure $96. A. put in a certain sum, B. put in twice as much as A., and C. as much as A. and B. both. What was each man's share of the gain?

Ans. { A.'s, $16. B.'s, 32. C.'s, 48.

9. Two men trade in company. Their joint stock is $800, of which B. put in $\frac{1}{2}$ of $\frac{2}{3}$ of $\frac{3}{4}$ of $\frac{6}{8}$ of 4 times the whole. What is each man's stock?

Ans. { A.'s, $200.
B.'s, 600.

10. A., B., and C. trade in company. A.'s stock is $250, B.'s $300, C.'s $550. They lose 5 per cent. by trading. What is each man's share of the loss?

Ans. { A.'s loss, $12.50.
B.'s loss, 15.00.
C.'s loss, 27.50.

11. A man by his will left his estate to his children, as follows: to A. he gave $5000, to B. $4500, to C. $4500, and to D. $4000; but his whole estate amounted to but $12000. How much did each receive?

Ans. { A. received 3333.33\frac{1}{3}$.
B. " 3000.00.
C. " 3000.00.
D. " 2666.66$\frac{2}{3}$.

ASSESSMENT OF TAXES.

Art. 222.—In order to the assessment of taxes on a town, the following facts should be known:

1. The amount of tax assessed by the Legislature of the State.

2. The inventory of all the rateable property in the town.

RULE.

I. *From the tax raised by the town, deduct the amount of poll taxes.*

II. *Find the tax on a dollar, and multiply each man's inventory by it, and to the product add his poll tax.*

EXAMPLES.

1. A town inventoried at $160,000, raises a tax of $3400. There are 400 rateable polls, taxed 50 cents each. What is the tax on a dollar, and what is A.'s tax, whose *real* and *personal* estate is inventoried at $1683, and who pays for one poll?

QUESTIONS.—1. What facts should be known, in order to the assessment of taxes? 2. What is the rule?

First deduct the amount of poll tax for 400 polls, at 50 cents each, 400×.50=$200, amount of poll tax; then $3400−200 =$3200 to be assessed on the whole property. Secondly, find the tax on a dollar.

Operation.

What tax.	1 $
$160,000	3200 $
Ans.	.02 cts. on $1.

Then to find A.'s tax, multiply the amount of his inventory by the tax on a dollar, and to the product add his poll tax—thus, $1683×.02=$33.66+.50=$34.16, A.'s tax.

Or, having found the tax on a dollar, a table may be formed, containing the tax on 1, 2, 3, or to 20 dollars; then on 30, 40, &c., to 100 dollars; then on 110, 120, &c., to 1000 dollars. Then, having the inventory of the property of an individual, his tax may be readily made out.

TABLE.

Tax on	$1	is	$.02	Tax on	$17	is	$.34	Tax on	$150	is	$3.00
"	2	"	.04	"	18	"	.36	"	160	"	3.20
"	3	"	.06	"	19	"	.38	"	170	"	3.40
"	4	"	.08	"	20	"	.40	"	180	"	3.60
"	5	"	.10	"	30	"	.60	"	190	"	3.80
"	6	"	.12	"	40	"	.80	"	200	"	4.00
"	7	"	.14	"	50	"	1.00	"	300	"	6.00
"	8	"	.16	"	60	"	1.20	"	400	"	8.00
"	9	"	.18	"	70	"	1.40	"	500	"	10.00
"	10	"	.20	"	80	"	1.60	"	600	"	12.00
"	11	"	.22	"	90	"	1.80	"	700	"	14.00
"	12	"	.24	"	100	"	2.00	"	800	"	16.00
"	13	"	.26	"	110	"	2.20	"	900	"	18.00
"	14	"	.28	"	120	"	2.40	"	1000	"	20.00
"	15	"	.30	"	130	"	2.60				
"	16	"	.32	"	140	"	2.80				

We find by the table, the tax on	$1000	to be	$20.00
" " " "	600	"	12.00
" " " "	80	"	1.60
" " " "	3	"	.06
A.'s inventory,	$1683		$33.66
Poll tax,			50
Amount of A.'s tax,			$34.16

2. The inventory of *real* and *personal* estate in the town of ———, for the year 1849, is $800,000. The amount assessed on each rateable poll is $1. The number of polls is 400. The amount of town tax voted to be raised for the year 1849 is $2000. The proportion of state tax for said town for that year is $400; county tax $200; the amount of school tax is $800; the highway tax is $1200. How much is A.'s town, state, county, school, and highway tax, whose whole estate is inventoried at $5000, and who pays for one poll?

Ans.			
	Town tax,	$11.847.	Total, $27.25.
	State tax,	2.369.	
	County tax,	1.189.	
	School tax,	4.739.	
	Highway tax,	7.108.	

DOUBLE FELLOWSHIP.

Art. 223.—1. Two men, A. and B., hire a pasture for $36. A. put in 8 oxen 6 weeks, and B. 12 oxen 8 weeks. How much must each pay?

It is evident that the pasturage of 8 oxen, 6 weeks, is the same as of 1 ox 48 weeks; and the pasturage of 12 oxen, 8 weeks, is the same as of 1 ox 96 weeks. The shares of A. and B. are the same as though A. had put in 1 ox 48 weeks, and B. 1 ox 96 weeks; $96+48=144$ weeks. Then A.'s share of the rent will be $\frac{48}{144}=\frac{1}{3}$ of $36=$12, and B.'s share will be $\frac{96}{144}=\frac{2}{3}$ of $36=$24. Hence the

RULE.

Multiply each man's stock by the time it is continued in trade, and consider the product his share of the joint stock, and proceed as in Single Fellowship

Operation.			*Operation.*	
How many $	48 weeks.		How many $	96 weeks.
Weeks. 144	36 $		Weeks. 144	36 $
	$12, A.'s share.			$24, B.'s share.

2. A. and B. trade in company. A. put in $3000, for 6 months; B. put in $4000, for 10 months; and C. put in

$2500 for 12 months. They gained $880. What is each man's share of the gain?

Ans. { A.'s share, $180. B.'s share, 400. C.'s share, 300. }

3. Three men trade in company. A. put in $4000 for 12 months, B. put in $3000 for 15 months, and C. put in $5000 for 6 months. The whole gain was $615. What was their respective shares?

Ans. { A.'s gain, $240. B.'s gain, 225. C.'s gain, 150. }

4. A., B., and C. made a stock for 2 years. A. put in at first $1000. At the end of 6 months he put in $500 more. B. put in $1600, and after 8 months took out $400. C. put in $2000 for 20 months, and then took out $1500. They gain $1000. What is each man's share?

5. A., B., and C. lost in trade $263.90. A.'s stock was $580, for $6\frac{1}{2}$ months; B.'s stock was $580, for $9\frac{1}{2}$ months; C.'s stock was $870, for $8\frac{2}{3}$ months. What is each man's share of the loss?

Ans. { A.'s loss, $59.15. B.'s loss, 86.45. C.'s loss, 118.30. }

6. A. commenced business on the first of January, with a capital of $3800; on the first of May he took in B. as a partner, with a capital of $2700; on the first of August, they admit C. as a partner, with a capital of $4000: at the end of the year they dissolve partnership; each took his share of the stock and gain, the gain being $4360. How much did each take?

Ans. { A. took $6080. B. " 3780. C. " 5000. }

INVOLUTION.

Art. 224.—INVOLUTION is multiplying a number into itself. The product is called a power; the number so multiplied is called a *root*, or the *first* power. The product of any number multiplied into itself is called the *second* power, or *square*. If the square be multiplied by the *first* power, the product is called the *cube*, or *third* power.

QUESTIONS.—1. What is Involution? 2. What is a power?

The power is sometimes denoted by a small figure, called the *index*, or *exponent*, of the power, placed above the given number at the right hand—Thus, 3^2 denotes that the second power of 3 is required, or it shows how many times 3 is to be involved or multiplied. This may be illustrated by the following:

$2^2=2\times2=4$, the second power of 2.
$2^3=2\times2\times2=8$, the third power of 2.
$2^4=2\times2\times2\times2=16$, the fourth power of 2.
$2^5=2\times2\times2\times2\times2=32$, the fifth power, or sursolid.
$2^6=2\times2\times2\times2\times2\times2=64$, the sixth power, or square cubes.

The product of any two powers is always that power whose index is the sum of the indices, or exponents, of the power multiplied, thus:

0	1	2	3	4	5	6	7	8	9	10
1	2	4	8	16	32	64	128	256	512	1024

If 16, which is the 4th power of 2, be multiplied into 64, the 6th power of 2, we shall have 1024, the power indicated by the multiplication of $2^4\times2^6=2^{6+4}=2^{10}=1024$.

EXAMPLES.

1. What is the square, or second power, of 25 ? *Ans.* 625.
2. What is the square, or second power, of 145 ? *Ans.* 21025.
3. What is the cube, or third power, of 23 ? *Ans.* 12167.
4. What is the cube, or third power, of 159 ? *Ans.* 4019679.
5. What is the biquadrate, or fourth power, of 29 ? *Ans.* 707281.
6. What is the fifth power of 134 ?
7. What is the square of 1 ? 8. What is the cube of 1 ? *Ans.* 1 ; 1.
9. What is the square of $\frac{1}{2}$? 10. What is the cube of $\frac{1}{2}$? *Ans.* $\frac{1}{4}$; $\frac{1}{8}$.
11. What is the cube of 1.5 ? 12. What is the cube of 2.25 ?
13. What is the square of $2\frac{2}{5}$? *Ans.* $\frac{144}{25}=5.76$.

QUESTIONS.—3. What is a root ? 4. What is the second power ?—the third power ? 5. How is a power denoted ?

OBS.—Mixed numbers may be reduced to improper fractions, before involving: Thus, $2\frac{2}{5}=\frac{12}{5}$; or they may be reduced to decimal: Thus, $2\frac{2}{5}=2.4$.

The powers of the nine digits, from the first to the ninth, may be seen by the following

TABLE.

Roots	1	2	3	4	5	6	7	8	9
Squares	1	4	9	16	25	36	49	64	81
Cubes	1	8	27	64	125	216	343	512	729
4th power ..	1	16	81	256	625	1296	2401	4096	6561
5th power ..	1	32	243	1024	3125	7776	16807	32768	59049
6th power ..	1	64	729	4096	15625	46656	117649	262144	531441
7th power ..	1	128	2187	16384	78125	279936	823543	2097152	4782969
8th power ..	1	256	6561	65536	390625	1679616	5764801	16777216	43046721
9th power ..	1	512	19683	262144	1953125	10077696	40353607	134217728	387420489

EVOLUTION

Art. 225.—EVOLUTION, the opposite of Involution, is the extracting of the root of any number, or the finding of such a number as, when multiplied into itself a certain number of times, will produce a given number. Thus, 3 is the square root of 9, because $3\times3=9$; also, 3 is the cube root of 27, because $3\times3\times3=27$.

Any given power may be found by a continued multiplication of the number into itself; yet there are numbers whose precise root can never be found; but, by the use of decimals, we can arrive sufficiently near for all practical purposes.

A number whose precise root cannot be found, is called a *surd*, or *irrational number*, and its root a *surd* root.

The square root may be denoted by this character, $\surd$, called the *radical sign*, placed before the power; and the other roots by the same sign, with the index of the root placed over it, or by the fractional indices placed on the right hand. Thus, the quare root of 9 is expressed, $\sqrt{9}$, or $9^{\frac{1}{2}}$, and the cube root of 27 thus: $\sqrt[3]{27}$, or $27^{\frac{1}{3}}$.

QUESTIONS.—1. What is Evolution? 2. How may any given power be found? 3. Can the precise roots of all powers be found? 4. How can we approximate sufficiently near for practical purposes? 5. What is a number called whose precise root cannot be found? 6. What is the advantage of denoting roots by the fractional indices?

The method of denoting roots by the fractional indices is preferable, as, by it, not the root only is denoted, but the power. The numerator of the index denotes the power, and the denominator the root of the number over which it is placed.

If the power is expressed by several numbers, with the sign + or − between them, a line, or *vinculum*, is drawn from the top of the sign over all the numbers. Thus, the square root of 12+4 is $\sqrt{12+4}=4$, and the cube root of 357−14 is $\sqrt[3]{357-14}=7$.

EXTRACTION OF THE SQUARE ROOT.

Formation of the Square, and Extraction of the Square Root.

Art. 226.—It has been shown, that to obtain the square of any number, whether entire or fractional, we have only to multiply that number into itself. Therefore, *To extract the square root, is to find a number, which, multiplied into itself once, will produce a given number.*

The principle applied in the extraction of the square root, will be better understood by attending, first, to the formation of the square.

The square of any number expressed by a single figure, will contain no figure of a higher denomination than tens. (See *Table of Powers.*)

Numbers which are produced by the multiplication of a number into itself, are called *perfect squares.*

There are but nine perfect squares among all the numbers, which can be expressed by one or two figures. The square roots of all other numbers, expressed by one or two figures, will be found between two whole numbers differing from each other by unity. Thus, 37, which is comprised between 36 and 49, has for its square root a number between 6 and 7; and 95, which is comprised between 81 and 100, has for its square root a number between 9 and 10.

What is the square of 32?

$$\begin{array}{r} \text{tens. units.} \\ 32=3+2 \\ 3+2 \\ \hline 6+4 \\ 9+\ 6 \\ \hline 9+12+4=1024 \end{array}$$

Thus, it appears, that the square of a number made up of tens and units, contains the *square of the tens, plus twice the products of the tens into the units, plus the square of the units.*

What is the square root of 1024?

It is evident, that the root will contain more than one figure, since the number is composed of more than two places; and it will contain no more than two, for 1024 is less than 10,000, the square of 100. It will also be perceived, from the foregoing process, that the square of the tens, the first figure of the root, must be found in the two left-hand figures, which we will separate from the others by a point; thus, 10̇24. The two parts, of two figures each, are called *periods*. The period 10 is comprised between the squares, 9 and 16, whose roots are 3 and 4; hence, 3 is the tens, or the first figure of the root sought.

```
          1̇024(32
           9
          ------
 3×2=6)12|4
62×2=   124
```

We write 3, the first figure of the root, on the right of the given number, and its square, 9, we subtract from 10, the left-hand period, and to the remainder we bring down the next period. Having subtracted the square of the tens from the given number, the remainder, 124, contains twice the product of the tens into the units, *plus* the square of the units; but since tens into units cannot give a product of less name than tens, it follows that the right-hand figure, 4, can form no part of the double product of the tens into the units; therefore, if we divide 12, twice the product of tens into the units, by twice 3, the tens of the quotient, we shall obtain the unit figure of the root. We will now write this quotient figure on the right of the other, and multiply 62 by 2, the last quotient figure. We thus obtain, 1st, the square of the units; 2d, twice the product of the tens into the units; hence 32 is the required root.

What is the square root of 572?

Operation.

```
   5̇72̇(23
   4
   ---
43)172
   129
   ---
    43
```

In this example the remainder, 43, shows that 572 is not a *perfect square;* but 23 is the greatest square contained in 572; that is, it is the entire part of the root. This may be shown, thus: *The difference between the squares of two consecutive numbers, is equal to twice the less number, plus* 1. The difference between the squares of 8 and 9 is

$17=\overline{8\times2}+1$, and $\overline{23\times2}+1=47$, which is greater than 43, the remainder, which shows that 23 is the entire part of the root.

The foregoing rule may now be applied to finding the length of one side of a square surface, whose area is expressed by the given number.

EXAMPLES.

Art. 227.—1. What is the length of one side of a square garden, containing 576 square rods, or what is the square root of 576?

We first distinguish* the number whose root is to be found, into periods of two figures each, denoted by the index of the root. By the number of periods, we perceive that the root will consist of two figures, a unit and ten. As the second power of ten cannot be less than a hundred, we look for the square of tens in the second, or left-hand period, which is 5. We find the nearest square in 5 to be 4, and its root 2, or 2 tens, which we place in the quotient as the first figure of the root; and its square 4, or 400, under the period, and subtracting it, we have a remainder of 1, or 100, to which we add 76, the next period. Had the garden contained but 400 square rods, we should now have obtained the length of one side, 2 tens=20, and $20\times20=400$; consequently, 400 rods would be disposed of in the form of a square. (See Fig. 1.) But we have a remainder of 176 rods, to be added to the square, and in such a manner that its form shall not be altered. We must, therefore, make an equal addition on two sides. Then $20+20=40$, the length of the whole addition. To find the width of the addition, we place the double of the root

$$\begin{array}{l} \dot{5}7\dot{6}(2 \\ \underline{4} \\ 176 \end{array}$$

Fig. 1.

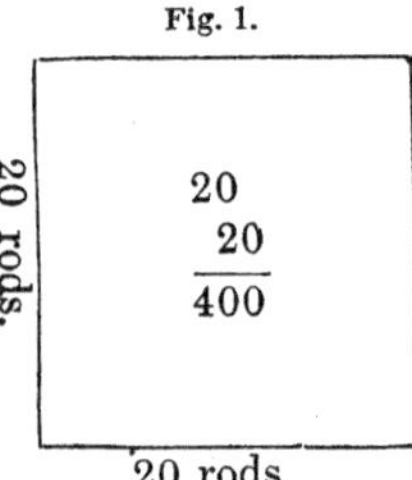

* It is distinguished into periods of two figures each, because the second power can never have more than twice as many figures as its root, and never but one less than twice as many. The third power can never have more than three times as many figures as its root, and never but two less than three times as many. Distinguish, therefore, any number into periods of as many figures as are denoted by the index of the root.

already found, on the left hand of the dividend, for a divisor.

```
    5̇76̇(24
 44)176
    176
Proof: 24×24=576
```

If we divide 176, the number of rods to be added, by 40, the length of the addition, (or 17 by 4, rejecting the unit figure of the dividend and divisor,) we have 4 rods, the width of the addition. Then $40 \times 4 = 160$, the number of rods added on the two sides; still there is a remainder of 16 rods. As the additions made are no longer than the sides of the square, there will be a deficiency in the corner, (see Fig. 2,) of a square whose sides are equal to the width of the addition, $4 \times 4 = 16$ rods. We therefore place 4, the last quotient figure, on the right of the divisor, because its square is necessary to supply this deficiency. The whole divisor now multiplied by the last quotient figure, equals 176, the number of rods which were to be added to the square. We have now obtained 24, the root of 576, or the length of one side of a square garden containing 576 square rods. Proof by Involution: $24 \times 24 = 576$.

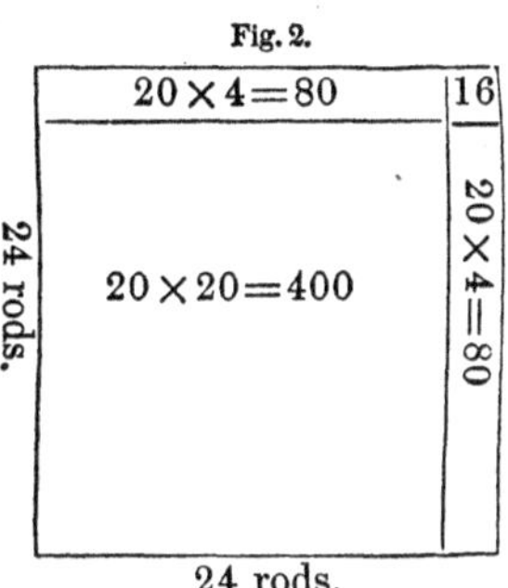

Fig. 2.

From the preceding example and illustration we derive the following

RULE.

I. *Distinguish the given number into periods of two figures each, by putting a dot over the units, and another over the hundreds, and so on. The dots show the number of figures of which the root will consist.*

II. *Find the root of the greatest square number in the left-hand period, and place it as a quotient in division. Place the square of the root found, under said period, and subtract it therefrom, and to the remainder bring down the next period, for a dividend.*

III. *Double the root already found, for a divisor; see how often the divisor is contained in the dividend, (excepting the right-hand figure,) and place the result for the next figure in the root, and also on the right hand of the divisor.*

IV. *Multiply the divisor by the figure in the root last found, and subtract the product from the dividend. To the remainder, bring down the next period, for a new dividend. Double the root now found, for a new divisor, and proceed in the operation as before, until all the periods are brought down.*

Obs.—Doubling the right-hand figure of the last divisor, observing to add 1 to the place of tens, when the double of the unit figure is over ten, is the same as doubling the root, or quotient.

EXAMPLES.

Art. 228.—2. What is the square root of 119716?

Operation.

```
 119716(346 Ans.
  9
64)297
   256
686)4116
    4116
```

3. What is the square root of 1444? *Ans.* 38.
4. What is the square root of 59536? *Ans.* 244.
5. What is the square root of 124896? *Ans.* 353.4+.

Obs. 1.—When there is a remainder, after all the figures are brought down, ciphers may be annexed, and the operation continued to any assigned degree of exactness.

6. What is the square root of 67321? *Ans.* 259.46+.
7. What is the square root of 25289? *Ans.* 159.02+.
8. What is the square root of 21027? *Ans.* 145.006+.
9. What is the square root of 6842.723400? *Ans.* 82.7207+.

Obs. 2.—When there are whole numbers and decimals in the given sum, point off both ways from the units' place; if the decimals be an odd number, annex ciphers, and make them even.

10. What is the square root of 10.4976? *Ans.* 3.24.
11. What is the square root of 336.234? *Ans.* 18.333+.
12. What is the square root of .108241? *Ans.* .329.
13. What is the length of a square field containing 7744 square rods? *Ans.* 88 rods.

14. What is the square root of $\frac{9}{81}$? *Ans.* $\frac{3}{9}$.

OBS. 3.—The square root of a fraction may be found by extracting the root of the numerator and denominator.

15. What is the square root of $\frac{16}{25}$? *Ans.* $\frac{4}{5}$.

16. What is the square root of $\frac{4}{49}$? *Ans.* $\frac{2}{7}$.

17. What is the square root of $\frac{49}{64}$? *Ans.* $\frac{7}{8}$.

18. What is the square root of $\frac{1}{2}$? *Ans.* .707+.

OBS. 4.—When the numerator and denominator are *surd* numbers, reduce the fraction to a decimal, and extract the root as above directed.

19. What is the square root of $\frac{3}{4}$? *Ans.* .866+.

20. What is the square root of $\frac{7}{8}$? *Ans.* .9355+.

21. How many rows on one side of a square cornfield, containing 15376 hills? *Ans.* 124.

22. An army of 242064 men are drawn up in a solid body, in the form of a square. What is the number of men in rank and file? *Ans.* 492.

23. A man has 841 peach-trees, which he wishes to plant in the form of a square. How many must be planted in each row? *Ans.* 29.

24. There is a circular pond, containing 110889 square rods. What will be the length of a square field containing the same number of rods? *Ans.* 333 rods.

25. A number of men gave £22 1*s.* for a charitable purpose, each giving as many shillings as there were men. What was the number of men? *Ans.* 21.

26. What is the length of one side of a square acre of land? *Ans.* 12.64+.

27. The diameter of a circle is 6 inches. What is the diameter of a circle 4 times as large? *Ans.* 12.

OBS. 5.—Circles are to one another as the squares of their diameter; therefore, to find the required diameter, square the *given* diameter, multiply the square by the given ratio, and the square root of the product will be the diameter required.

28. The diameter of a circle is 24 feet. What is the diameter of a circle one-fourth as large? *Ans.* 12 feet.

29. In the right-angled triangle ABC, the side AC is 9 feet, and the side BC 12 feet. What is the length of the side AB?

In every right-angled triangle, the square of the hypotenuse is equal to the sum of the squares of the base and perpendicular; therefore, the square root of the sum of the squares of the base and perpendicular, will be the hypotenuse, and the square root of the difference of the square of the hypotenuse, and either of the other sides, will be the remaining side.

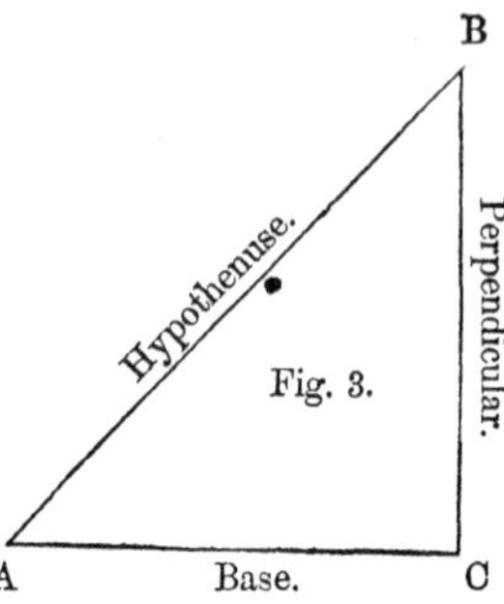

$$AC^2 = 9^2 = 81$$
$$BC^2 = 12^2 = 144$$
$$AB^2 = 225$$
$AB = \sqrt{225} = 15$ feet, *Ans.*

30. What is the distance between the opposite corners of a room, 20 feet in length and 15 in width? *Ans.* 25 feet.

31. If the distance between the opposite corners of a room be 25 feet, and the width of the room be 15 feet, what is the length? *Ans.* 20 feet.

32. If a room be 20 feet in length, and 25 feet between the opposite corners, what is the width? *Ans.* 15 feet.

33. Two men owning a pasture 32 rods in width, and 50 rods between the opposite corners, agreed to divide said pasture into two equal parts by a wall running through it lengthwise. Suppose they pay 50 cents a rod for building the wall, what does it cost them? *Ans.* $19.209.

34. Suppose a ladder 50 feet long, to be so placed as to reach a window 30 feet from the ground on one side of the street, and without moving it at the foot, will reach a window 20 feet high, on the other side; what is the width of the street? *Ans.* 85.825+ feet.

35. Two men travel from the same place—one due east, the other due north. One travels 40 miles the first day, the other 30. What is the nearest distance between them at night? *Ans.* 50 miles.

36. A. and B. set out together, and travel in the same direction on parallel courses, which are 20 miles apart. A. travels 45 miles, and B. 25. What is the distance between them at night? *Ans.* 28+ miles.

37. Suppose a pine-tree to stand 25 feet from the end of a house 40 feet in length, the foot of the tree being on a level with the foundation of the chimney, which stands in the centre of the house, and a line reaching from the foot of the tree to the top of the chimney, be 75 feet, what is the height of the chimney? and if the height of the tree be $\frac{1}{3}$ of $\frac{3}{4}$ of $\frac{4}{7}$ of 14 of the height of the chimney, what will be the length of a line reaching from the top of the chimney to the top of the tree?

Ans. { 60 feet, height of the chimney.
75 feet, length of the line.

Art. 229.—To find a mean proportional between two num bers.

RULE.

Multiply the given numbers together, and the square root of their product is the mean proportion sought.

1. What is the mean proportional between 3 and 12?

Operation.

$3 \times 12 = 36$, and $\sqrt{36} = 6$ *Ans.*

It is evident, that the ratio of 3 to 6 is the same as the ratio of 6 to 12; for $\frac{3}{6} = \frac{1}{2}$, and $\frac{6}{12} = \frac{1}{2}$.

2. What is the mean proportional between 12 and 48? *Ans.* 24
3. What is the mean proportional between 9 and 81? *Ans.* 27.
4. What is the mean proportional between 25 and 625? *Ans.* 125.

EXTRACTION OF THE CUBE ROOT.

Formation of the Cube, and Extraction of the Cube Root.

Art. 230.—The *third power*, or *cube* of any number, is the product of that number multiplied into its square; and the *cube root* is a number which, multiplied into its square, will produce the given number.

Roots and powers are *correlative* terms; that is, if 3 is the cube root of 27, then 27 is the third power, or cube, of 3.

There are but nine *perfect cubes* among numbers expressed by one, two, or three figures; each of the other numbers has for its cube root a whole number, *plus* a fraction. Thus 64 is the cube of 4, and 27 is the cube of 3; therefore, the cube root of each number between 27 and 64 must be 3 *plus* a fraction.

What is the cube of 24?

$$
\begin{array}{r}
\text{tens. units.} \\
24=2+\ 4 \\
2+\ 4 \\
\hline
8+16 \\
4+\ 8\qquad \\
\hline
4+16+16 \\
2+\ 4 \\
\hline
16+64+64 \\
8+32+32\qquad\quad \\
\hline
8+48+96+64=13824
\end{array}
$$

It will be perceived, from the above process, that the cube of a number composed of tens and units, is made up of four parts, viz: 1. The cube of the tens, (8 thousands.) 2. Three times the product of the square of the tens into the units, (48 hundreds.) 3. Three times the product of the tens into the square of the units, (96 tens.) 4. The cube of the units, (64 units.)

To extract the cube root is to find a number which, multiplied into its square, will produce the given number.

What is the cube root of 13824?

Operation.

$$
\begin{array}{r}
1\dot{3}82\dot{4}(24 \\
8\quad\ \ \\
\hline
2^2\times3=12)58|24
\end{array}
$$

As this number is greater than 1000, which is the cube of 10, but less than 1,000,000, its root will consist of two figures, tens and units; but the cube of tens cannot be less than thousands; therefore, the three figures, 824, on the right, cannot form a part of it. Hence we separate these from 13 by a point, and look for the cube of tens in 13, the left-hand period. The root of the greatest cube contained in 13 is 2, which is the tens in the required root; for the cube of 20, which is 8000, is less, and the cube of 30, which is 27000, is greater than the given number;

therefore, the required root is composed of 2 tens, *plus* a certain number of units less than *ten.*

We now subtract 8, the cube of the tens, from 13, and bring down the next period, 824. We have now 5824, which contains the three remaining parts of the cube, viz: *Three times the product of the square of the tens into the units, plus three times the product of the tens into the square of the units, plus the cube of the units.* Now, as the square of tens gives hundreds, it follows, that three times the square of the tens into units must be contained in 58, which we separate from 24 by a line. If we now divide 58 by three times the square of the tens, we shall obtain the units of the required root. • We may ascertain whether the unit figure be right, by cubing the quotient, or by applying the following principle: *The difference between the cubes of two consecutive numbers is equal to three times the square of the least number, plus three times this number, plus* 1. Thus, the difference between the cube of 3 and the cube of 4, is equal to $\overline{9\times3}+\overline{3\times3}+1=37$, which is the difference between the cube of 3 and the cube of 4. Therefore, had we written 3 in the unit's place, the remainder would have been equal to 3 times the square of 23, *plus* three times 23, *plus* 1, which would show that the unit figure must be increased.

Thus far the illustration has been general,—applied to *numbers merely*—numbers in the abstract. We may now apply it to solid bodies. Numbers which represent, or stand for things, are called *concrete,* as question first below.

EXAMPLES.

Art. 231.—1. What is the length of one side of a solid block containing 13824 solid inches, or what is the cube root of 13824 ?

OBS.—The foregoing operation can be better understood by blocks prepared for the purpose. It is necessary to have one cubical block, of a convenient size, to represent the greatest cube in the left-hand period, and three other blocks, equal to the sides of the first block, but of indefinite thickness, to represent the additions upon the sides. Then three other blocks, equal in length to the sides of the cube, and their other dimensions equal to the thickness of the additions on the sides of the cube. Lastly: a small cubic block, of dimensions equal to the thickness of the additions, to fill the deficiency at the corner. By placing these blocks as above described, the several steps in the operation may be easily understood. It may be observed, however, that this illustration would serve only for concrete numbers, as in the above question.

Having distinguished the given number into periods of three figures each, denoted by the index of the root, we perceive, by the number of periods, that the root will consist of two figures. As the cube of ten cannot be less than a thousand, $10\times10\times10=1000$, we look for the cube of tens in the second, or left-hand period. We find, by trial, the greatest cube in 13, or 13000, to be 8, or 8000, and its root, 2 or 2 tens, (the length of one side of the cube, Fig. 4,) which we place in the quotient, as the first figure of the root, and its cube, $20\times20\times20=8000$, under that period; and, subtracting it, we have a remainder of 5, or 5000—to which we bring down the next period. Had the cube contained but 8000 solid inches, we should now have found its root, or the length of one side. But we have 5824 inches to be added to the cube, and in such a manner that its cubic form shall not be altered. It is obvious, that an equal addition must be made on three sides. As each side is 20 inches square, we have $20\times20\times3=1200$; or, which is the same thing, multiply the square of the quotient by 300. $2\times2\times300=1200$ inches surface, to which the additions are to be made. It will be seen (Fig. 5) that there are three deficiencies along the sides, *a a a*, where the additions meet, 20 inches in length, $20\times3=60$, or multiply the quotient by 30; $2\times30=60$. We have, then, $1200+60=1260$, which may be considered the points where the additions are to be made. Then $5824\div1260=4$ inches, the thickness of the addition, or the

Operation.

$$
\begin{array}{rr}
 & 13\dot{8}2\dot{4}(24 \text{ root.} \\
2^3=2\times2\times2=8 & \\
2^2\times300+60=1260) & 5824 \\
1200\times4= & 4800 \\
60\times4\times4= & 960 \\
4\times4\times4= & 64 \\
 & 5824
\end{array}
$$

Fig. 4.

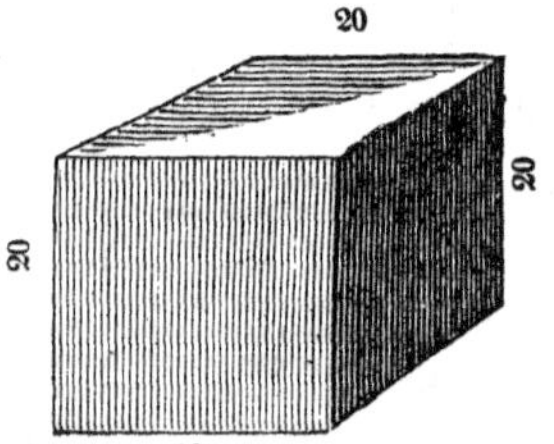

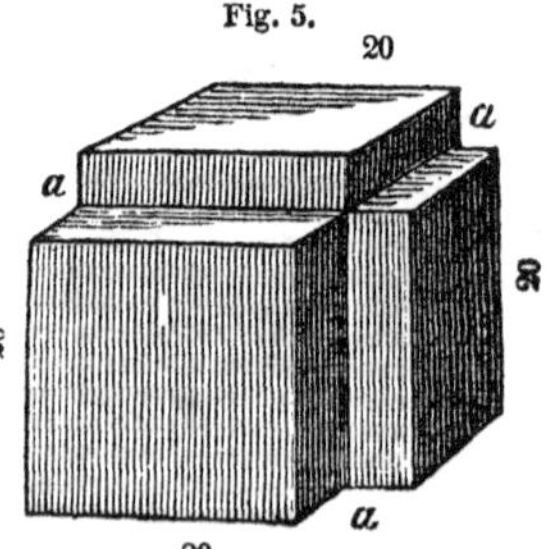

Fig. 5.

second figure of the root. The area of the sides multiplied by the thickness, $1200 \times 4 = 4800$ inches, the amount of the addition upon the sides. Then the number of inches necessary to fill the deficiencies where the additions on the sides meet, is $60 \times 4 \times 4 = 960$ inches. Still there is a deficiency of a small cube in the corner, (Fig. 6,) whose dimensions are equal to the thickness of the additions: $4 \times 4 \times 4 = 64$ inches. This supplied, and the cube is completed. (Fig. 7.) The sum of all the additions will be a subtrahend equal to the dividend; $4800 + 960 + 64 = 5824$. We have now found the length of one side of the cube to be 24 inches. Proof by Involution:

Fig. 6.

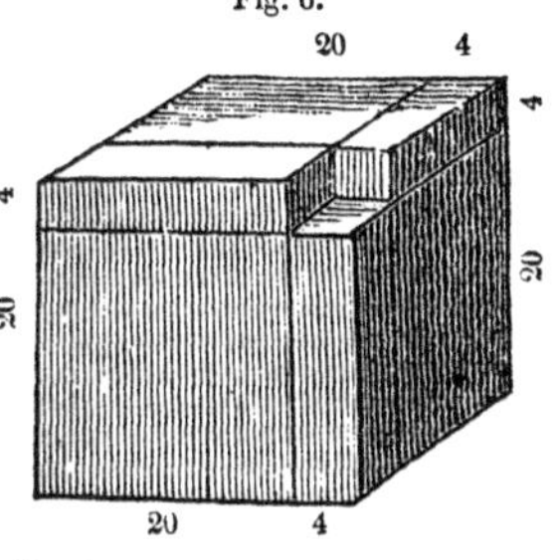

$$24 \times 24 \times 24 = 13824.$$

Art. 232.—Hence it appears, that a cube is a solid body, having six equal sides, and its cube root is the length of one of those sides.

From the foregoing example and illustration we derive the following

Fig. 7.

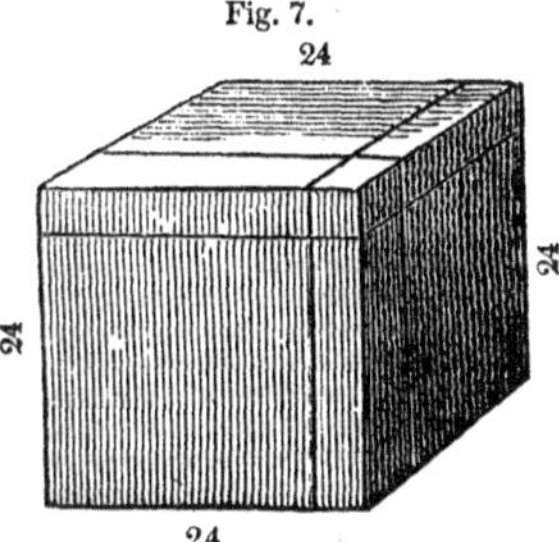

RULE.

I. *Distinguish the given number into periods of three figures each, beginning at the right hand.*

II. *Find the greatest cube in the left-hand period, and place its root as a quotient in division.*

III. *Subtract the cube from said period, and to the remainder bring down the next period, for a dividend.*

IV. *Multiply the square of the quotient by* 300, *calling it the triple square, and the quotient by* 30, *calling it the triple quotient, and the sum of these call the divisor.*

OBS.—The triple quotient is not indispensable in forming the divisor.

V. *Seek how many times the divisor is contained in the dividend, and place the result in the quotient, for the second figure of the root.*

VI. *Multiply the triple square by the last quotient figure, and write the product under the dividend; multiply the triple quotient by the square of the last quotient figure, and place this product under the last; under these write the cube of the last quotient figure, and call their sum the subtrahend. Subtract the subtrahend from the dividend, and to the remainder bring down the next period, for a new dividend, and proceed as before, till the work is finished.*

EXAMPLES.

2. What is the cube root of 1906624?

Operation.

$$
\begin{array}{rl}
1\times1\times300=300 & \dot{1}90\dot{6}62\dot{4}(124\ \textit{Ans.} \\
1\times\ 30=\ \ 30 & 1 \\
\text{Divisor,}\ \ 330 & \ \ 906\ \text{dividend.} \\
300\times2\ =600 & \\
30\times2^2=120 & \\
2^3=\ \ \ 8 & \\
 & 728\ \text{subtrahend.}
\end{array}
$$

$$
\begin{array}{r}
\overline{12^2}\times300+\overline{12\times30}=43560)178624 \\
43200\times4\ =172800 \\
360\times4^2=\ \ \ 5760 \\
4^3=\ \ \ \ \ \ 64 \\
\text{Subtrahend, }178624
\end{array}
$$

3. What is the cube root of 941192? *Ans.* 98.
4. What is the cube root of 6331625? *Ans.* 185.
5. What is the cube root of 11543176000? *Ans.* 2260.
6. What is the cube root of 34.328125? *Ans.* 3.25.
7. What is the cube root of .000729? *Ans.* .09.
8. What is the cube root of .003375? *Ans.* .15.
9. What is the cube root of 5? of 3?
10. What is the cube root of $\frac{8}{125}$? *Ans.* $\frac{2}{5}$.
11. What is the cube root of $\frac{216}{343}$? *Ans.* $\frac{6}{7}$.
12. What is the cube root of $\frac{1728}{2197}$? *Ans.* $\frac{12}{13}$.
13. What is the cube root of $\frac{1}{27}$? *Ans.* $\frac{1}{3}$.
14. A certain hill contains 11543176 cubical feet. What is the length of one side of a cubical mound, containing an equal number of feet? *Ans.* 226 feet.

15. The contents of an oblong cellar is 9261 cubical feet. What is the length of one side of a cubical cellar, of the same capacity? *Ans.* 21 feet.

16. A merchant bought cloth to the amount of $393.04, but forgets the number of pieces, and also the number of yards in each piece, and what the cloth cost per yard; but remembers that he paid as many cents per yard as there were yards in each piece, and that there were as many in each piece as there were pieces. What did he pay per yard? *Ans.* 34 cents.

17. What is the width of a cubical vessel, containing 75 wine gallons, each 231 cubic inches?

18. Required the side of a cubic box that shall contain a bushel? *Ans.* 12.9+inches.

Art. 233.—*Solids of the same form are to one another as the cubes of their similar sides, or diameters.*

EXAMPLES.

1. If a bullet, weighing 72 lbs., be 8 inches in diameter, what is the diameter of a bullet weighing 9 lbs.? *Ans.* 4 inches.

Statement.

$8^3=512$ $72 : 9 :: 512 : 64^{\frac{1}{3}}$. Or thus:

	9
8 72	512 $64^{\frac{1}{3}}$
	4 *Ans.*

2. A bullet, 2 inches in diameter, weighs 4 lbs. What is the weight of a bullet 5 inches in diameter? *Ans.* $62\frac{1}{2}$ lbs.

3. If a silver ball, 9 inches in diameter, be worth $400, what is the worth of another ball, 12 inches in diameter? *Ans.* $948.148+.

Art. 234.—To find two mean proportionals between two numbers.

RULE.

Divide the greater by the less, and extract the cube root of the quotient: multiply the lesser number by this root, and the product will be the lesser mean; multiply this mean by the same root, and the product will be the greater mean.

EXAMPLES.

1. What are the two mean proportionals between 4 and 256?

$256 \div 4 = 64$; then $\sqrt{64} = 4$, and $4 \times 4 = 16$, the lesser, and $16 \times 4 = 64$, the greater. *Proof,* 4 : 16 :: 64 : 256.

2. What are the two mean proportionals between 5 and 625?
Ans. 25 and 125.

3. What are the two mean proportionals between 7 and 2401?
Ans. 49 and 343.

EXTRACTION OF ROOTS IN GENERAL.

RULE.

Art. 235.—I. *Point the given number into periods of as many figures as the index of the root directs. Thus, for the square root, two figures; cube root, three; fourth root, four, etc.*

II. *Find, by trial, the greatest root in the left-hand period, and subtract its power from that period, and to the remainder bring down the first figure of the next period, for a dividend.*

III. *Involve the root, already found, to the next inferior power to that which is given, and multiply it by the number denoting the given power, for a divisor, by which find the second figure of the root.*

IV. *Involve the whole root now found to the given power; subtract it from the given number, as before, and bring down the first figure of the next period to the remainder, for a new dividend, and proceed as before, till the work is finished.*

Obs.—The roots of most of the powers may be found by repeated extractions of the square and cube root—Thus:

For the 4th root, take the square root of the square root.
For the 6th " take the square root of the cube root.

Questions.—1. Rule for finding a mean proportional between two numbers? 2. What is a cube? 3. What is a cube root? 4. What is it to extract the cube root? 5. What is the rule? 6. Why do you distinguish the given number into periods of three figures each? 7. Why do you multiply the square of the quotient by 300? 8. Why the quotient by 30? 9. Why the triple square by the last quotient figure? 10. Why the triple quotient by the square of the last quotient figure? 11. Explain the process of illustrating this rule by blocks. 12. What proportion have solids to one another? 13. Rule for finding two mean proportionals between two numbers? 14. Rule for extracting roots in general?

For the 8th root, take the square root of the 4th root.
For the 9th " take the cube root of the cube root.
For the 12th " take the cube root of the 4th root.

EXAMPLES.

1. What is the square root of 7569?

Operation.

```
               7569(87
 8×  8=        64  =square, or 2d power, of the quotient.
 8×  2=16, 16)116=dividend.
87×87=         7569=square of the quotient.
               0000
```

2. What is the fifth root of 4084101?

Operation.

```
                     4084101(21
 2×2×2×2×2=          32    =  5th power of the quotient.
 2×2×2×2×5=80)88    =  1st dividend.
                     4084101
21×21×21×21×21=4084101=5th power of the quotient.
```

3. What is the fourth root of 140283207936? *Ans.* 612.

4. What is the seventh root of 4586471424? *Ans.* 24.

5. What is the ninth root of 1352605460594688? *Ans.* 48.

ARITHMETICAL PROGRESSION.

Art. 236.—ARITHMETICAL PROGRESSION is when a series of numbers increases by a common excess, or decreases by a common difference.

When numbers increase by a common excess, they form the *ascending series,* as 2, 4, 6, 8, 10, 12, etc.

QUESTION.—1. What is Arithmetical Progression?

When numbers decrease by a common difference, they form the *descending series*, as 12, 10, 8, 6, 4, 2, etc.

The numbers forming the series are called the terms; the first and last terms are called the *extremes*, and the other terms the *means*.

When any even number of terms differs by Arithmetical Progression, the sum of the two extremes will equal the sum of any two means equally distant from the extremes; as 2, 4, 6, 8, 10, 12. The two extremes, $2+12=6+8$, the two means. When the number of terms is odd, the double of the mean will equal the sum of the two extremes, or the sum of any two numbers equally distant from the extremes; as 1, 2, 3, 4, 5. The double of the mean $3\times2=5+1=6$.

In Arithmetical Progression, five things are to be considered, viz.: the first and last terms, the number, common difference, and sum of all the terms; any three of which being given, the other two may be found.

1. If I buy 4 books, giving 2 cents for the first, 4 for the second, and so on, with a common difference of 2, what do I pay for the last book?

It is evident, that if we add 2 cents, the common difference, to the price of the first book, we shall have the price of the second, and so on to the last; thus, $2+2=4$, $4+2=6$, $6+2=8$ cents, the answer. It will be seen that 2, the common difference, is added to every term but the last. If, then, we multiply the number of terms, less 1, by the common difference, we have the difference between the cost of the first book and the last; thus, $3\times2=6$, and $6+2=8$, as before. Therefore,

Art. 237.—When the first term, the number of terms, and common difference are given, to find the last term:

RULE.

Multiply the number of terms, less 1, *by the common difference, and to the product add the first term, and the sum will be the last term.*

2. If the first term of a series be 5, the number of terms 35, and the common difference 3, what is the last term?

Ans. $\overline{35-1}\times3=102+5=107$.

QUESTIONS.—2. When is the series ascending? 3. When descending? 4. What is meant by the terms? 5. What is meant by the extremes? 6. By the means?

3. If I buy 80 yards of cloth, giving 6 cents for the first, 10 for the second, and so on, with a common difference of 4, what do I pay for the last yard? *Ans.* 322 cents.

4. Suppose a man purchase 40 sheep, paying 3 pence for the first, 10 for the second, and so on, with a common difference of 7, what does he pay for the last sheep?

Ans. 276 pence.

5. If 96 acres of land be sold at the rate of 10 cents for the first acre, 19 for the second, and so on, with a common difference of 9, for how much is the last acre sold?

Ans. 865 cents.

6. If I buy 4 books, the prices of which are, in Arithmetical Progression, giving 2 cents for the first, and 8 for the last, what is the common difference in the prices of the books?

This question is the reverse of question 1st. $8-2=6$, $6\div3=2$, the common difference. It is plain, that the difference between the price of the first and last book, is the whole addition made to the price of the first book; and as the addition is made equally to the three books, it is equally plain that the whole addition, divided by the number of additions, will be the addition made to the price of each book. Therefore—

Art. 238.—When the extremes and number of terms are given, to find the common difference, we have this

RULE.

Divide the difference of the extremes by the number of terms less 1, and the quotient will be the common difference.

7. If the first term of a series be 3, the last term 276, and the number of terms 40, what is the common difference?

Ans. 7.

8. A man on a journey travels the first day 2 miles, and increases his travel daily by an equal excess for 15 days, so that the last day he travels 72 miles. What was the daily increase? *Ans.* 5 miles.

9. Bought books, paying 2 cents for the first, and 8 cents for the last, with a common difference of 2. What number of books did I buy?

As the difference of the extremes, divided by the number of the terms less 1, will give the common difference, it is evident that the difference of the extremes divided by the common difference, will give the number of the terms less 1. Then, 8−2=6, the difference of the extremes, and 6÷2=3, which is one less than the number of terms; then 3+1=4, the number of books purchased. Therefore—

Art. 239.—When the first and last terms, and the common difference are given, to find the number of terms—

RULE.

Divide the difference of the extremes by the common difference, and the quotient will be 1 *less than the number of terms.*

10. If the first term of a series be 2, and the last term 72, the common difference 5, what is the number of terms? *Ans.* 15.

11. A man bought sheep, paying at the rate of 3 pence for the first, and 276 for the last, with a common difference of 7. What number did he buy? *Ans.* 40.

12. A man has a number of sons, the common difference of whose ages is 4 years; the youngest is 8, the eldest 40 years old. How many sons has he? *Ans.* 9.

13. If I buy 4 books, paying 2 cents for the first, and 8 cents for the last, how many cents do I pay in all?

If the price of the first book is 2 cents, and the price of the last is 8 cents, it is evident that the average price of the books is half way between 2 cents, the price of the first, and 8 cents, the price of the last book: 2+8÷2=5. Then 5, the average price, multiplied by the number of books, will give the whole cost: 5×4=20 cents. The same may also be shown by writing the double series, thus:

$$\begin{array}{cccc} 2 + & 4 + & 6 + & 8 \\ 8 + & 6 + & 4 + & 2 \\ \hline 10 & 10 & 10 & 10 \end{array}$$

It will be seen by this formula, that the sum of any two corresponding terms in the double series is equal to the sum of the two extremes in the simple series; if, therefore, we multiply the sum of the extremes by the number of terms, we shall obtain a sum twice too large. Therefore—

Art. 240.—When the first and last terms, and the number of terms are given, to find the sum of the series—

RULE.

Multiply half the sum of the extremes by the number of terms. The product will be the sum of the series.

14. A man has 9 sons; the youngest is 8, the eldest 40 years old. What is the sum of their ages?

Ans. 216 years.

15. How many times will a clock strike in a day, if constructed like the clocks of Venice, to run till 24 o'clock?

Ans. 300.

16. If a triangular piece of land, 60 rods in length, be 1 rod wide at one end, and 60 at the other, what number of square rods does it contain? *Ans.* 1830.

GEOMETRICAL PROGRESSION.

Art. 241.—A GEOMETRICAL PROGRESSION is a series of terms, which increase by a uniform multiplier, or decrease by a uniform divisor; as 3, 6, 12, 24, etc., increasing by a uniform multiplier, 2; or 54, 18, 6, 2, $\frac{2}{3}$, etc., decreasing by a uniform divisor, 3.

The multiplier, or divisor, which produces the series, is called the *ratio.*

1. A man bought 5 yards of cloth, paying 3 cents for the first yard, 6 for the second, and so on, doubling the price to the last. What was the price of the last yard?

$3 \times 2 \times 2 \times 2 \times 2 = 48$, the cost of the last yard.

From the above operation it will be seen that the cost of the second yard is the product of the ratio multiplied by the cost of the first yard; and that the cost of the third yard, is the product of the second power of the ratio multiplied by the cost of the first yard, or the first term; and finally, that the cost of the fifth yard, or the last term, is the product of the fourth power of the ratio, multiplied by the cost of the first yard. It appears, also, that any term in the series may be found by in-

QUESTIONS.—1. What is Geometrical Progression? 2. What is an ascending series? 3. What a descending? 4. What is the ratio? 5. When the first term and ratio are given, how do you find the last term? 6. When the first and last terms, and the ratio are given, how do you find the sum of the series?

volving the ratio to a power less 1 than the number of terms, and multiplying that power by the first term.

OBS.—The process of involving the ratio to a high power, may be shortened by multiplying together those lower powers whose indices added equal the index of the power sought. To find the fifth power of 3, we may multiply together the second and third powers, for the index of the second power of 3, and the index of the third power added,

equal 5, $3 \overset{2+3=5}{} 3$ and $\overline{3^2 \times 3^3} = \overline{9 \times 27} = 243$, the 5th power of 3.

Art. 242.—When the first term and ratio are given, to find the last term—

RULE.

Involve the ratio to a power whose index is 1 less than the number of terms, and multiply this power by the first term. The product will be the answer, if the series is increasing; but if it is decreasing, divide the first term by the ratio.

2. If I hire a man for 12 months, and agree to pay him 1 dollar for the first month, 3 for the second, and so in a tripl proportion, what must I pay him for the last month?

$\overset{1}{3}\ \overset{2}{9}\ \overset{3}{27}\ \overset{4}{81}\ \overset{5}{243}\ \overset{6}{729},\ \overset{6}{729} \overset{+}{\times} \overset{5}{243} \overset{=}{=} \overset{11}{177147} \times 1 = \$177147$ *Ans.*

It will be seen that the sum of the indices of the fifth and sixth powers added, equal 11, which is 1 less than the number of terms; and the fifth and sixth powers multiplied together, equal the 11th power of the ratio, which multiplied by the first term, gives the answer, or the last term.

3. A man bought 20 cows, paying 2 farthings for the first, 10 for the second, and so on, in a five-fold ratio. What was the price of the last cow?

Ans. £39736429850 5*s.* 2*d.* 2*qrs.*

4. A man bought 5 yards of cloth, giving 2 cents for the first, and 32 for the last; the prices forming a geometrical series, the ratio of which was 2. What was the whole cost of the cloth?

The price of the cloth would be the sum of the following numbers: $2+4+8+16+32=62$, the whole cost. It will be seen, that the whole cost is the same as the difference between the two extremes divided by the ratio less 1 added to the greater extreme: Thus, $32-2=30$, and $30 \div 1 = 30$, and $30+32=62$.

Again, if any term of a corresponding series be multiplied by the ratio, the product will be the succeeding term. We will now form a new series, and write it one step farther to the right of that from which it is formed; if we now subtract the first series from the second, we find that all the terms but the first in the first series and the last in the second, disappear, thus:

2	4	8	16	32	
	4	8	16	32	64
−2					64=64− 2=62

OBS.—If the ratio were 3 we should have double the first series, if it were 4 we should have triple; hence we divide by the ratio less one.

Art. 243.—Hence, when the first term, last term, and ratio, are given, to find the sum of the series, we have the following

RULES.

I. *Multiply the last term by the ratio, and from the product subtract the first term, and divide the remainder by the ratio less* 1; *the quotient will be the answer.*

II. *Divide the difference between the two extremes by the ratio less* 1, *and add the quotient to the greater term; their sum will be the answer.*

5. The extremes of a Geometrical Progression are 3 and 18673, and the ratio 11. What is the sum of the series?

Operation 1*st.*

$$\overline{\overline{\overline{18673\times 11}-3}\div 10}=20540\ Ans.$$

Operation 2*d.*

$$\frac{18673-3}{11-1}+18673=20540\ Ans.$$

6. If I discharge a debt by paying 1 dollar the first month, 4 the second, and so on, in a four-fold ratio, the last payment being 65536 dollars; what was the whole debt?

Ans. $87381.

7. The first term in a geometrical series is 2, the number of terms 10, and the ratio 3. What is the sum of the series?

Ans. 59048.

OBS.—The last term may be found by rule first, or the two processes of finding the last term and the sum of the series may be reduced to one, thus:

Art. 244.—When the first term, the number of terms, and the ratio are given, to find the sum of the series—

RULE.

Involve the ratio to a power whose index is equal to the number of terms, from which subtract 1; *divide the remainder by the ratio less* 1, *and the quotient, multiplied by the first term, will be the answer.*

8. A man sold 15 yards of cloth; the first yard for 1 shilling, the second for 2, the third for 4, and so on, doubling the price of each succeeding yard. For how much did he sell the whole?

Operation.

$$2^{15}=32768, \text{ and } \frac{32768-1}{2-1}\times 1=32767s.$$

2|0)3276|7

£1638 7*s. Ans.*

9. A man bought 20 yards of cloth, agreeing to pay 3 pence for the first yard, 9 pence for the second, and so on in a triple proportion to the last. What did he pay for the whole? *Ans.* £21792402 10*s.*

10. A gentleman bought a horse, agreeing to pay what his shoes would amount to, at 1 cent for the first nail, 2 for the second, 4 for the third, and so on, doubling the price of each succeeding nail to the last. The number of nails was 32; what was the price of the horse? *Ans.* $42949672.95.

11. A laborer wrought 20 days, and received for the first day's labor 4 grains of rye, for the second 12, for the third 36, &c. How much did his wages amount to, allowing 7680 grains to make a pint, and the whole to be disposed of at $1 per bushel? *Ans.* $14187+.

COMPOUND INTEREST BY PROGRESSION.

Art. 245.—1. What is the amount of $6, for four years, at 6 per cent., compound interest?

Operation.

```
        $6   1st term.
      1.06
      ----
       636   2d.
      1.06
      ----
      3816
     636
    ------
    6.7416   3d.
      1.06
    ------
    404496
   67416
  --------
  7.146096   4th.
      1.06
  --------
  42876576
 7146096
----------
7.57486176   5th.
```

It will be seen, that this question may be solved by the rule after Example 1st in Progression. The principal is the first term, the amount of $1 for one year the ratio, and the number of years, 1 less than the number of terms. The question may be thus stated:—If the first term be 6, the number of terms 5, and the ratio 1.06, what is the last term?

$$1.06^4 = 1.262 \times 6 = 7.572 \text{ dollars.}$$

The amount of £1, or $1, at 5 or 6 per cent., may be found by the table for Compound Interest, (see Art. 211.)

2. What is the amount of $30, for 7 years, at 5 per cent., compound interest? *Ans.* $42.213.

3. What is the amount of $7, for 4 years, at 9 per cent., compound interest? *Ans.* $9.881.

4. If the amount of a certain sum for 6 years, at 6 per cent., compound interest, be $56.74040, what is that sum, or principal?

It will be seen that this question is the reverse of the preceding. If the amount be the product of that power of the ratio denoted by the number of years and the principal, then the amount divided by that power of the ratio will be the principal.

$$\frac{56.74040}{1.06^6} = \$40 \text{ } Ans.$$

5. If the amount of $40, for 6 years, compound interest, be $56.74040, what is the rate per cent.?

$$\frac{56.74040}{40}=1.41851=\text{the 6th power of the ratio};$$

then, by extracting the 6th root, we have 1.06 for the ratio.
Ans. 6 per cent.

6. If the amount of $40, at 6 per cent., compound interest, be $56.74040, what is the time?

$$\frac{56.74040}{40}=1.41851=1.06 \text{ raised to a power whose index}$$

is equal to the time; therefore, if we divide 1.41851 by 1.06 until there is no remainder, it is plain that the number of divisions will be the time required; or, having found the power of the ratio, we may look in the table under the given rate per cent., and against the power we shall find the number of years.
Ans. 6 years.

7. In what time will $60 amount to $75.74820, at 6 per cent., compound interest? *Ans.* 4 years.

ANNUITIES AT COMPOUND INTEREST.

Art. 246.—An annuity is a sum of money payable yearly for a certain number of years, or forever.

When annuities are not paid at the time they become due, they are said to be in arrears.

The sum of all the annuities remaining unpaid, together with the interest on each, for the time they have remained due, is called the *amount*.

EXAMPLES.

1. What is the amount of an annual pension of $100, which has remained unpaid 5 years, allowing 6 per cent., compound interest?

The last year's pension will be $100, without interest, because it is paid as soon as due; the last but one will be $106, the amount of $100 for one year; the last but two, $112.36, the amount of $100 for two years at compound interest, and so on, forming a geometrical progression. The sum of these amounts will be the sum of the series, or the amount due.

Art. 247.—Hence, when the annuity, time, and rate per cent. are given, to find the amount, we have the following

RULE.

Involve the ratio to a power denoted by the number of years; from this power subtract 1; divide the remainder by the ratio, less 1, and the quotient, multiplied by the annuity, will be the amount.

The above example may be stated thus: If the first term be 100, the number of terms 5, and the ratio 1.06, what is the sum of all the terms?

$$\frac{1.06^5-1}{1.06\ -1}\times 100=563.7.\qquad \textit{Ans.}\ \$563.7.$$

2. What is the amount of an annuity of $70, to continue 5 years, allowing 6 per cent., compound interest?

Ans. $394.59.

3. What is the amount of an annuity of $160, to continue 10 years, at 5 per cent., compound interest?

Ans. $2012.448.

4. If a yearly rent of $75 be in arrears 4 years, to what does it amount, at 6 per cent., compound interest?

Ans. $328.087.

5. A salary of $600 remains unpaid 5 years. To what does it amount, allowing 6 per cent., compound interest?

Ans. $3382.255.

Art. 248.—The annuity, time, and rate being given, to find the present worth.

RULE.

Find the amount of the annuity in arrears for the whole time; this amount, divided by that power of the ratio denoted by the number of years, will give the present worth.

6. What is the present worth of an annual pension of $96, to continue 4 years, allowing 6 per cent., compound interest?

Ans. $332.643.

7. What is the present worth of an annual salary of $400, to continue 5 years, allowing 5 per cent., compound interest?

Ans. $1731.792.

QUESTIONS.—1. What is an annuity? 2. When are annuities said to be in arrears? 3. What is the *amount?* 4. What is the rule, when the annuity, time, and rate per cent. are given, to find the amount? 5. Rule, when the amount, time, and rate are given, to find the present worth?

TABLE,

Showing the present worth of $1 *or* £1 *annuity, at* 5 *and* 6 *per cent., compound interest, for any number of years from* 1 *to* 40.

Years.	5 per cent.	6 per cent.	Years.	5 per cent.	6 per cent.
1	0.95238	0.94339	21	12.82115	11.76407
2	1.85941	1.83339	22	13.163	12.04158
3	2.72325	2.67301	23	13.48807	12.30338
4	3.54595	3.4651	24	13.79864	12.55035
5	4.32948	4.21236	25	14.09394	12.78335
6	5.07569	4.91732	26	14.37518	13.00316
7	5.78637	5.58238	27	14.64303	13.21053
8	6.46321	6.20979	28	14.89813	13.40616
9	7.10782	6.80169	29	15.14107	13.59072
10	7.72173	7.36008	30	15.37245	13.76483
11	8.30641	7.88687	31	15.59281	13.92908
12	8.86325	8.38384	32	15.80268	14.08398
13	9.39357	8.85268	33	16.00255	14.22917
14	9.89864	9.29498	34	16.1929	14.36613
15	10.37966	9.71225	35	16.37419	14.49825
16	10.83777	10.10589	36	16.54685	14.62098
17	11.27407	10.47726	37	16.71129	14.73678
18	11.68958	10.8276	38	16.36789	14.84602
19	12.08532	11.15811	39	17.01704	14.94907
20	12.46221	11.46992	40	17.15909	14.92640

Obs.—To find the present worth of any annuity, at 5 or 6 per cent., by the above table :—First find the present worth of $1 or £1 annuity; then multiply it by the given annuity, and the product will be the present worth.

8. What is the present worth of an annuity of $300, to continue 25 years, at 6 per cent., compound interest?

The present worth of $1 annuity, by the table, for 25 years, is 12.78335. Then, 12.78335 × 300 = $3835.005 *Ans.*

9. What ready money will purchase an annuity of $250, to continue 40 years, at 6 per cent., compound interest?

Ans. $3731.6.

Annuities taken in reversion at compound interest.

Art. 249.—Annuities taken in reversion are certain sums of money, payable yearly for a limited period, but not to commence until after the expiration of a certain time.

Questions.—6. What are annuities taken in reversion? 7. Rule?

RULE.

Find the present worth, to commence immediately, and this sum, divided by the power of the ratio denoted by the time in reversion, will give the answer.

10. What is the present worth of a reversion of a lease of $40 per annum, to continue 6 years, but not to commence until the end of three years, allowing 6 per cent. to the purchaser?

Present worth, . . 196.69280 / Third power of the ratio, 1.19101 = 165.147.

Ans. $165.147.

The same result may be obtained by finding the present worth of the annuity to commence immediately, and to continue the whole time. Thus, 3+6=9 years, and from the present worth for this time subtract the present worth of the annuity for the time of reversion, 3 years. Or, by the table, find the present worth of $1 for the whole time; from the sum subtract the present worth of $1 for the time of reversion, and multiply the difference by the given annuity. Thus,

The whole time, ...	6.80169
The time of reversion, . . .	2.67301
Difference, .	4.12868
	40
	$165.14720 *Ans.*

11. What is the present worth of $50, payable yearly for 4 years, but not to commence until 2 years, at 6 per cent.?

Ans. $154.1965.

12. What is the present worth of the reversion of a lease of $70 per annum, to continue 20 years, but not to commence until the end of 8 years, allowing 6 per cent. to the purchaser?

Ans. $503.7459.

13. What is the present worth of a lease of $200, to continue 30 years, but not to commence until the end of 10 years, allowing 6 per cent.? *Ans.* $1513.264.

Art. 250.—To find the present worth of an annuity to continue forever.

RULE.

Divide the annuity by the rate per cent., and the quotient will be the present worth.

14. What is the present worth of a freehold estate whose yearly rent is $60, allowing 6 per cent. to the purchaser?

$\frac{60}{.06}=\$1000$ *Ans.* It is evident that the estate is worth as much money as, at the given rate per cent., would give interest equal to the rent.

15. What is $300 annuity worth, to continue forever, allowing 5 per cent. to the purchaser? *Ans.* $6000.

Art. 251.—To find the present worth of a freehold estate, in reversion, at compound interest.

RULE

Find the value, as though it were to be entered on immediately, by the foregoing rule, and divide this value by that power of the ratio denoted by the time of reversion; and the quotient will be the present worth of the estate in reversion.

16. Suppose a freehold estate, of $48 per annum, to commence two years hence, be put on sale. What is the value, allowing 6 per cent to the purchaser?

$$\frac{48}{.06}=800,\ \frac{800}{1.06^2}=\frac{800}{1.1236}=\$711.997\ \textit{Ans.}$$

17. Which is the more valuable, a term of 16 years, in an estate of $100 per annum, or the reversion of such an estate forever after 16 years, computing at the rate of 5 per cent., compound interest?

Ans. The term of 16 years, by $167.551+.

PERMUTATION.

Art. 252.—PERMUTATION is the method of finding how many changes may be made upon the order of any given number of things.

1. How many changes can be made of the first three letters of the alphabet?

QUESTIONS.—8. Rule for finding the present worth of an annuity to continue forever? 9. Rule for finding the present worth of a freehold estate in reversion at compound interest? 1. What is permutation of quantities?

The letter *a* can occupy but 1 position; *a* and *b* can change places, and occupy 2 positions, *ab* and *ba*, $1\times2=2$. The three letters, *a*, *b*, and *c*, can, any two of them, leaving out the third, have two positions, $1\times2=2$; consequently, when the third is taken in, there can be $1\times2\times3=6$ positions, which may be expressed thus: *abc, acb, bac, bca, cba, cab*. The same may be shown of any number of things. Hence, to find the number of changes which can be made of any given number of different things—

RULE.

Multiply all the terms of the natural series of numbers, from 1 *up to the given number, continually together, and the last product will be the answer required.*

2. Christ Church, in Boston, has 8 bells. How many changes can be rung upon them?

$1\times2\times3\times4\times5\times6\times7\times8=40320$ *Ans.*

3. Six men met at a public house, and agreed to remain so long as they could occupy different situations at the dinner-table. How long did they remain, and what was the price of their board, at 25 cents for each dinner?

Ans. { 720 days.
$1080 board.

POSITION.

Art. 253.—Position teaches to find the true number by the use of false, or supposed numbers. It is of two kinds, Single and Double.

Art. 254.—Single Position is so called, because the true number is obtained by the use of *one* supposed number.

1. A., B., and C. travelled. C. paid a certain part of the expense; B. paid double, and A. treble the sum which C. paid. The amount of their expenses was $60. What did each one pay?

Suppose C.'s expense was $8; then, by the conditions of the question, B.'s expense was $8\times2=\$16$; and A.'s $8\times3=\$24$;

Questions.—2. Rule for finding the number of permutations? 3. What is Position? 4. What is Single Position?

and the sum of their expenses \$8+\$16+\$24=\$48. As the ratios, in the true and supposed, are the same, it follows, that the true sum of their expenses will have the same ratio to the true expense of each individual, that the sum of their supposed expenses has to the supposed expenses of each individual. Thus:

48 : 8 : : 60 : 10, C.'s expense;
48 : 16 : : 60 : 20, B.'s expense; and
48 : 24 : : 60 : 30, A.'s expense.

RULE.

Suppose any number, and proceed in the operation as though it were the true; then, as the result of the operation, or sum of the errors, is to the supposed number, so is the given number to the true number required.

EXAMPLES.

2. A person, after spending $\frac{1}{2}$ and $\frac{1}{3}$ of his income, had \$30 left. What had he at first?

Suppose \$60
$\frac{1}{2}=30$
$\frac{1}{3}=20$
60—50=10 income left:
Then 10 : 60 : : 30 : 180 *Ans.*

Or by fractions: $\frac{1}{2}=\frac{3}{6}$, and $\frac{1}{3}=\frac{2}{6}$; then $\frac{3}{6}+\frac{2}{6}=\frac{5}{6}$, the income spent, and $\frac{1}{6}$ remains=\$30; then $\frac{6}{6}=30\times6=$\$180, as before.

3. A certain sum of money is to be divided between 5 men, in such a manner that A. shall have $\frac{1}{4}$, B. $\frac{1}{5}$, C. $\frac{1}{10}$, D. $\frac{1}{20}$, and E. the remainder, which is \$40. What is the sum? *Ans.* \$100.

4. A schoolmaster being asked how many scholars he had, replied, if he had as many more, $\frac{1}{2}$ and $\frac{1}{4}$ as many more, he would have 11 less than 99. How many had he? *Ans.* 32.

5. A man bought a horse, chaise, and harness for \$216. The horse cost twice as much as the harness, and the harness one third as much as the chaise. What was the cost of the chaise? *Ans.* \$108.

6. What number is that whose $\frac{1}{2}$, $\frac{1}{3}$, $\frac{1}{10}$, $\frac{1}{9}$, $\frac{1}{5}$, and $\frac{1}{6}$ make 127? *Ans.* 90.

7. A man being asked his age, said, If you add to its double $\frac{1}{3}$, $\frac{1}{5}$, $\frac{1}{9}$, and $\frac{1}{15}$ of my age, it will be 122. What was his age? *Ans.* 45.

8. A certain sum of money is to be divided among 4 persons, in such a manner, that the first shall have $\frac{3}{4}$ of $\frac{1}{2}$ of $\frac{2}{3}$; the second $\frac{1}{11}$ of $\frac{11}{12}$ of 2; the third $\frac{1}{45}$ of $\frac{45}{8}$; the fourth has $110. What is the sum divided? *Ans.* $240.

9. A. and B. having found a purse of money, disputed who should have it. A. said that $\frac{1}{5}$, $\frac{1}{10}$, and $\frac{1}{20}$ of it amounted to $35, and if B. would tell him how much was in it, he should have the whole; otherwise he should have nothing. How much did the purse contain? *Ans.* $100.

DOUBLE POSITION.

Art. 255.—DOUBLE POSITION teaches to discover the true, by the use of *two* supposed numbers.

RULE.

I. *Suppose two numbers, and proceed with each according to the conditions of the question, as in Single Position, noting the error. The difference between the result and the given sum is the error.*

II. *Multiply the first supposition by the second error, and the second supposition by the first error.*

III. *If the errors are alike—that is, both too great or both too small, divide the difference of the products by the difference of the errors.*

IV. *If the errors are unlike—that is, one too large, and the other too small, divide the sum of the products by the sum of the errors.*

OBS.—This rule is founded on the supposition that the first error is to the second as the difference between the true and first supposed number, is to the difference between the true and second supposed number. When this is not the case, the exact number cannot be obtained by this rule.

QUESTIONS.—5. What is Double Position? 6. On what supposition is this rule founded? 7. Verify the principle.

EXAMPLES.

1. A man being asked what his carriage cost, replied, If it had cost twice as much as it did, and \$20 more, it would have cost \$370. What was the cost of the carriage?

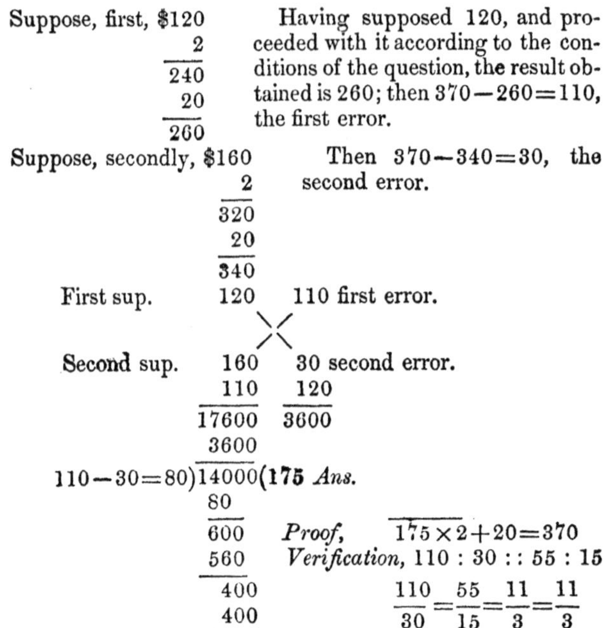

Suppose, first, \$120
2
240
20
260

Having supposed 120, and proceeded with it according to the conditions of the question, the result obtained is 260; then 370—260=110, the first error.

Suppose, secondly, \$160
2
320
20
340

Then 370—340=30, the second error.

First sup. 120 — 110 first error.

Second sup. 160 — 30 second error.

160 × 110 = 17600; 30 × 120 = 3600

17600 − 3600 = 14000

110—30=80)14000(**175** *Ans.*
80
600
560
400
400

Proof, $\overline{175}\times 2+20=370$

Verification, 110 : 30 : : 55 : 15

$$\frac{110}{30}=\frac{55}{15}=\frac{11}{3}=\frac{11}{3}$$

The foregoing question may be thus solved: Let x equal the cost of the carriage; then by the conditions of the question, $2x+20=370$.

Solution.

x=cost of carriage
$2x+20=370$
$2x=370-20=350$
$2x=350=$; then $x=\frac{350}{2}=175$ *Ans.*

2. A., B., and C. built a house, which cost \$228. B. paid

$30 more than A., and C paid as much as A. and B. What did each pay?

Ans. { A. paid $42. B. paid $72. C. paid $114.

3. A. and B. have the same income. A. saves $\frac{1}{4}$ of his annually, but B., by spending $120 per annum more than A., at the end of 6 years, finds himself $120 in debt. What is their income, and what does each spend annually?

Ans. { Their income $400. A. spends $300, and B. spends $420.

4. A man has two silver cups of unequal weight, having one cover to both, weighing 5 oz. When the cover is put on the less cup, it weighs double the greater; when put upon the greater cup, it weighs three times the less. What is the weight of each cup?

Ans. { The less, 3 oz. The greater, 4 oz.

5. There is a fish whose head is 3 feet long, his tail is as long as his head and half the length of his body, and his body is as long as his head and tail. What is the length of the fish?

Ans. 24 feet.

6. A man being asked, in the afternoon, what o'clock it was, answered, that the time passed from noon was equal to $\frac{1}{8}$ of the time to midnight. Required the time.

Ans. 20 minutes past 1 o'clock.

7. A gentleman has two horses, and one carriage which is worth $100. If the first horse be harnessed into the carriage, he and the carriage together will be worth three times as much as the second horse; but if the second be harnessed into the carriage, they will be worth seven times as much as the first horse. What is the value of each horse?

Ans. $20 and $40.

8. A laborer was hired 60 days upon this condition, that for every day he wrought he should receive 3*s.* 4*d.*, and for every day he was idle he should forfeit 1*s.* 8*d.* At the expiration of the time he received £3 15*s.* How many days did he work, and how many days was he idle?

Ans. { He was employed 35 days, and was idle 25.

ALLIGATION.

Art. 260.—Alligation is the method of mixing two or more simples, of different qualities, so that the composition may be of a mean, or middle quality.

When the quantities and prices of the simples are given, to find the mean price of the mixture compounded of them, the process is called

ALLIGATION MEDIAL.

Art. 261.—1. If I mix 8 lbs. of sugar, worth 10 cents a pound, with 10 lbs., worth 15 cents a pound, what is 1 lb. of the mixture worth?

Eight pounds, at 10 cents a pound, are worth $10\times8=80$ cents, and 10 pounds, at 15 cents, are worth $15\times10=150$ cents; then, $80+150=230$ cents, the price of the whole mixture, and $8+10=18$ pounds, the whole mixture; then $\$2.30\div18$ lbs.$=12\frac{7}{9}$ cts., the worth of 1 pound of the mixture. Hence the

RULE.

Multiply each quantity by its price, and divide the sum of the products by the sum of the quantities. The quotient will be the rate of the compound required.

EXAMPLES.

2. A grocer mixes sugar, 5 lbs. at 6 cts., 8 lbs. at 5 cts., and 7 lbs. at 10 cts. a lb. What is 1 lb. of the mixture worth?
Ans. 7 cts.

3. A farmer mixes 12 bushels of wheat at $1.75 a bushel, 8 bushels of rye at $1, and 6 bushels of corn at 80 cts. a bushel. What is a bushel of the mixture worth? *Ans.* $1.30.

4. A goldsmith melted together 12 lbs. of gold, 21 carats fine, 8 lbs. 20 carats fine, 9 lbs. 22 carats fine, and 7 lbs. 18 carats fine. Of what fineness is the mixture?
Ans. $20\frac{4}{9}$ carats fine.

5. A merchant mixed 8 gallons of wine, at 4*s.* 2*d.* per gal-

lon, with 10 gallons at 6*s.* 5*d.*, and 12 gallons at 8*s.* 4*d.* per gallon. What is a gallon of the mixture worth?

Ans. 6*s.* 7*d.*

6. If 4 lbs. of tea, at 6*s.* per lb., 8 lbs. at 5*s.*, and 6 lbs. at 3*s.*, be mixed together, what is 1 lb. of the mixture worth?

Ans. $4\frac{5}{9}$ shillings.

ALLIGATION ALTERNATE.

Art. 262.—Alligation Alternate is when the prices of the simples to be mixed, and the mean rate, are given, to find what quantity of each is to be taken at a given rate.

1. I have corn at 50 cents a bushel, and oats at 30 cents a bushel, which I would mix, so that the mixture may be worth 40 cents a bushel. What quantity of each must be taken?

It is evident that equal quantities of each must be taken, for the price of the corn exceeds the mean rate as much as the price of the oats falls short of it, which is 10 cents in each case. We find, also, that the whole mixture, which is 20 bushels, at the mean rate, 40 cents a bushel, equals the price of 10 bushels of oats at 30, and 10 bushels of corn at 50 cents a bushel.

RULE.

I. *Reduce the rates of all the simples to the same denomination, and write them in a column under each other, and the mean rate on the left hand.*

II. *Connect the rate of each simple, which is less than the rate of the compound, with one that is greater, and each that is greater with one that is less.*

III. *Write the difference between each rate, and that of the compound against the number with which it is connected. Then, if only one difference stand against any rate, it will express the relative quantity to be taken of that rate; but if more than one, their sum will express that quantity.*

EXAMPLES.

Art. 263.—2. A farmer has wheat at \$1.50, rye at \$1.00, corn at 90, and oats at 40 cents a bushel, which he mixes so that the mixture is worth 95 cents a bushel. What quantity of each does he take?

Operation.

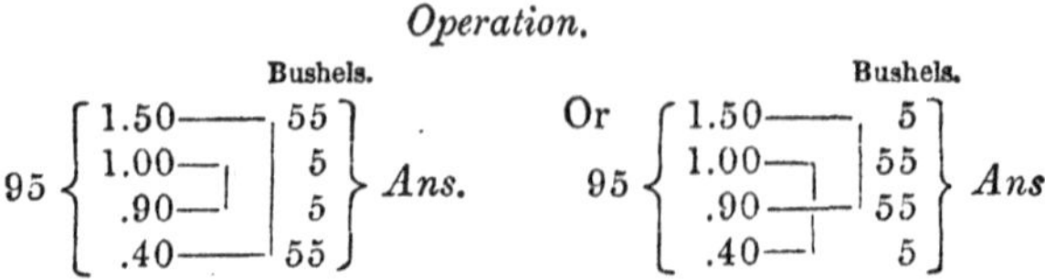

By linking the price of the different simples, as above, their quantities are mutually mixed, and the portion taken of each depends upon the manner of linking them. In the first operation, the price of the wheat, which is greater than the mean price, is linked with the price of the oats, which is less. The price of the wheat is found to be as much greater than the mean rate, as the price of the oats is less; therefore an equal quantity of each is taken. The same is true of the corn and oats. In the second operation, the price of the wheat is linked with the price of the corn. The difference between the price of the wheat and the mean rate, is 55, and the difference between the price of the corn and the mean rate, is 5. Hence, it appears that the less the difference between the price of a simple and the mean rate, the greater will be the quantity taken of that simple; and the greater the difference the less the quantity.

3. A merchant has teas at 72 cents, at 62 cents, and 57 cents a pound, which he would mix, so that the mixture may be worth 67 cents per lb. What quantity of each must be taken?

Operation. lbs.

		lbs.	
67	72	10+5=15	*Ans.*
	62	5= 5	
	57	5= 5	

The correctness of the above operation may be ascertained thus: The cost of 15 lbs. at 72 cents, is $10.80, and the cost of 5 lbs. at 62 cents, is $3.10, and the cost of 5 lbs. at 57 cents, is $2.85. Then the whole cost is $10.80+$3.10+$2.85=$16.75, which, divided by 25 lbs., gives the mean price, $16.75÷25=67 cents. Hence, it appears that Alligation Alternate is the reverse of Alligation Medial, and may be proved by it.

4. A grocer mixes wines at 29*s*., 24*s*., 22*s*., and 17*s*. a gallon, so that the mixture is worth 23*s*. per gallon. How much of each sort does he take?

1st *Ans.* { 1 gal. at 29*s*. / 6 gal. at 24*s*. / 6 gal. at 22*s*. / 1 gal. at 17*s*.

2d *Ans.* { 6 gal. at 29*s*. / 1 " at 24*s*. / 1 " at 22*s*. / 6 " at 17*s*.

3d *Ans.* { 7 gal. at 29*s*. / 6 " at 24*s*. / 6 " at 22*s*. / 7 " at 17*s*.

As many different answers may be obtained to questions in this rule as there are modes of linking the prices of the simples. Let the number of simples be what it may, and with how many soever each one is linked, since the price of one that is less than the mean rate, is always linked with one that is greater, there will always be an equal balance of loss and gain between the two, and consequently an equal balance on the whole.

5. It is required to mix brandy at 12*s*., wine at 9*s*., cider at 2*s*., beer at 1*s*., and water at 0*s*., per gallon, so that the mixture may be worth 7*s*. per gallon. What quantity of each must be taken?

Ans. { 13 gals. brandy. / 5 " wine. / 2 " cider. / 5 " beer. / 5 " water.

Art. 264.—When the composition is limited in quantity.

RULE.

Find the proportion of each quantity as before; then say, As the sum of the quantities is to the given quantity, so is each of the differences to the required quantity.

EXAMPLES.

6. Suppose a mass of pure gold, a mass of pure silver, and a mass which is a mixture of gold and silver, each weighing 9 oz.; by immersing them in water, it is found that the quantity of water displaced by the gold is 5; by the silver 8, and by

23

the mixture 7. What part of the mixture is gold, and what part silver?

$$7 \begin{cases} 5 — 1 \\ 8 — 2 \end{cases}$$

$$\overline{3} : 9 :: \begin{cases} 1 : 3 \text{ gold.} \\ 2 : 6 \text{ silver.} \end{cases}$$

By a similar problem, Archimedes detected the fraud of the artist employed by Hiero, king of Syracuse, to make him a crown of pure gold.

7. A druggist has medicines at 6*d.*, 3*d.*, 9*d.*, and 4*d.* per oz., and would form a compound of 15 oz., worth 5*d.* per oz. How much of each sort must he take?

Ans. $\begin{cases} 1\frac{7}{8} \text{ oz. at } 6d. \\ 7\frac{1}{2} \quad \text{"} \quad 3d. \\ 3\frac{3}{4} \quad \text{"} \quad 9d. \\ 1\frac{7}{8} \quad \text{"} \quad 4d. \end{cases}$

8. A goldsmith would melt together gold of 13, of 14, of 15, and of 21 carats fine, to form a composition of 35 oz. 18 carats fine. What proportion of each must he take?

Ans. $\left\{ \begin{matrix} 5 \text{ of } 13 \\ 5 \text{ " } 14 \\ 5 \text{ " } 15 \\ 20 \text{ " } 21 \end{matrix} \right\}$ carats fine.

9. How many gallons of water, worth 0*s.* per gal., must be mixed with wine worth 12*s.* per gal., so as to fill a cask of 20 gallons, and that a gallon of the mixture may be afforded at 9*s.* per gallon? *Ans.* $\begin{cases} 5 \text{ gal. water.} \\ 15 \text{ gal. wine.} \end{cases}$

Art. 265.—When one of the simples is limited to a certain quantity.

RULE.

Find the proportional quantities, or differences, as before; then say, As the difference standing against the given quantity is to the given quantity, so are the other differences severally to the several quantities required.

EXAMPLES.

10. A grocer mixes sugar at 9 cts., 12 cts., and 14 cts., with 16 lbs. at 15 cts. How much of each sort must he take, that the mixture may be worth 13 cts. per lb.?

$$13\begin{cases} 9 \text{———} 2 \\ 12 \text{—} \;\; 1 \\ 14 \text{—} \;\; 1 \\ 15 \text{———} 4 \end{cases}$$ against the given quantity.

$$4 : 16 :: \begin{cases} 2 : 8 \text{ lbs. at } 9s. \\ 1 : 4 \text{ " " } 12s. \\ 1 : 4 \text{ " " } 14s. \end{cases} \textit{Ans.}$$

11. A grocer would mix flour, at $6, $5, $12 a barrel, with 10 barrels at $11 a barrel. How much of each kind must he take, that the mixture may be worth $10 a barrel?

Ans. { 4 at $6, 2 at $5, 8 at $12 } per barrel.

12. How much water must be mixed with 100 gallons of brandy, worth 7*s.* 6*d.* per gallon, to reduce it to 6*s.* 3*d.* per gallon? *Ans.* 20 gallons.

13. A farmer would mix barley at 50 cents, oats at 30, with 20 bushels of rye at 60 cents a bushel. How much of each sort must he take, that the provender may be worth 40 cents per bushel? *Ans.* { 60 bushels of oats, and 20 bushels of barley.

DUODECIMALS.

Art. 266.—This rule is principally used in measuring surfaces and solids. Calculations are generally made in feet, inches, or primes, seconds, thirds, fourths, and so on. These subdivisions of the foot are made by a common divisor, 12, an inch being $\frac{1}{12}$ of a foot, a second $\frac{1}{12}$ of an inch, or $\frac{1}{144}$ of a foot, thus forming a descending series of a geometrical progression, whose first term is 1, and the ratio 12. Hence the term *duodecimal*. It is derived from the Latin word *duodecim*, which signifies twelve. Duodecimals, then, are fractions of a foot, as may be seen by the following:

Questions.—1. What are Duodecimals? 2. For what is the rule chiefly used? 3. What are the divisions of the foot?

TABLE I.

1′	inch, or prime, is................	$\frac{1}{12}$	of a foot.
1″	second is $\frac{1}{12}$ of $\frac{1}{12}$	$\frac{1}{144}$	"
1‴	third is $\frac{1}{12}$ of $\frac{1}{12}$ of $\frac{1}{12}$	$\frac{1}{1728}$	"
1⁗	fourth is $\frac{1}{12}$ of $\frac{1}{12}$ of $\frac{1}{12}$ of $\frac{1}{12}$	$\frac{1}{20736}$	"

The marks ′, ″, ‴, ⁗, placed over numbers, denote the denomination, and are called *indices.* In Multiplication of Duodecimals, the denomination of the product is denoted by the sum of the indices. That is, if inches be multiplied by inches, the product will be seconds; thus, $2' \times 2' = 4''$. If inches be multiplied by seconds, the product will be thirds; thus, $2' \times 2'' = 4'''$, etc.

TABLE II.

12⁗	fourths make..........	1‴	third.
12‴	thirds................	1″	second.
12″	seconds	1′	inch, or prime.
12′	inches, or primes	1	foot.

MULTIPLICATION OF DUODECIMALS.

Art. 267.—1. How many square feet in a board, 11 feet 2 inches long, and 1 foot 4 inches wide?

Operation.

11	2′		
1	4′		
11	2′		
3	8′	8″	
14	10′	8″	*Ans.*

Having written numbers of the same denomination under each other, as in Multiplication of Compound Numbers, we commence with the feet in the multiplier, and say, once $\frac{2}{12}$ is $\frac{2}{12}$, and 11 feet multiplied by 1 is 11. Proceeding to the next figure in the multiplier, which is 4 inches, or $\frac{4}{12}$, we say, 4 times 2 are 8, but 4 is $\frac{4}{12}$, and 2 is $\frac{2}{12}$; therefore, $\frac{4}{12} \times \frac{2}{12} = \frac{8}{144}$ of a foot, or 8″ seconds, which being less than 1 prime, we write it in the place of seconds, and proceed to the next figure in the multiplicand, which is 11. Multiplying 11 feet by $\frac{4}{12}$, we have $11 \times \frac{4}{12} = \frac{44}{12} = 3$ feet 8′ inches. Having written 8 in the place of inches, and 3 in the place of feet, we add the several partial products, and obtain 14 feet 10′ 8″, the answer. By examining the foregoing operation, it will be seen, that the first product, being the product of inches by feet, is inches. The second product is the product of feet, and consequently is feet. The third product is the product of inches by inches; the sum of the indices being two, it is 12ths of an inch. The fourth and last is the product of feet by inches, and is 12ths of a foot. Therefore, to multiply feet, inches, etc., by numbers of corresponding denominations, we have the following

RULE.

I. *Write the several denominations of the multiplier under the corresponding denominations of the multiplicand.*

II. *Multiply first the lowest denomination in the multiplicand by the highest in the multiplier, observing to carry* 1 *for every* 12 *from a lower to a higher denomination.*

It is to be remembered that the denomination of the product of two numbers is denoted by the sum of the indices.

EXAMPLES.

Art. 268.—2. How many square feet in a marble slab, 5 feet 7 inches in length, and 4 feet 8 inches in breadth?

Operation.

$$\begin{array}{rrrrl} & 5 & 7' & & \\ & 4 & 8' & & \\ \hline & 22 & 4' & & \\ & 3 & 8' & 8'' & \\ \hline \text{ft.} & 26 & 0' & 8'' & \textit{Ans.} \end{array}$$

Duodecimals may also be written as decimal fractions, observing to carry for 12 instead of 10. Thus,

$$\begin{array}{ll} \text{ft.} & \\ 5 \quad 7' = & 5.7 \\ 4 \quad 8' = & 4.8 \\ \hline & 388 \\ & 224 \\ \hline & 26.08 = 26 \text{ ft. } 0'\ 8'' \end{array}$$

3. How many square feet in a room 15 feet 8 inches in length, and 14 feet 9 inches in breadth? *Ans.* 231 ft. 1 in.

4. What is the product of 15 ft. 2′ 6″ × 20 ft. 3′ 7″?

5. How many solid feet in a block 4 ft. 8′ long, 3 ft. 6′ wide, and 2 ft. 9′ thick? *Ans.* 44 ft. 11′.

OBS. 1.–The solid contents may be found by multiplying the length by the breadth, and that product by the thickness.

6. How much wood in a load 9 ft. 8′ long, 8 ft. 7′ wide, and 3 ft. high? *Ans.* 1 cord, 120 ft. 11′.

7. How many square feet in a stock of 20 boards, 13 ft. 11′ long, and 1 ft. 7′ wide? *Ans.* 440 ft. 8′ 4″.

8. How many feet of flooring in a room 30 feet 6 inches long, and 19 feet 5 inches in width? *Ans.* 592 ft. 2′ 6″.

9. How much wood in a cubic pile, 12 feet 3 inches on each side, and what will it be worth at $4 25 per cord?
Ans. to the last, $61.035.

10. How many square yards in the walls of a room, 14 feet 8 inches long, 11 ft. 6 inches wide, and 7 ft. 11 inches high?
Ans. 46 yds. 0 ft. 0′ 4″ 10‴ 8⁗.

11. How many cord-feet in a pile of wood 38 feet long, 7 feet 2′ wide, and 4 feet 7′ in height?
Ans. 78 cord-feet, 0′ 1″ 9‴.

12. How many cord-feet in a load of wood 8 feet long, 3 feet 6 inches high, and 4 feet 5 inches wide?
Ans. 7 cord-feet, 11 solid feet 8 in.

13. How much wood in a load 9 feet long, 3 feet 4 inches wide, and 2 feet 6 inches high?
Ans. 4 cord-feet, 11 solid feet.

OBS. 2.–Many mechanics and surveyors take dimensions in feet and *decimal* parts. This method is preferable, inasmuch as by it the calculations of the artificer are rendered more simple and easy. For such, it is convenient to have a rule, or scale, four feet long, divided into feet, and each foot into ten equal parts. One foot, on one end of the rule, should be divided into one hundred equal parts. The former division will be 10ths, and the latter 100ths of a foot. Dimensions taken by this rule are calculated the same as other decimal fractions.

14. How many square feet in a board 20.5 in length, and 1.8 in width? *Ans.* $36\frac{9}{10}$.

15. How much wood in a pile 40.5 in length, 5.4 in width, and 6.2 in height; and what will it be worth at $3.75 a cord?
Ans. to the last, $39.724+.

16. What will be the cost of a marble slab, 13.9 feet in length, and 2.1 feet in width, at $1.18 per square foot?
Ans. $34.444.

MISCELLANEOUS RULES.

MENSURATION.

MENSURATION OF SURFACES.

DEFINITIONS.

1. A point is a small dot; or, mathematically considered, is that which has no parts, being of itself indivisible.

2. A line has length, but no breadth.

3. A superficies, or surface, called also area, has length and breadth, but no thickness.

4. A solid has length, breadth, and thickness.

5. A right line is the shortest that can be drawn between two points: as AB.

A —————— B

6. The inclination of two lines meeting one another, or the opening between them, is called an angle: as ABC; B the angular point.

7. If a right line fall upon another righ. line, so as to incline to neither side, but make the angles on each side equal, then those angles are called right angles, and the line is said to be perpendicular to the other line: as ABC, right angle.

8. An obtuse angle is greater than a right angle: as LBG.

9. An acute angle is less than a right angle: as ABL.

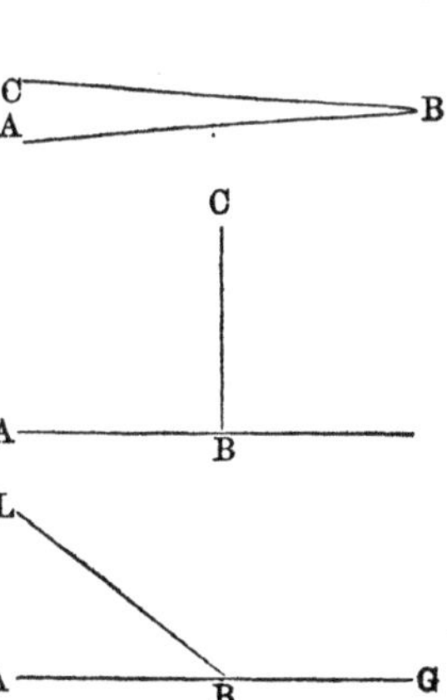

10. A circle is a round figure bounded by a single line, in every part equally distant from some point, which is called the centre.

11. The circumference or periphery of a circle, is the bounding line.

12. The radius of a circle (AO) is a line drawn from the centre to the circumference. Therefore, all radii of the same circle are equal.

13. The diameter of a circle (AC) is a right line drawn from one side of the circumference to the other, passing through the centre ; and it divides the circle into two equal parts called semicircles.

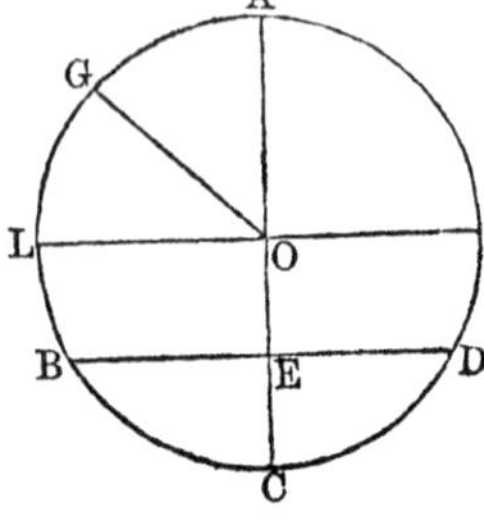

14. The circumference of every circle is supposed to be divided into 360 equal parts, called degrees; and each degree into 60 equal parts, called minutes; and each minute into 60 equal parts, called seconds; and these into thirds, etc. Hence a semicircle contains 180 degrees, and a quadrant 90 degrees.

15. An arc of a circle (BCD) is any part of the circumference.

16. A chord (BD) is a right line drawn from one end of an arc to another, and is the measure of the arc. The chord of an arc of 60 degrees is equal in length to the radius of the circle of which the arc is a part.

17. The segment of a circle is a part of a circle cut off by a chord.

18. A sector of a circle (LOG) is a space contained between two radii and an arc less than a semicircle.

19. Parallel lines (LO and BE) are such as are equally distant from each other.

20. A triangle is a figure bounded by three lines.

21. An equilateral triangle (ABC) has its three sides equal in length to each other. (CE is the perpendicular height.)

22. An isosceles triangle has two of its sides equal.

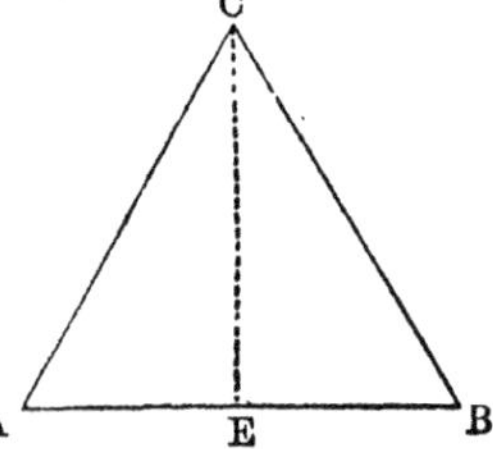

23. A scalene triangle has three unequal sides.

24. A right-angled triangle has one right angle.

25. An obtuse-angled triangle has one obtuse angle.

26. An acute-angled triangle has all its angles acute.

27. Acute and obtuse angled triangles, are called oblique-angled triangles, or simply oblique triangles; in which the lower side is called the base, and the other two, legs.

28. In a right-angled triangle the longest side is called the hypotenuse, and the other two, legs, or base and perpendicular.

OBS.—The three angles of every triangle being added together, will amount to 180 degrees; consequently, the two acute angles of a right-angled triangle amount to 90 degrees, the right angle being also 90.

29. The perpendicular height of a triangle is a line drawn from one end of the angles perpendicular to its opposite side.

30. A square (ABCD) is a figure bounded by four equal sides, and containing four right angles.

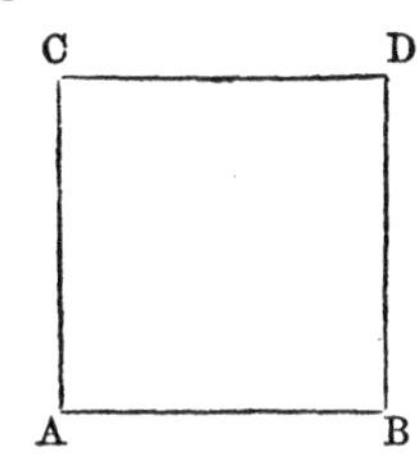

31. A parallelogram (EFGH) is a figure bounded by four sides, the opposite ones being equal, and the angles right.

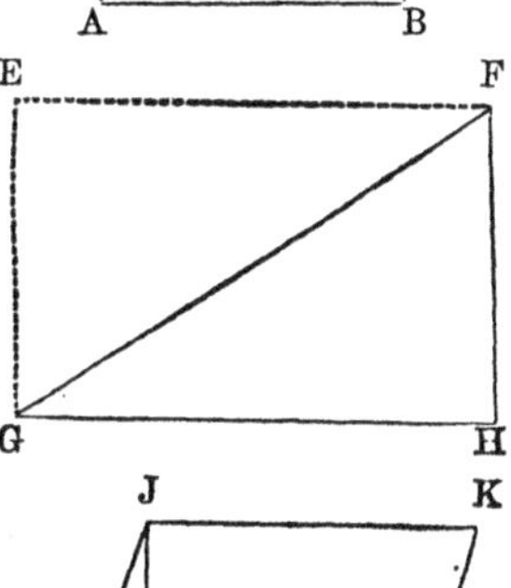

32. A rhombus (JKLM) is a figure bounded by four equal sides, but has its angles oblique.
JN perpendicular height of a rhombus.

J K

L N M

33. A rhomboid (EFGH) is a figure bounded by four sides, the opposite ones being equal, but the angles oblique.

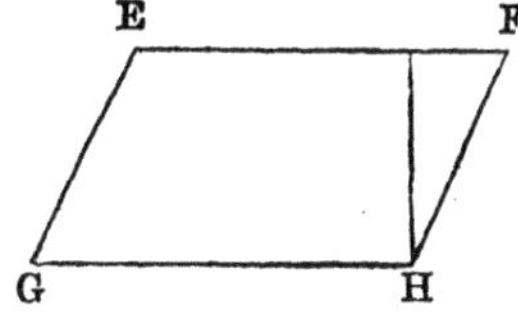

34. The perpendicular height of a rhombus, or rhomboides, is a line drawn from one of the angles to its opposite side.

35. A trapezoid (ABCD) is a figure bounded by four sides, two of which are parallel, though of unequal lengths.

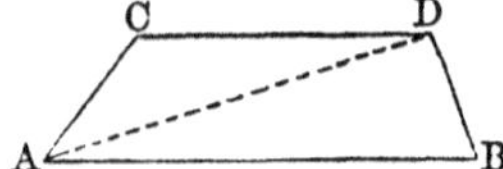

36. A trapeze, or trapezium, is a figure bounded by four unequal sides.

37. A diagonal is a line drawn between two opposite angles.

38. Figures which consist of more than four sides are called polygons; if the sides are equal to each other, they are called regular polygons, and are sometimes named from the number of their sides, as pentagon, or hexagon, a figure of five or six sides, etc. If the sides are unequal, they are called irregular polygons.

An irregular plane figure.

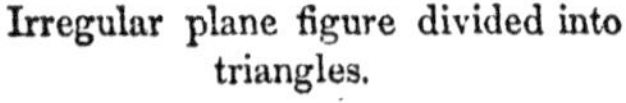
Irregular plane figure divided into triangles.

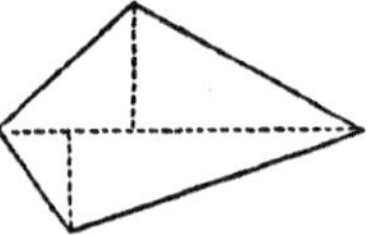

Heptagon.

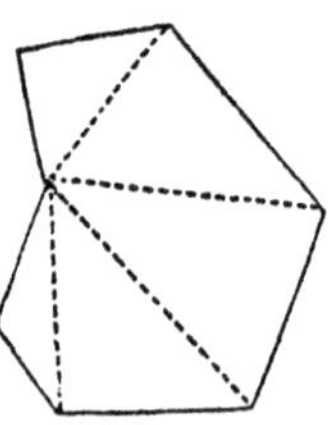

(The dotted lines represent a division into triangles.)

39. The area of a figure is the space contained between the bounding lines of its surface, without regard to thickness. The area is reckoned so many square inches, square feet, square yards, or square rods, etc.

Art. 269.—To find the area of a square, or parallelogram.

RULE.

Multiply the length by the breadth, or perpendicular height, and the product will be the area.

1. How many square rods in a field 28 rods on each side?
$28 \times 28 = 784$ rods, *Ans.*

2. What is the area of a square field, one side of which is 25.35 chains? *Ans.* 642.622 chains.

3. What is the area of a field 30.5 chains in length, and 24.5 in width? *Ans.* 747.25 chains.

4. How many square feet in a board 18.8 feet long, and 2.7 feet wide? *Ans.* 50.76.

5. How many acres in a rectangular piece of ground, 64 rods long and 24 rods wide? *Ans.* $9\frac{3}{5}$.

Art. 370.—To find the area of a triangle.

RULE.

Multiply the perpendicular by the base, and one-half the product will be the area; or, multiply the base by half the perpendicular height, and the product will be the area.

1. What is the area of a triangle whose base is 20 feet, and whose height is 18 feet?
$18 \times 20 = 360 \div 2 = 180$ feet, *Ans.*

2. What is the area of a triangle whose base is 55 rods, and its height 24.6 rods? *Ans.* 676.5.

3 How many feet of boards will it take to cover the gable end of a barn, 38 feet wide, the height from the beam to the top being 12.5 feet? *Ans.* 237.5.

When the three sides of a triangle are known, the area may be found by the following

RULE.

Add together the three sides, and from half their sum subtract each side separately; multiply the half sum and the remainders together continually, and the square root of the product will be the area.

4. What is the area of a triangle whose three sides are 14, 12, and 8 rods?

14+12+8=34÷2=17, the half sum: then,

17 17 17
14 12 8

Rem. 3 × 5 × 9 × 17=2295: then $\sqrt{2295}$=47.9+ rds.

5. The three sides of a triangle are 6, 8, and 10 chains. What is the area? *Ans.* 24 chains.

Art. 271.—To find the area of a trapezoid.

RULE.

Multiply half the sum of the two parallel sides by the perpendicular distance between them: the product will be the area.

1. What is the area of a piece of land that is 30 chains long, 20 chains wide at one end, and 18 chains at the other?

$\frac{20+18}{2}$=19, the half sum of the two sides:
then 19 × 30=570 chains, *Ans.*

2. What is the content of a board, 10 feet long, 10 inches wide at one end, and 2 feet ten inches at the other? *Ans.* 18$\frac{1}{3}$ feet.

3. What is the area of a hall, 40 feet long, and at one end 30 feet, and at the other 24 feet wide? *Ans.* 1080 feet.

4. How many acres in a farm 300 rods long, 80 rods wide at one end, and 60 at the other? *Ans.* 131 acres, 40 rods.

Art. 272.—To measure any irregular plane figure.

RULE.

The whole may be divided into triangles, and measured separately. The sum of the area of the triangles will be the area of the whole.

OF THE CIRCLE.

The circumference of a circle is found by calculation to be about 3$\frac{1}{7}$ times the diameter; or more accurately, by decimals, as 1 is to 3.1416, or as 113 is to 355, so is the diameter to the circumference. Hence, if the diameter is given, to find the circumference, it may be found by multiplying the diameter by 3$\frac{1}{7}$, or by 3.1416; or as 113 is to 355, so is the diameter to the circumference.

1. What is the circumference of a circle whose diameter is 42 feet?

$42 \times 3\frac{1}{7} = 132$ ft.; or $42 \times 3.1416 = 131.9472$ feet;
or 113 : 355 : : 42 : 131.946+ feet.

By reversing the foregoing, the diameter may be found, the circumference being given.

2. If the circumference of a circle be 132 feet, what is the diameter? $132 \div 3\frac{1}{7} = 42$ feet, *Ans.*

3. Suppose the diameter of a circular pond to be 121 rods, what is the circumference? *Ans.* 380.28+ rods.

4. If the circumference of a circular field be 198 rods, what is the diameter? *Ans.* 63 rods.

5. What is the diameter of a tree, whose circumference is $9\frac{3}{7}$ feet? *Ans.* 3 feet.

6. If the circumference of the earth is 25000.8528 miles, what is the diameter? *Ans.* 7958 miles.

7. The diameter of the earth being 7958 miles, what is the circumference? *Ans.* 25000.8528.

Art. 273.—To find the area of a circle.

RULE.

Multiply half the diameter by half the circumference; the product will be the area.

1. What is the area of a circular grove, whose diameter is 147 rods, and circumference 462 rods?

Ans. $\overline{462 \div 2} \times \overline{147 \div 2} = 16978\frac{1}{2}$ rods.

2. What is the area of a circle whose diameter is 28, and the circumference 88 rods? *Ans.* 616 rods.

3. How many square rods in a circle whose circumference is 63, and the diameter 20 rods? *Ans.* 315.

Art. 274.—The diameter given, to find the area.

RULE.

Multiply the square of the diameter by .7854, and the proauct will be the area.

1. What is the area of a circle whose diameter is 28 rods?

$28 \times 28 \times .7854 = 615.7536$ rods, *Ans.*

2. What is the area of a circle whose diameter is 59 rods?

Ans. 2733.9774 rods.

Art. 275.—The circumference given, to find the area.

RULE.

Multiply the square of the circumference by .07958, and the product will be the area.

1. What is the area of a circle whose circumference is 46 rods? 46×46×.07958=168.39128 rods, *Ans.*

2. What is the area of a circle whose circumference is 44 rods? *Ans.* 154.06688 rods.

Art. 276.—To find the area of an oval, or ellipsis.

RULE.

Multiply the longest and shortest diameters together, and the product by .7854. The last product will be the area.

1. What is the area of an oval, whose longest diameter is 7 ft., and the shortest 5 ft.? 7×5×.7854=27489 ft. *Ans.*

2. What is the area of an oval, whose longest diameter is 15, and the shortest 13 feet? *Ans.* 153.153 ft.

Art. 277.—To find the area of a globe, or sphere.

RULE

Multiply the circumference by the diameter. The product will be the area.

1. What is the area of a globe, whose circumference is 44 feet, and the diameter 14? 44×14=616 ft. *Ans.*

2. How many square inches in the surface of a ball, 1 inch in diameter? *Ans.* 3.1416.

3. How many square miles in the surface of the earth, allowing its circumference to be 25000 and its diameter 8000 miles? *Ans.* 200000000.

Art. 278.—Given the chord of an arc, and its height, to find the diameter of a circle, of which the arc is a part.

RULE.

Divide the square of half the chord by the height, and the quotient, added to the height, will be the diameter required.

1. Given the chord, BD, 287, and the height, CE, 78 feet, to find the diameter, AC.

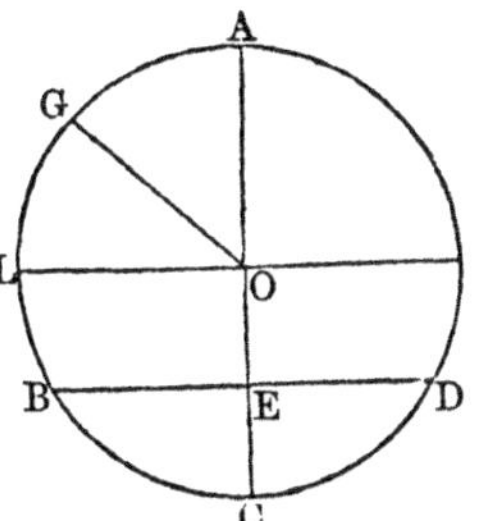

Operation.

$287 \div 2 = 143.5$, and $\overline{143.5^2} =$ 20592.25, and $20592.25 \div 78 = 264$, and $264 + 78 = 342$, the required diameter.

2. Given the chord 178, and the height 257 yards, to find the diameter. *Ans.* 287.821 yards.

3. Given the chord 843, height 648 links. *Ans.* 922.17 links.

4. Given the chord 40, height 12 yards, to find the diameter. *Ans.* $45\frac{1}{3}$ yards.

5. Given the chord 560, height 45 links, to find the diameter. *Ans.* $1787\frac{2}{9}$ links.

Art. 279.—Given the radius and number of degrees in an arc of a circle, to find the length of the arc.

RULE

Multiply the radius by the number of degrees in the arc, and by .0174533.

Or find the circumference, multiply it by the degrees, and divide by 360°.

1. Required the length of an arc, AC, of 57°, in a circle of which the radius, AB, is 38 feet.

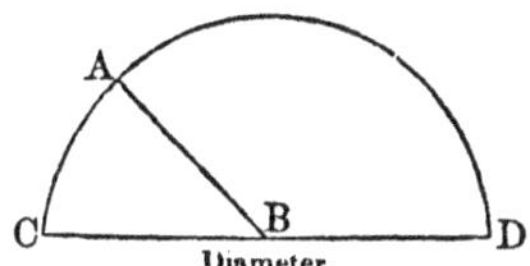

Operation.

$.0174533 \times 57 \times 38 = 37.8038478$ feet, *Ans.*

2. What is the length of an arc of 19° 37′, the radius being 98 yards? *Ans.* $0174533 \times 19°.617 \times 98 = 33.553$ yards.

3. What is the length of an arc of 83° 24′, radius 32 poles? *Ans.* 1 furlong, 6 poles, 3 yards, 6.72 inches.

4. What is the length of an arc of 150°, radius 19 ells? *Ans.* 49 ells, 27 inches.

5. What is the length of an arc of 17° 50′, radius 178 miles?
Ans. 55 miles, 3 furlongs, 8 poles, $4\frac{1}{2}$ yards.

Art. 280.—To find the area of a sector of a circle.

RULE.

I. *If the length of the arc be known, multiply half the arc by the radius, or the arc by half the radius.*

II. *If the angle of the sector be given, find the length of the arc, and proceed as before.*

1. What is the area of a sector, of which the arc is 79, and the radius of the circle 47 yards?
Ans. $\frac{79}{2}=39.5$ and $39.5\times47=1856.5$ square yards.

2. What is the area of a sector, of which the arc is 17 feet 5 inches, and the radius 22 feet?
Ans. 191.583 square feet=21 yards, 2.583 feet.

3. What is the area of a sector, of which the angle is 127° 16′, the radius 133 feet?
Ans. 1 rood, 32 poles, 4.845 yards.

Obs.—The area of the circle is 55571.63245; and this multiplied by $127\frac{4}{15}$, and divided by 360=19645.60+.

4. What is the area of a sector, of which the angle is 27 degrees, the radius 97 miles? *Ans.* 2216.95 miles.

5. What is the area of a sector, of which the angle is 137° 20′, the radius 456 links?
Ans. 2 acres, 1 rood, 38 poles, 21.9 yards.

6. What is the area of a sector, of which the arc is 156 feet, the radius 478 feet?
Ans. 3 roods, 16 poles, 28 yards, 6 feet.

Art. 281.—To find the area of a ring contained by two concentric circles.

Obs.—*Concentric* means having the same centre.

RULE.

Multiply the sum of the diameters by their difference, and that product by .7854.

1. Required the area of the ring ABC—DEF, of which the diameters are 10 and 6, or OC, 5, and OF, 3 feet.

Ans. $10+6=16$, and $10-6=4$, then $16\times4\times.7854=50.2656$ feet.

2. What is the area of a ring, of which the diameters are 72 and 48 ft.?

Ans. 2261.952 square ft. = 8 rods, $9\frac{1}{3}$ yards.

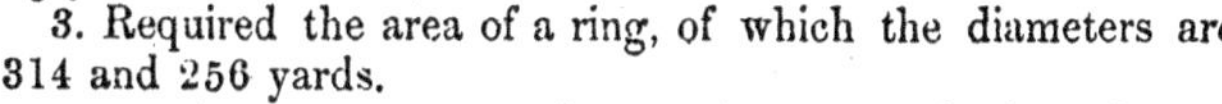

3. Required the area of a ring, of which the diameters are 314 and 256 yards.

Ans. 5 acres, 1 rood, 18 poles, 10 yards, 7.42 feet.

4. What is the area of a ring, of which the diameters are 246 and 228 inches? *Ans.* 46 feet, 77 inches.

Art. 282.—To find the area of a space bounded on one side by a curve line.

RULE.

Let perpendiculars be erected upon the base, so numerous that the part of the curve between any two nearest to one another shall differ but little from a straight line. Then add the perpendiculars at the extremities of the base, if there are any, and to half the sum add the remaining perpendiculars. Multiply the sum by the base, and divide the product by the number of parts into which the base is divided by the perpendiculars: the quotient will be the area, nearly.

1. Suppose the perpendiculars at the extremities of the base to be 10 and 16, and the others to be 11, 14, 16, and the base to be 20 feet.

Operation.

$\frac{10+16}{2}=13$, and $13+11+14+16=54$; and $\overline{54\times20}\div4=270$ square feet, the area.

2. A. curve-lined space meets the base at one of its extremities, and the perpendicular at the other extremity is 96; the other perpendiculars are 83, 70, 64, 51, 38, 25, and the base 325 links. What is the area? *Ans.* $17596\frac{3}{7}$ square links.

3. Perpendiculars were raised from the base to a curve; those at the ends were 364 and 578, the others were 396, 418, 453, 512, 554 links, and the base 1260 links. What is the area?

Ans. 5 acres, 3 roods, 22 poles, 4 yards, 3.2 feet.

4. A curve meets the base at one extremity; the base is 2364; the perpendicular, at the other extremity, 758, and the others are 642, 587, 524, 432, 417, and 335 links. What is the area?

Ans. 1119860$\frac{4}{7}$ links=11 acres, 31 poles, 23.5 yards.

MENSURATION OF SOLIDS.

Art. 283.—The Mensuration of Solids includes the mensuration of all bodies which have length, breadth, and thickness.

DEFINITIONS.

1. Solids are figures, having length, breadth, and thickness.

2. A prism is a solid, whose ends are any plane figures, which are equal and similar, and its sides are parallelograms.

Obs.—A prism is called a triangular prism, when its ends are triangles; a square prism, when its ends are squares; a pentagonal prism, when its ends are pentagons; and so on.

3. A cube is a square prism, having six sides, which are all squares.

4. A parallelopiped is a solid, having six rectangular sides, every opposite pair of which are equal, and parallel.

5. A cylinder is a round prism, having circles for its ends.

6. A pyramid is a solid, having any plane figure for a base, and its sides are triangles, whose vertices meet in a point at the top, called the vertex of the pyramid.

7. A cone is a round pyramid, having a circular base.

8. A sphere is a solid, bounded by one continued convex surface, every point of which is equally distant from a point within, called the *centre*. The sphere may be conceived to be formed by the revolution of a semicircle about its diameter, which remains fixed.

A hemisphere is half a sphere.

9. The segment of a pyramid, sphere, or any other solid, is a part cut off the top by a plane, parallel to the base of that figure.

10. A frustum is the part that remains at the bottom after the segment is cut off.

11. The sector of a sphere is composed of a segment less than a hemisphere, and of a cone, having the same base with the segment, and its vertex in the centre of the sphere.

12. The axis of a solid is a line drawn from the middle of one end to the middle of the opposite end; as between the opposite ends of a prism. The axis of a sphere is the same as a diameter, or a line passing through the centre, and terminating at the surface on both sides.

13. The height, or altitude of a solid, is a line drawn from its vertex, or top, perpendicular to its base.

Art. 284.—To find the solidity of a cube.

RULE.

Multiply the length, breadth, and thickness together, and the product will be the area.

1. If the length of one side of a cubical block be 14 inches, what is its solidity? $14\times14\times14=2744$ inches, *Ans.*

2. How many cubical feet in a mound, each side of which is 25.5 feet? *Ans.* 16581.375 feet.

Art. 285.—To find the solidity of a prism, or cylinder.

RULE.

Find the area of the end, and multiply it by the length. The product will be the area.

What is the solidity of a prism, the area of whose end is 2.6 feet, and whose length is 16 feet?

$2.6\times16=41.6$ feet, *Ans.*

Art. 286.—To find the side of the largest stick of timber that can be hewn from a round log

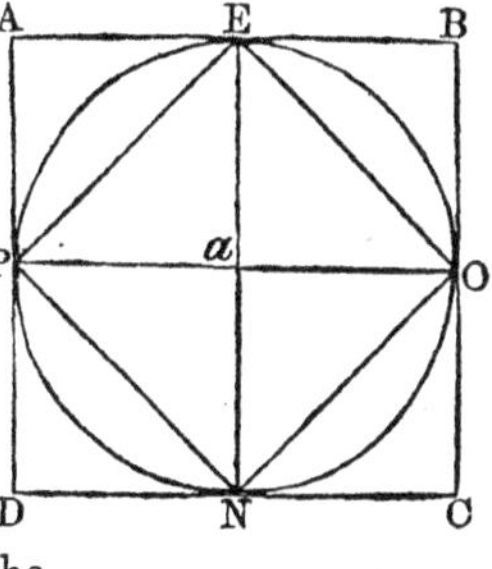

The circle, PEON, represents the end of a round stick of timber; ABCD, a circumscribed square, and PEON, an inscribed square. It will be perceived that the square ABCD is double the square PEON. But the square ABCD is equal to the square of PO, the diameter of the circle; but PO is equal to $Pa+aO=Pa+aE$. Now $\overline{Pa+aE}^2=PE^2$, the side of the largest inscribed square. Hence the

RULE.

Extract the square root of double the square of half the diameter at the smallest end of the stick, for the side of the stick when squared.

1. What will be the side of the largest stick of square timber which can be hewn from a round log, 18 inches in diameter at the smallest end?

$\sqrt{9\times9\times2}=12.727+$ inches, *Ans.*

2. The diameter of a log at the smallest end is 24 inches. What will be the side of the largest stick of timber that can be hewn from it? *Ans.* 16.97+ inches

Art. 287.—To find the solidity of a pyramid, or cone.

RULE.

Multiply the area of the base by one third of the height, and the product will be the area.

1. What is the contents of a cone, whose height is 21 feet, and the diameter of the base 9.5 feet?

$\overline{9.5 \times 9.5 \times .7854 \times 21} \div 3 = 496.176$ feet, *Ans.*

2. How many solid feet in a cone, whose height is 48 feet, and whose diameter at the base is 13 feet?

Ans. 2123.7216 feet.

Art. 288.—To find the solid contents of a globe, or sphere.

RULE.

Multiply the cube of the diameter by .5236, *or multiply the square of the diameter by one sixth of the circumference.*

1. What is the solidity of a ball, 9 inches in diameter?

$9 \times 9 \times 9 \times .5236 = 381.7044$, *Ans.*

2. What is the solidity of a globe, whose diameter is 13 inches? *Ans.* 1150.3492 inches.

Art. 289.—To find the solid contents of the segment of a sphere, the height and base of the segment being given.

RULE.

To three times the square of the radius of the base of the segment, add the square of the height, and multiply this sum by the height of the segment, and this product by .5236.

How many cubic feet are there in a coal-pit, the diameter of whose base is 103 feet, and whose height is 9 feet?

Ans. 37877.0931.

GAUGING.

Art. 290.—Gauging is the art of measuring all kinds of vessels, such as pipes, hogsheads, barrels, etc.

RULE.

Add the square of the head diameter to the square of the bung diameter ; multiply the sum by the length, and the product by .0014 for ale gallons, or by .0017 for wine gallons.

1. What is the contents of a cask, whose diameters are 18 and 26 inches, and its length 38 inches ?

$\overline{26\times26}+\overline{18\times18}\times38=38000$; then $38000\times.0017=64.6$ wine measure. $38000\times.0014=53.2$ gallons, beer measure.

2. How many wine gallons will fill a cask 50 inches in length, bung diameter 38, head diameter 30 inches ?

Ans. 199.24 gallons.

MEASURING GRAIN, ETC.

Art. 291.—When the grain is heaped in the form of a cone.

RULE.

Measure the perpendicular height of the heap, and also the slanting height, from the top to the floor, in inches ; then multiply the difference of the squares of those two heights by the perpendicular height, and this product by .0005. The last product will be the contents in bushels.

1. How many bushels in a parcel of wheat heaped in the form of a cone ; the perpendicular height being 40 inches, and the slanting height 90 inches ?

The square of 90=8100
The square of 40=1600
6500 difference of squares.
40 perpendicular height.
260000
.0005
130.0000 bushels, *Ans.*

2. What number of bushels in a conical heap of rye, the perpendicular height being 35 inches and the slanting height 65 inches? *Ans.* 52.5 bushels.

Art. 292.—When grain is heaped against the side of the barn.

RULE.

Multiply the difference of the squares of the heights by one half of the perpendicular height, and this product by .0005. The result will be the contents in bushels.

1. How many bushels of oats are in a heap, the perpendicular height being 30 inches, and the slanting height 60 inches? *Ans.* 20.25 bushels.

2. How many bushels of beans are in a heap, the perpendicular height being 25 inches, and the slanting height 50 inches?

Art. 293.—When grain is heaped in the corner of the barn.

RULE.

Multiply the difference of the squares of the heights by one fourth of the perpendicular height, and this product by .0005. The result will be the contents in bushels.

1. Required the number of bushels of grain heaped in the corner of the barn; the perpendicular height being 40 inches, and the slanting height 70 inches? *Ans.* 16.5.

2. How many bushels of barley in the corner of a box, the perpendicular height being 24 inches, and the slanting height 36 inches? *Ans.* 2.16 bushels.

TONNAGE OF VESSELS.

CARPENTERS' RULE.

Art. 294.—*For single-decked vessels, multiply the length and breadth at the main beam, and depth in the hold, together, and divide the product by 95, and the quotient is the tons. But for a double-decked vessel, take half of the breadth of the main beam for the depth of the hold, and proceed as before.*

1. What is the tonnage of a single-decked vessel, whose length is 67 feet, breadth 24 feet, and depth 12 feet? *Ans.* $203\frac{11}{95}$ tons.

2. What is the tonnage of a double-decked vessel, whose length is 80 feet, and breadth 30 feet? *Ans.* $378\frac{18}{19}$ tons.

GOVERNMENT RULE.

"*If the vessel be double decked, take the length thereof from the fore part of the main stem to the after part of the stern-post, above the upper deck; the breadth thereof at the broadest part above the main wales, half of which breadth shall be accounted the depth of such vessel, and then deduct from the length three fifths of the breadth; multiply the remainder by the breadth, and the product by the depth, and divide this last product by 95, the quotient whereof shall be deemed the true contents, or tonnage of such ship or vessel; and if such ship or vessel be single decked, take the length and breadth, as above directed, deduct from the length three fifths of the breadth, and take the depth from the under side of the deck plank to the ceiling in the hold, and then multiply and divide as aforesaid, and the quotient shall be deemed the tonnage."*

1. What is the government tonnage of a single-decked vessel, whose length is 90 feet, breadth 40 feet, and depth in the hold 12 feet? *Ans.* $333\frac{9}{19}$ tons.

MECHANICAL POWERS.

Art. 295.—That body which communicates motion to another, is called a *power:* the body which receives the motion is called the *weight.*

The mechanical powers are six: the Lever, the Wheel and Axle, the Pulley, the Screw, the Inclined Plane, and the Wedge.

OF THE LEVER.

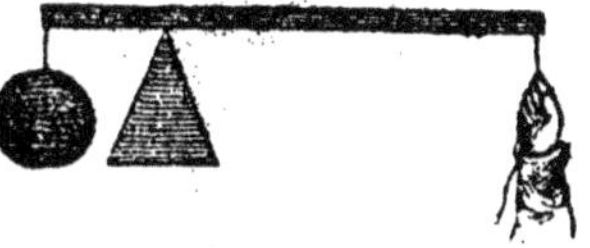

Art. 296.—The lever is a bar, moveable about a fixed point, called its fulcrum, or prop. It is, in theory, considered an inflexible line, without weight. There are several kinds of lever used in mechanics. The more common kind is that which is here shown.

It is a principle in mechanics, that the power is to the weight as the velocity of the weight is to the velocity of the power.

Art. 297.—To find what weight may be balanced by a given power.

RULE

As the distance between the body to be raised, or balanced, and the fulcrum, or prop, is to the distance between the prop and the point where the power is applied, so is the power to the weight which it will balance.

1. If a man, weighing 160 lbs., rest on a lever 10 feet long, what weight will he balance on the other end, supposing the prop to be 1 foot from the weight?

1 : 9 : : 160 : 1440 lbs. *Ans.*

2. If a weight of 1440 lbs. were to be raised by a lever 10 feet long, the prop being 1 foot from the weight, what power must be applied to the other end, to balance the weight?

Ans. 160 lbs.

3. At what distance from the prop must a power of 160 lbs. be applied, to balance 1440 lbs., 1 foot from the prop?

Ans. 9 feet.

4. At what distance from a weight of 1440 lbs. must a prop be placed, so that a power of 160 lbs., applied 9 feet from the prop, may balance it? *Ans.* 1 foot.

OF THE WHEEL AND AXLE.

Art. 298.—The wheel and axle are here represented with the weight attached to the circumference of the axle, and the power applied to the circumference of the wheel. The principle of the lever is employed in the wheel and axle.

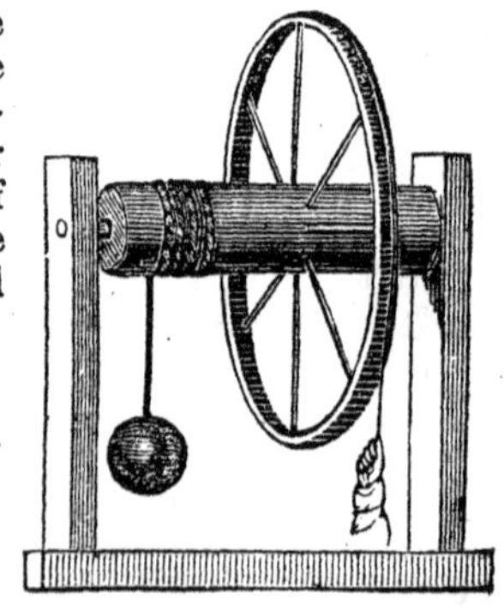

RULE.

As the diameter of the axle is to the diameter of the wheel, so is the power applied to the wheel, to the weight suspended on the axle.

1. If the diameter of the axle be 6 inches, and that of the

wheel be 60 incnes, what weight applied to the wheel will balance 10 lbs. on the axle? 60 : 6 : : 10 : 1 lb. *Ans.*

2. If the diameter of the wheel be 60 inches, what must be the diameter of the axle, so that 1 lb. on the wheel may balance 10 lbs. on the axle? *Ans.* 6 inches.

3. If the diameter of the axle be 6 inches, what must be the diameter of the wheel, so that 10 lbs. on the axle may balance 1 lb. on the wheel? *Ans.* 60 inches.

THE PULLEY.

Art. 299.—The pulley is a small wheel, moveable about its axis by means of a cord, which passes over it.

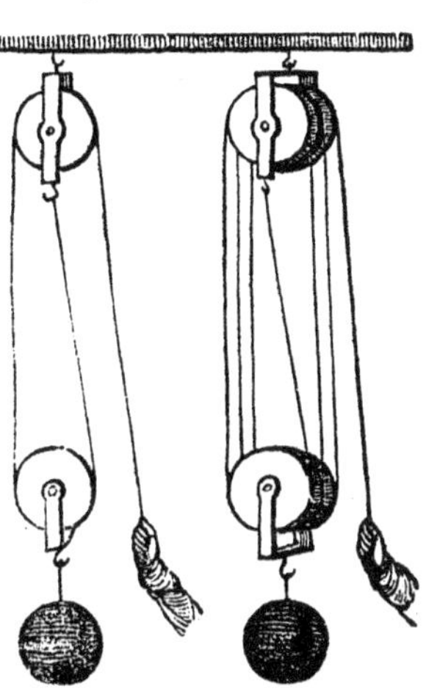

When the axis of a pulley is fixed, the pulley only changes the direction of the power; if moveable pulleys are used, an equilibrium is produced, when the power is to the weight as one to the number of ropes applied to them. If each moveable pulley has its own rope, each pulley will be double the power.

Art. 300.—The number of moveable pulleys and the power given, to find what weight may be raised.

RULE.

As 1 *is to twice the number of moveable pulleys, so is the power to the weight.*

OBS.—Reverse the rule, to find the power.

1. What weight would balance a power of 45 lbs., applied to a cord that runs over 3 moveable pulleys?

2. If a cord, which runs over 2 moveable pulleys, be attached to an axle 3 inches in diameter, the wheel of the axle being 28 inches in diameter, and a power of 10 lbs. be exerted

at the circumference of the wheel, what weight would be raised under the pulleys?

Thus, $3 : 28 :: \overline{10 \times 3 \times 2} : 560$ lbs. *Ans.*

OF THE SCREW.

Art. 301.—The screw is a spiral thread, or groove, cut round a cylinder, and everywhere making the same angle with the length of the cylinder.

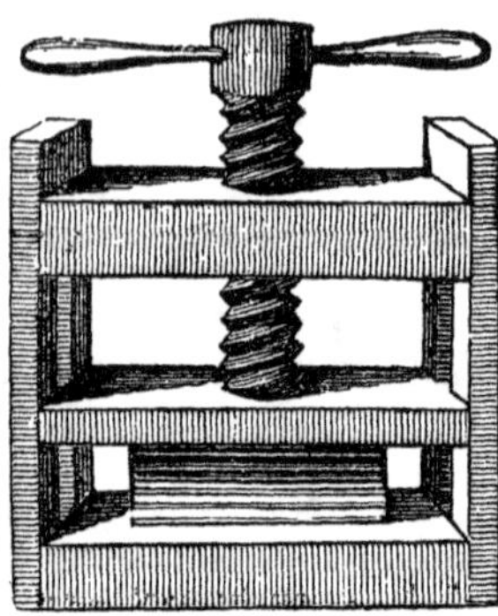

The power is to the weight which is to be raised, as the distance between two contiguous threads of the screw is to the circumference of a circle, described by the power applied at the end of the lever.

RULE.

Multiply twice the length of the lever by 3.1416, *which will give the circumference of the circle; then say, as the circumference is to the distance between the threads of the screw, so is the weight to be raised to the power which will raise it.*

1. The threads of a screw are 1 inch asunder; the lever, by which it is turned, is 30 inches long, and the weight to be raised is 1 ton=2240 lbs. What power must be applied to turn the screw?

$30 \times 2 = 60$, and $60 \times 3.1416 = 188.496$ inches, the circumference. Then $188.496 : 1 :: 2240 : 11.88$ lbs. *Ans.*

2. If the lever be 30 inches, the circumference of the circle described by the power 188.496, the threads of the screw 1 inch asunder, and the power 11.88 lbs., what weight will be raised? *Ans.* 2240 lbs.

3. If the weight be 2240 lbs., the power 11.88 lbs., and the lever 30 inches in length, what is the distance between the threads of the screw? *Ans.* 1 inch.

4. If the power be 11.88 lbs., the weight 2240 lbs., and the threads 1 inch asunder, what is the length of the lever?

Ans. 30 inches, nearly.

INCLINED PLANE.

Art. 302.—An inclined plane is a plane which makes an acute angle with the horizon.

To find the power that will draw a weight up an inclined plane.

RULE.

As the length of the plane is to the perpendicular height of the plane, so is the weight to the power.

1. An inclined plane is 40 feet in length, and 8 feet in perpendicular height. What power is sufficient to balance a weight of 2000 pounds? *Ans.* 400 lbs.

2. A certain railroad, 200 rods in length, has a perpendicular elevation of 20 feet. What power is sufficient to sustain a train of cars weighing 100,000 pounds? *Ans.* $606\frac{2}{33}$.

THE WEDGE.

Art. 303.—The wedge is composed of two inclined planes, whose bases are joined.

When the resisting forces, and the power which acts on the wedge, are in equilibrium, the weight will be to the power as the height of the wedge to a line drawn from the middle of the base to one side, and parallel to the direction in which the resisting force acts on that side.

Art. 304.—To find the force of the wedge.

RULE.

As the breadth, or thickness, of the head of the wedge, is to one of its slanting sides, so is the power which acts against its head, to the force produced at its side.

Suppose 100 lbs. to be applied to the head of a wedge, 2 inches broad, and 20 inches long, what force would be effected on each side? *Ans.* 1000 lbs.

MATHEMATICAL PROBLEMS.

Art. 305.—PROB. I. The sum and difference of two numbers given, to find those numbers.

RULE.

Subtract the difference from the sum, and divide the remainder by 2. The quotient will be the smaller number. Then add the given difference to the smaller number, and this sum will be the larger number.

EXAMPLE.

An assembly of 344 persons is convened in two rooms, one of which has 142 persons more in it than the other. How many are in each?

Operation.

344—142=202; then 202÷2=101 persons, in one room; then 101+142=243, in the other.

Art. 306.—PROB. II. The sum of two numbers, and the difference of their squares given, to find those numbers.

RULE.

Divide the difference of their squares by the sum of the numbers, and the quotient will be their difference. We then have their sum and difference, to find each number, by Prob. I.

EXAMPLE.

A. and B. played at marbles, having at first 14 each; but after playing several games, B. having lost some of his, would not play any longer, and it was found that the difference of the squares of what each then had, was 336. How many did B. lose?

Thus $336 \div \overline{14+14}=12$ difference; 14=half sum, and 12÷2=6=half difference. Then 14+6=20, A. retired with; and 14—6=8, B. retired with: then 14—8=6, B. lost.

Art. 307.—PROB. III. The difference of two numbers, and the difference of their squares given, to find those numbers.

RULE.

Divide the difference of their squares by the difference of their numbers, and the quotient will be their sum; then proceed by Prob. I.

EXAMPLE.

Said William to John, Father gave me $12 more than he gave Charles; and the difference of the squares of our separate parcels is 288. How much did he give each? Thus, $288 \div 12 = 24$, the sum: then $\overline{24 - 12} \div 2 = 6$; then $12 + 6 =$ $18, William's share; and $12 - 6 =$ $6, Charles had given him.

Art. 308.—PROB. IV. The sum of two numbers and their quotient given, to find those numbers.

RULE.

Add 1 *to the quotient, and by this sum divide the sum of the two numbers: this will give the less number. Subtract the less number from the sum, and you will obtain the greater number.*

EXAMPLES.

1. Divide 100 into two such parts, that if the greater be divided by the less, the quotient will be just 30.

Operation.

Thus $100 \div \overline{30 + 1} = 3\frac{7}{31}$, the less part; then $100 - 3\frac{7}{31} = 96\frac{24}{31}$, greater.

2. The sum of A. and B.'s ages is 45, and if you divide A.'s by B.'s the quotient will be 4. What is the age of each?

Ans. A.'s 36 years; B.'s 9 years.

Art. 309.—PROB. V. The difference of two numbers, and the quotient given, to find those numbers.

RULE.

The difference of the two numbers divided by the quotient less 1, *will be the less number. Add the less number to the difference, and you will have the greater number.*

A greyhound, in pursuit of a hare, ran three times as fast as the hare, and when he overtook the hare he had run 30 rods more than she. How many rods did each run?

Operation.

$30 \div \overline{3 - 1} = 15$ rods, the hare ran; then $15 + 30 = 45$ rods, the greyhound ran.

Art. 310.—PROB. VI. To find the true weight of any quantity when weighed in each scale of a balance, whose beam is unequally divided.

RULE.

Take the square root of the product of the different weights, for the true weight.

A parcel of sugar weighs in one scale 25 lbs.; in the other 30 lbs. What was its true weight?

$$\sqrt{25\times30}=27.856.$$

Art. 311.—Prob. VII. The base and perpendicular given, to find the hypotenuse.

RULE.

The square root of the sum of the squares of the base and perpendicular will be the hypotenuse.

This rule is illustrated by the following figure.

If the base of a right-angled triangle be 9 feet, and the perpendicular 12, what is the hypotenuse?

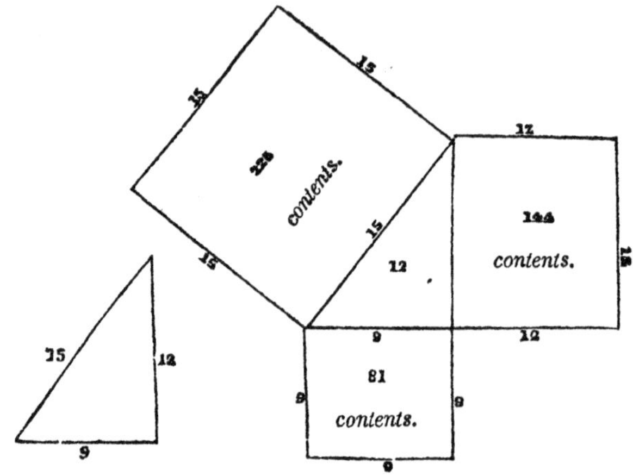

Art. 312.—Prob. VIII. Given the base and sum of the perpendicular and hypotenuse of a right-angled triangle, to find the perpendicular.

RULE.

From the square of the sum subtract the square of the base, and divide the remainder by twice the sum, and the quotient will be the perpendicular.

A tree, 100 feet in height, is broken off—the top of the tree reaches the ground 30 feet from the bottom, while the part broken off rests on the stump. How high from the ground was it broken off? *Ans.* $45\frac{1}{2}$ feet.

Art. 313.—Prob. IX. Given the base and the difference of the hypotenuse and perpendicular of a right-angled triangle, to find the perpendicular.

RULE.

From the square of the base subtract the square of the given difference, and divide the remainder by twice the difference.

EXAMPLE.

If the base of a right-angled triangle be 30 feet, and the difference of the other two sides 6 feet, what is the length of the perpendicular? *Ans.* 72 feet.

Art. 314.—Prob. X. To find the diameter of the earth, from the known height of a distant mountain, whose summit is just visible in the horizon.

RULE.

From the square of the distance, divided by the height, subtract the height.

The highest point of the Andes is about 4 miles above the bed of the ocean. If a straight line from this touch the surface of the water at the distance of $178\frac{1}{4}$ miles, what is the diameter of the earth? *Ans.* 7940.

Art. 315.—Prob. XI. To find the greatest distance at which a given object can be seen on the surface of the earth.

RULE.

To the product of the height of the object into the diameter of the earth, add the square of the height; and extract the square root of the sum.

1. If the diameter of the earth be 7940 miles, and Mount Etna 2 miles high, how far can it be seen at sea? *Ans.* 126+ miles.

$$\sqrt{7940\times2+2^2}=126.$$

Obs.—The actual distance at which an object can be seen is increased by the refraction of the air.

2. A man standing on a level with the ocean, has his eye raised $5\frac{1}{2}$ feet above the water. To what distance can he see the surface? *Ans.* $2\frac{7}{8}$ miles.

Art. 316.—Prob. XII. To find the height of an object at sea, or on the surface of the earth, having only the distance given.

RULE

From the given distance, take the distance which the elevation of the eye above the surface will give, found by the last problem; then divide the square of the remainder by the diameter of the earth, and the quotient will be the height required.

Art. 317.—Prob. XIII. To find the contents of squared timber.

RULE.

Multiply the mean breadth by the mean thickness: the product, multiplied by the length, will give the contents.

Required the contents of a log, the length 24 feet 6 inches, mean breadth 1 foot 1 inch, and mean thickness 1 foot 1 inch. *Ans.* 28 feet 9 inches 6'''

Art. 318.—Prob. XIV. To find the contents of round timber.

COMMON RULE.

Take one fourth of the mean girt, and square it, and multiply it by the length, for the contents.

Obs.—1. Tapering timber should be divided into pieces of eight or ten feet long, and these parts should be computed separately, and added.

2. In order to reduce the tree to such a circumference as it would have without its bark, a deduction is generally made of $\frac{1}{2}$ or $\frac{3}{4}$ of an inch for every foot of quarter-girt for young oak, ash, beech, etc.; but 1, or even $1\frac{1}{2}$ inch, must be allowed for old oak, for every foot of quarter-girt.

3. The common rule gives the contents too small, by 3 feet on every 11 feet of contents; yet it is universally used in practice, being originally introduced in order to compensate the purchaser of round timber for the waste occasioned by squaring it.

Rule II.—*Take one fifth of the girt, and square it, and multiply by twice the length, for the contents.*

1. Required the contents of a tree 24 feet long, and its girts at the ends 14 and 2 feet?

Ans. 96 feet, by the common rule; the true content is 122.88 feet.

2. How much timber in a tree 18 feet long, and its mean girt 5 feet 8 inches?

Ans. Common rule, 36 feet $1\frac{1}{2}$ inch; true content, 46 ft. 2 inches 10″ 6‴.

LEVELLING.

Art. 319.—PROB. I. To find the difference in the height of two places, by levelling rods.

RULE.

Set up the levelling rods perpendicular to the horizon, and at equal distances from the spirit level; observe the points where the line of level strikes the rods before and behind, and measure the heights of these points above the ground; level in the same manner, from the second station to the third, from the third to the fourth, etc. The difference between the sum of the heights at the back stations, and at the forward stations, will be the difference between the height of the first station and the last.

If the stations are numerous, it will be expedient to place the back and forward heights in separate columns in a table, as in the following example.

	Back heights.		Fore heights.	
	Feet.	Inches.	Feet.	Inches.
1st observation, . .	5	2	7	5
2d " . .	2	8	6	3
3d " . .	3	6	5	9
4th " . .	4	5	3	2
5th " . .	8	7	1	7
	24	4	24	2
	24	2		
Difference, . .	0	2		

If the sum of the forward heights is less than the sum of the back heights, it is evident that the last station must be higher than the first.

Art. 320.—PROB. II. To find the difference between the true and apparent lèvel, for any given distance.

OBS.—1. The *true level* is a *curve*, which either coincides with, or is parallel to, the surface of water at rest.

2. The *apparent level* is a *straight line*, which is a *tangent* to the true level, at the point where the observation is made.
3. The difference between the true and the apparent level is nearly equal to the square of the distance, divided by the diameter of the earth.

1. What is the difference between the true and apparent level, for a distance of one English mile, supposing the earth to be 7940 miles in diameter?

Ans. 7.98 inches, or 8 inches, nearly.

2. A tangent to a certain point on the ocean strikes the top of a mountain 23 miles distant. What is the height of the mountain? *Ans.* 352 feet.

PHILOSOPHICAL PROBLEMS.

Art. 321.—Prob. I. To find the time in which pendulums of different lengths would vibrate, that which vibrates seconds being 39.2 inches.

The time of the vibrations of pendulums are to each other, as the square roots of their lengths; or, their lengths are as the squares of their times of vibrations.

RULE.

As the square of one second is to the square of the time in seconds in which a pendulum would vibrate, so is 39.2 inches to the length of the required pendulum.

EXAMPLES.

1. Required the length of a pendulum that vibrates once in 8 seconds. $1^2 : 8^2 :: 39.2$ in. : 2508.8 in. $= 209\frac{1}{15}$ ft. *Ans.*

2. How often will a pendulum vibrate, whose length is 100 feet? *Ans.* 5.53+ seconds.

Art. 322.—Prob. II. By having the height of a tide on the earth given, to find the height of one at the moon.

RULE.

As the cube of the moon's diameter, multiplied by its density, is to the cube of the earth's diameter, multiplied by its density, so is the height of a tide on the earth, to the height of one at the moon.

EXAMPLE.

The moon's diameter is 2180 miles, and its density 494; the earth's diameter is 7964 miles, and its density 400. If, then, by the attraction of the moon, a tide of 6 feet is raised at the earth, what will be the height of a tide raised by the attraction of the earth at the moon? *Ans.* 236.8+ feet.

I. If the diameters of two globes be equal, and their densities different, the weight of a body on their surfaces will be as their densities.

II. If their densities be equal, and their diameters different, the weight of a body will be as $\frac{1}{3}$ of their circumferences.

III. If their diameters and densities be both different, the weight will be as $\frac{2}{3}$ of their semidiameters multiplied by their densities.

TABLE.

	Density.	Diameter.	Semidiameter.	$\frac{2}{3}$ semidiameter.
Sun	100	883246	441623	294415
Jupiter	94.5	89170	44585	29723
Saturn	67	79042	39521	26347
Earth	400	7964	3982	2654
Moon	494	2180	1090	726

Art. 323.—PROB. III. To find how far a heavy body will fall in a given time, near the surface of the earth.

OBS.—Heavy bodies, near the surface of the earth, fall 16 feet in 1 second of time; and the velocities they acquire in falling are as the squares of the times; therefore, to find the distance any body will fall in a given time, we adopt the following

RULE.

As 1 second is to the square of the time in seconds that the body is falling, so is 16 feet to the distance in feet, that the body will fall in the given time.

How far will a leaden bullet fall in 8 seconds?

$1^2 : 8^2 :: 16$ ft. : 1024 ft. := *Ans.*

Art. 324.—PROB. IV. The velocity given, to find the space fallen through, to acquire that velocity.

RULE.

Divide the velocity by 8, and the square of the quotient will be the distance fallen through to acquire that velocity.

1. The velocity of a cannon-ball is 424 feet per second. From what height must it fall to acquire that velocity? *Ans.* 2809 feet.

2. At what distance must a body have fallen to acquire the velocity of 1024 feet per second? *Ans.* 3 miles, 544 feet.

Art. 325.—PROB. V. The velocity given per second, to find the time.

RULE.

Divide the velocity by 8, *and a fourth part of the quotient will be the time in seconds.*

1. How long must a body be falling to acquire a velocity of 304 feet per second? *Ans.* $9\frac{1}{2}$ seconds.

2. How long must a body be falling to acquire a velocity of 864 feet per second? *Ans.* 27 seconds.

Art. 326.—PROB. VI. The space through which a body has fallen, given, to find the time it has been falling.

RULE.

Divide the square root of the space fallen through by 4, *and the quotient will be the time in which it was falling.*

How long would a ball be falling from the top of a tower, 900 feet high, to the earth? *Ans.* $7\frac{1}{2}$ seconds.

Art. 327.—PROB. VII. The time given, to find the space fallen through.

RULE.

Multiply the time by 4, *and the square of the product will be the space fallen through in the given time.*

1. What is the difference between the depth of two wells, into each of which, should a stone be dropped at the same instant, one would reach the bottom in 5 seconds, and the other in 3?

$5\times4=20$, and $20\times20=400$: then $3\times4=12$, and $12\times12=144$: then $400-144=256$ ft. *Ans.*

2. A ball was seen to fall half the way from the top of a tower in the last second of time. How long was it in descending, and what was its height before its descent? *Ans.* 186.486+ feet.

Art. 328.—Prob. VIII. To find the velocity, per second, with which a heavy body will begin to descend, at any distance from the earth's surface.

RULE.

As the square of the earth's semidiameter is to 16 feet, so is the square of any other distance from the earth's centre, inversely, to the velocity with which it begins to descend per second.

With what velocity per second will a ball begin to descend, if raised 3000 miles above the earth's surface?

As $\overline{4000\times 4000}:16::\overline{4000+3000}\times\overline{4000\times 3000}:5.22449$ feet, *Ans.*

And if the height is required, and the velocity given, thus, as $16:\overline{4000\times 4000}::5.22449:49000000$, and $\sqrt{49000000}-4000=3000$ miles, *Ans.*

Art. 329.—Prob. IX. The weight of a body, and the space fallen through, given, to find the force with which it will strike.

RULE.

Multiply the space fallen through by 64; then multiply the square root of this product by the weight, and the product is the momentum, or force with which it will strike.

There is a monument 64 feet high. Supposing a stone, weighing 4 tons, should fall from its top to the earth, what would be its force, or momentum? *Ans.* 573440 lbs.

That is, it would strike the earth with more force than the weight of two hundred and fifty tons.

Art. 330.—Prob. X. To find the magnitude of any thing, when the weight is known.

RULE.

Divide the weight by the specific gravity found in the table, and the quotient will be the magnitude sought.

What is the magnitude of several fragments of clear glass, whose weight is 13 ounces?

$13\div 2600=.005$ of a cubic foot; and $.005\times 1728=8.640$ cubic inches, *Ans.*

TABLE,

Showing the specific gravity of several solid and fluid bodies, in Avoirdupois Weight.

The specific gravity of a body is its weight compared with pure, or distilled water.

A Cubic foot of	Ounce.	A Cubic foot of	Ounce.
Platina, rendered malleable and hammered	20170	Brick	2000
Very fine Gold	19637	Live Sulphur	2000
Standard Gold	18888	Nitre	1900
Moidore Gold	17140	Alabaster	1875
Guinea Gold	17793	Dry Ivory	1885
Quicksilver	13600	Brimstone	1800
Lead	11325	Solid subs. Gunpowder	1745
Fine Silver	11087	Alum	1714
Standard Silver	10535	Ebony	1717
Rose Copper	9000	Human Blood	1054
Copper	8843	Amber	1030
Plate Brass	8000	Cows' Milk	1030
Steel	7852	Sea Water	1030
Cast Brass	7850	Pure Water	1000
Iron	7645	Red Wine	993
Block Tin	7321	Oil of Amber	978
Cast Iron	7135	Proof Spirits	925
Lead Ore	6800	Dry Oak	925
Copper Ore	3775	Olive Oil	913
Diamond	3400	Loose Gunpowder	872
Crystal Glass	3150	Spirits of Turpentine	864
White Marble	2707	Alcohol, or pure spirits	850
Black "	2704	Elm and Ash	800
Rock Crystal	2658	Oil of Turpentine	772
Green Glass	2620	Dry Crab-tree	765
Clear Glass	2600	Ether	732
Stone, Flint	2582	White Pine	569
Stone, Paving	2570	Sassafras Wood	482
Stone, Cornelia	2568	Cork	240
Stone, Free	2352	Common Air	$1\frac{35}{100}$
		Inflammable Air	$0\frac{12}{100}$

Art. 331.—PROB. XI. The bulk and weight of any body given, to find its specific gravity.

RULE.

Divide the weight by the bulk, and the quotient is the specific gravity.

Suppose a piece of marble contains 8 cubic feet, and weighs $1353\frac{1}{2}$ pounds, or 21656 ounces. What is its specific gravity?

$21656 \div 8 = 2707$, the specific gravity, as required by the table.

ASTRONOMICAL PROBLEMS.

Art. 332.—PROB. I. To find the dominical letter for any year in the present century, and also to find on what day of the week January will begin.

RULE.

To the given year add its fourth part, rejecting the fractions; divide this sum by 7; if nothing remains, the dominical letter is A; but, if there be a remainder, subtract it from 8, and the residue will show the dominical letter, reckoning 0=A, 2=B, 3=C, 4=D, 5=E, 6=F, 7=G. These letters will also show on what day of the week January commences. For, when A is the dominical letter, January begins on Sunday; when B is the dominical letter, January begins on Saturday; C begins it on Friday; D begins it on Thursday; E on Wednesday; F on Tuesday; G on Monday.

1. Required the dominical letter for 1825.

4)1825 8−6=2=B, dominical letter.
456
7)2281
325−6

As B is the dominical letter, January will begin on Saturday, and the second day will be the Sabbath.

2. Required the dominical letter for 1842. *Ans.* B.
3. Required the dominical letter for 1837. *Ans.* A.
4. What is the dominical letter for 1801? *Ans.* D.
5. What is the dominical letter for 1845? *Ans.* E.

Art. 333.—PROB. II. To find on what day of the week any given day of the month will happen.

RULE.

Find by the last problem the dominical letter for the given year, and on what day in January will be the first Sabbath; and the corresponding days in the succeeding months will be as follows: Wednesday for February; Wednesday for March; Saturday for April; Monday for May; Thursday for June; Saturday for July; Tuesday for August; Friday for September; Sunday for October; Wednesday for November; Friday for December. Having found the day of the week for any day in the month, any other day may be easily obtained, as may be seen in the following example.

1. Let it be required to ascertain on what day of the week will be the 25th day of September, 1842.

The domirtical letter for 1842 is B; therefore, the 2d of January will be the Sabbath; and, by the above rule, the 2d of February will be Wednesday; the 2d of March will be Wednesday; the 2d of April will be Saturday; the 2d of May will be Monday; the 2d of June will be Thursday; the 2d of July will be Saturday; the 2d of August will be Tuesday; the 2d of September will be Friday. If the 2d be Friday, the 9th, 16th, and 23d will be Fridays. And if the 23d be Friday, the 24th will be Saturday, the 25th will be the Sabbath, the day required.

2. On what day of the week will be December 8, 1849?
Ans. Saturday.

3. On what day of the week will happen July 4, 1857?
Ans. Saturday.

4. On what day of the week were you born?

OF BALLS AND SHELLS.

Art. 334.—An iron ball, 4 inches in diameter, weighs 9 lbs., nearly; and a leaden one, $4\frac{1}{4}$, weighs about 17 lbs., and a pound of gunpowder measures about 30 cubic inches.

Given the diameter of an iron ball, to find the weight, and the converse.

RULE.

Divide the cube of the diameter by $7\frac{1}{9}$; the quotient will be the weight in pounds. Multiply the weight by $7\frac{1}{9}$. The cube root of the product will be the diameter.

1. What is the weight of an iron ball, of which the diameter is $3\frac{1}{2}$ inches? *Ans.* $3.5 \div 7\frac{1}{9} = 6.0293$ lbs.

2. What is the diameter of an iron ball which weighs 24 lbs.?
Ans. $\sqrt[3]{24 \times 7\frac{1}{9}} = \sqrt[3]{170.6} = 5.547$ inches diameter.

3. What does an iron ball weigh, whose diameter is 5.5 inches? *Ans.* 23.3965 lbs.

4. What is the diameter of an iron ball weighing 48 lbs?
Ans. 6.988 inches.

5. What does an iron ball weigh whose diameter is 4.6 inches? *Ans.* 13.688 lbs.

6. What is the diameter of an iron ball which weighs 36 lbs? *Ans.* 6.349 inches.

Art. 335.—Given the diameter of a leaden ball, to find its weight, and the converse.

RULE.

Divide the cube of the diameter by $4\frac{1}{2}$: *the quotient will be the weight in lbs. Multiply the weight by* $4\frac{1}{2}$: *the* $\sqrt[3]{\ }$ *of the product will be the diameter in inches.*

1. What is the weight of a leaden ball, whose diameter is 4.25 inches? *Ans.* $\overline{4.25}^3 \div 4\frac{1}{2} = 17.059$ lbs.

2. What is the diameter of a leaden ball which weighs 36 lbs.? *Ans.* 5.45 inches.

3. What is the weight of a leaden ball, of 4.6 inches in diameter? *Ans.* 21.63 lbs.

4 What is the diameter of a leaden ball weighing 48 lbs.? *Ans.* 6 inches.

PILING OF BALLS.

Art. 336.—Balls and shells are piled up in horizontal courses, upon a base of the form of an equilateral triangle, or of a square, or of a rectangle. The number of balls in a row diminishes, till, in the two first forms, it ends in a single ball, and in the last in a single row. The number of rows is equal to the number of balls in the lesser side of the under row. The number in the top row of a rectangular pile is one more than the difference between the length and breadth of the bottom row.

Art. 337.—Prob. I. To find the number of balls in a triangular pile.

RULE.

Multiply the number of balls in a side of the bottom row by that number increased by 1, *and again by that number increased by* 2: *the product, divided by* 6, *will be the number of balls in the pile.*

Required the number of balls in a triangular pile, of which each side of the base contains 30 balls. *Ans.* 4960.

Art. 338.—PROB. II. To find the number of balls in a square pile.

RULE.

To twice the number of balls in a side of the bottom, add 1, *and multiply the sum by the number in that row, and by that number increased by* 1: *the product, divided by* 6, *will give the number of balls in the pile.*

Let the side of the bottom row of a square pile contain 20 balls. How many balls are in the pile? *Ans.* 2870.

Art. 339.—PROB. III. To find the number of balls in a rectangular pile.

RULE.

From 3 *times the number in the length of the bottom row, increased by* 1, *subtract the number in the breadth, and multiply the remainder by the breadth, and by the breadth increased by* 1: *the product, divided by* 6, *will give the number of balls in the pile.*

Suppose the number of balls in the length of a rectangular pile to be 59, and in the breadth 20, what is the number in the pile? *Ans.* 11060.

Art. 340.—PROB. IV. To find the number of balls in an incomplete pile.

RULE.

From the number of balls in the complete pile subtract the number in the pile that is wanting, both computed as before: the remainder is the number in the incomplete pile.

Required the number of balls in a rectangular pile of 15 courses, the numbers in the bottom row being 60 and 25.
Ans. 14590.

TO FIND THE WEIGHT OF CATTLE.

Art. 341.—TAKE the girt behind the shoulder, and the length from the fore part of the shoulder-blade to the buttock, both in feet. Multiply the square of the girt by 4 times the length, and divide by 21. Multiply this quotient by 16, and it will give the weight of the four quarters, *nearly.*

OBS.—The four quarters are little more than $\frac{1}{2}$ the whole weight. The skin weighs nearly $\frac{1}{18}$, and the tallow very nearly $\frac{1}{12}$.

What will the four quarters of an ox weigh, whose girt is 6 feet 6 inches, and length 5 feet 10 inches.

Ans. $\overline{6.5^2} \times 23\frac{1}{3} \div 21 \times 16 = 751 +$lbs.

MISCELLANEOUS QUESTIONS.

Art. 342.—1. What is the product of 2*s.* 6*d.*, multiplied by 2*s.* 6*d.*? *Ans.* £$\frac{1}{64}$.

2. Purchased a book for 15 cents, and sold it for 18. What did I gain per cent.? *Ans.* 20 per cent.

3. Sold a book for 18 cents, and gained 20 per cent. What did it cost me?

4. Purchased a book for 15 cents; sold it so as to gain 20 per cent. What did I get for it?

5. If $\frac{1}{2}$ of $\frac{3}{7}$ of $\frac{7}{40}$ of $\frac{3}{5}$ of $\frac{5}{9}$ of a vessel be worth £378, how many dollars is $\frac{3}{32}$ of it worth? *Ans.* $9450.

6. A person owning $\frac{2}{5}$ of a ship, sold $\frac{5}{8}$ of his share for $3750. What was the whole ship worth? *Ans.* $15000.

7. What is the sum of the third and half third of 3*s.* 4*d.*? *Ans.* 1*s.* 8*d.*

8. How many solid feet in a stick of timber 17 inches square, and 6 feet 5 inches long? *Ans.* 12 ft. 1517 in.

9. A man owning $\frac{2}{5}$ of a farm, sold $\frac{1}{2}$ of his share for $245. What was the value of the farm? *Ans.* $1225.

10. A. holds B.'s note for $2000, dated June 1st, 1825, on which are the following endorsements, viz:

Received Sept. 1st,	1825	$96.
Dec. 10th,	1825	15.
Apr. 20th,	1826	36.
July 1st,	1826	200.
Jan. 10th,	1827	20.
Mar. 25th,	1827	90.

How much remains due June 1st, 1827? *Ans.* $1767.228.

11. What principal, at 5 per cent., will amount to $725 in 9 years? *Ans.* $500.

12. What principal will gain $150, in 1 year, at 6 per cent.? *Ans.* $2500.

13. What is the present worth of $590, due 3 years hence, discounting at 6 per cent.? *Ans.* $500.

14. What is the present worth of $200; $100 payable in 2 months, $50 in 3 months, and $50 in 5 months, discounting at 4 per cent.? *Ans.* $198.01.

15. A note of $500 amounted to $725 in 9 years. What was the rate per cent.? *Ans.* 5 per cent.

16. In what time will £420 amount to £520 16*s.* at 3 per cent.? *Ans.* 8 years.

17. What will $1350 amount to in 3 years, at 5 per cent., compound interest? *Ans.* $1562.793.

18. A. has 150 gallons of wine, which he will sell at 7*s.* 3*d.* per gallon, ready money, but in barter he will have 8*s.* per gallon. B. has linen at 3*s.* 6*d.* per yard, ready money. How must B. sell his linen per yard, in proportion to the bartering price of A.'s wine, and how many yards will be equal to A.'s wine?

Ans. { Bartering price, 3*s.* $10\frac{10}{29}$*d.*, and 310 yards 2 qrs. 3+nails.

19. A merchant bought hats at 4*s.* each, and sold them at 4*s.* 9*d.* What was the gain in laying out £100? *Ans.* £18 15*s.*

20. A merchant bought 10 tons of iron for £200. The freight and duties amounted to £25, and his own charges to £8 6*s.* 8*d.* For how much per pound must he sell it to gain 20 per cent.? *Ans.* 3*d.* per lb.

21. Sold a watch for £50, and by so doing lost 17 per cent., whereas I ought to have cleared 20 per cent. How much was it sold under its real value? *Ans.* £22 5*s.* $9\frac{1}{4}$*d.*

22. Four men trade in company. A.'s stock was $560; B.'s $1040; C.'s $1200; their whole stock was $3200. They gained in two years a sum equal to twice their stock, and $160 more. What was D.'s stock, and what was each man's share of the gain?

Ans. {
D.'s stock, $400.
A.'s gain, $1148.
B.'s " $2132.
C.'s " $2460.
D.'s " $820.

23. A., B., and C. trade in company. A. put in $20, B. $30, and C. a sum unknown. The gain was $36, of which $16

was C.'s share. What was C.'s stock, and what was A.'s and B.'s gain?

Ans. { C.'s stock, \$40. A.'s gain, \$8. B.'s " \$12. }

24. What is the square root of $42\frac{1}{4}$? *Ans.* $6\frac{1}{2}$.

25. What is the mean proportional between 24 and 96? *Ans.* 48.

26. A certain general has an army of 5625. How many must he place in rank and file, to form them into a square? *Ans.* 75.

27. Arrange 10952 men in such a manner that the number in rank may be double the file.

Ans. 74 in file, and 148 in rank.

28. There is a circle, whose diameter is 4 inches. What is the diameter of a circle 3 times as large? *Ans.* 6.928+.

29. Two boats start on a river at the same time, from places 300 miles apart; the one proceeding up the stream is retarded by the current 2 miles per hour, while that moving down the stream is accelerated 2 miles per hour; both are propelled by by a steam engine, which would move them in still water 8 miles per hour. How far from each starting-place will the boats meet? *Ans.* $112\frac{1}{2}$ miles from the lower place, and $187\frac{1}{2}$ miles from the upper place.

30. There are 3 circular ponds; the diameter of the less is 100 feet; the area of the greater is 3 times the area of the less. What is its diameter? *Ans.* 173.2+.

31. What is the superficial contents of one side of a cubical stone, containing 474552 solid inches? *Ans.* 6084 inches.

32. If a cube of silver, whose side is 4 inches, be worth £50, what is the side of a cube of like silver, worth 4 times as much? *Ans.* 6.349 inches.

33. The height of a tree standing on the bank of a river is 75 feet; a line reaching from the opposite shore to the top of the tree is 256 feet long. What is the breadth of the river? *Ans.* 244.7+ feet.

34. If a pipe 6 inches in diameter will discharge a certain quantity of water in 4 hours, in what time will 3 pipes, each 4 inches in diameter, discharge double the quantity? *Ans.* 6 hours.

35. Two men start from the same place and travel, one south 76 leagues, the other east 45 leagues. How far will they be apart? *Ans.* 88.3+ leagues.

36. What is the side of a cubical mound, containing 5832 solid feet? *Ans.* 18 feet.

37. If a ball, 6 inches in diameter, weigh 32 lbs., how much will a ball of the same metal weigh, whose diameter is 3 inches? *Ans.* 4 lbs.

38. The side of a cubical box is 2 feet. What is the side of a box which will contain three times as much? *Ans.* 2 feet $10\frac{3}{5}$ inches.

39. A refiner mixed 3 lbs. of gold, 22 carats fine, with 3 lbs. 20 carats fine. What was the fineness of the mixture? *Ans.* 21 carats.

40. How many gallons of water must be put to wine, at 3*s.* a gallon, to fill a vessel of 100 gallons, so that a gallon of the mixture may be afforded at 2*s.* 6*d.* per gallon? *Ans.* $16\frac{2}{3}$ gallons.

41. If 12 bushels of oats, at 1*s.* 6*d.* per bushel, be mixed with barley at 2*s.* 6*d.*, rye at 3*s.*, and wheat at 4*s.* per bushel, how much barley, rye, and wheat must be mixed with the 12 bushels of oats, that the mixture may be worth 2*s.* 9*d.* per bushel? *Ans.* 12 bushels of each sort.

42. A man travelled 6 miles the first day, 9 the second, increasing each day's journey 3 miles. He travelled 61 days. How many miles did he travel the last day? *Ans.* 186 miles.

43. If 100 pears be placed in a right line, 1 yard asunder, how many miles will a man travel, to pick them and carry them 1 at a time to a basket placed 1 yard from the first pear? *Ans.* 5 miles, 1300 yards.

44. Suppose 550 men are in a garrison, with provision sufficient to last them 9 months. How many must depart, that their provision may last them 11 months? *Ans.* 100.

45. If a man labor for me 16 days, when the price of labor is $1.25, how long must I labor for him, to requite the favor, when the price of labor is 75 cents per day? *Ans.* $26\frac{2}{3}$ days.

46. The third part of an army were killed, the fourth part were taken prisoners, and 1000 fled. How many were in this army? *Ans.* 2400.

47. An ignorant fop, wanting to purchase an elegant house, met with a facetious gentleman, who told him he had one which he would sell him on the following very reasonable terms, viz: that he should give him 1 penny for the first door, 2 for the second, 4 for the third, and so on, doubling the price

of each door, which were 36 in number. "It is a bargain," exclaimed the simpleton, "and here is a guinea to bind it." What did the house cost him? *Ans.* £286331153 1*s.* 3*d.*

48. A., B., and C. would divide $100 between them, so that B. may have $3 more than A., and C. $4 more than B. What is each man's share?

Ans. { A.'s $30. B.'s $33. C.'s $37.

49. Divide £340 among 3 men, in such a manner that the first shall have 3 times as much as the second, and the third 4 times as much as the first?

Ans. { 1st, £63$\frac{3}{4}$. 2d, £21$\frac{1}{4}$. 3d, £255.

50. A person having a certain number of dollars, said, if a third, a fourth, and a sixth of them were added, their sum would be $45. How many dollars had he? *Ans.* $60.

51. What number is that which, being multiplied by $\frac{3}{4}$, the product will be 15$\frac{3}{4}$? *Ans.* 21.

52. What number is that from which, if you subtract $\frac{2}{5}$ of itself, the remainder will be 12? *Ans.* 20.

53. What part of 25 is $\frac{5}{8}$ of a unit? *Ans.* $\frac{1}{40}$.

54. What number is that which, being multiplied by $\frac{2}{3}$, the product will be $\frac{1}{4}$? *Ans.* $\frac{3}{8}$.

55. If $\frac{3}{8}$ of a farm be worth £3740, what is the whole worth? *Ans.* £9973 6*s.* 8*d.*

56. A father dying, left his son a fortune, $\frac{3}{16}$ of which he spent in 6 months; $\frac{2}{3}$ of the remainder lasted him 12 months longer, when he had £348 left. What sum did he receive? *Ans.* £1284 18*s.* 5$\frac{7}{13}$*d.*

57. A young man received $210, which was $\frac{2}{3}$ of his elder brother's fortune, and 3 times the elder brother's fortune was $\frac{1}{2}$ the father's estate. What was the value of the estate? *Ans.* $1890.

58. A man has 80 shillings to divide among his laborers, consisting of an equal number of men, women, and boys. To every boy he gives 6*d.*, to every woman 8*d.*, to every man 1*s.* 4*d.* How many were there of each? *Ans.* 32.

59. Suppose a man pay to his laborers, men, women, and boys, £7 17*s.* 6*d.*; to every boy he gave 6*d.*, to every woman 8*d.*, and to every man 16*d.*; for every boy there were 3 women, and for every woman there were 2 men. How many were there of each? *Ans.* 15 boys, 45 women, and 90 men.

60. A gentleman bought a horse, a chaise, and harness, for £60; the horse cost twice as much as the chaise, and the harness half as much as the horse. What was the cost of each?

Ans. { Horse, £30. Chaise, £15. Harness, £15.

61. Divide $1000 among 3 men, in such a manner that as often as the first has $3, the second shall have $5, and the third $8.

Ans. { 1st, $187.50. 2d, $312.50. 3d, $500.00.

62. A. can do a piece of work in 10 days; D. can do the same in 13 days. In what time will both working together do the same work? *Ans.* $5\frac{15}{23}$ days.

63. If 6 lbs. of pepper be worth 13 lbs. of ginger, and 19 lbs. of ginger be worth $4\frac{3}{4}$ lbs. of cloves, and 10 lbs. of cloves be worth 63 lbs. of sugar, at 10 cents per pound, what is the value of 1 cwt. of pepper? *Ans.* $38.22.

64. A tradesman increased his estate annually one third, less $960, which he spent in his family. At the end of $3\frac{1}{4}$ years he found that his estate amounted to $30284. What had he at first? *Ans.* $13551.75.

65. A person wants a cylindrical vessel, 3 feet deep, which shall hold twice as much as another 28 inches deep, and 46 inches in diameter. What must be the diameter of the required vessel? *Ans.* 57.373 inches.

66. How long must be the tether of a horse which will allow him to graze quite round an acre of ground?

Ans. $39\frac{1}{4}$ yards.

67. What number is that which, being increased by its $\frac{1}{2}$, $\frac{1}{4}$, and 5 more, the number will be doubled? *Ans.* 20.

68. A man, after having spent $\frac{1}{2}$ and $\frac{1}{3}$ of his money, had £$26\frac{2}{3}$ left. How much had he at first? *Ans.* £160.

69. A vessel has 3 pipes; the first will fill it in $\frac{1}{2}$ of an hour, the second in $\frac{1}{3}$ of an hour, and the third in $\frac{1}{4}$ of an hour. In what time will all running together fill it?

Ans. $\frac{1}{9}$ of an hour.

70. A. and B. employed equal sums in trade. A. gained a sum equal to $\frac{1}{4}$ of his stock; B. lost $225; then A.'s money was double that of B's. What was each man's stock?

Ans. $600.

71. There is a certain number, $\frac{7}{8}$ of which exceeds $\frac{4}{5}$ by 6. What is that number? *Ans.* 80.

72. If $60 be divided between 4 men in such a proportion that the first shall have $\frac{1}{3}$, the second $\frac{1}{4}$, the third, $\frac{1}{5}$, and the fourth $\frac{1}{6}$, what will each receive?

Ans. {
1st, $$21\frac{1}{19}$.
2d, $15\frac{15}{19}$.
3d, $12\frac{12}{19}$.
4th, $10\frac{10}{19}$.

73. A., B., C., and D. spent 35 shillings at a reckoning, and being a little *dipped,* agreed that A. should pay $\frac{2}{3}$, B. $\frac{1}{2}$, C. $\frac{1}{3}$, and D. $\frac{1}{4}$. What does each pay, in this proportion?

Ans. {
A. 13*s.* 4*d.*
B. 10*s.*
C. 6*s.* 8*d.*
D. 5*s.*

74. The wheels of a chaise, each 4 feet high, in turning within a ring, moved so that the outer wheel made two turns while the inner made one, and their distance from one another was 5 feet. What were the circumferences of the tracks described by them?

Ans. {
Outer, 62.8318 feet.
Inner, 31.4159 feet.

75. A., B., and C. traded in company, and gained £350, of which A. took a certain sum; B. took 4 times as much as A., and C. 8 times as much as B. What were their respective shares of the gain?

Ans. {
A.'s gain, £9 9*s.* 2*d.* $1\frac{3}{37}$*qr.*
B.'s gain, £37 16*s.* 9*d.* $0\frac{12}{37}$*qr.*
C.'s gain, £302 14*s.* 0*d.* $2\frac{22}{37}$*qrs.*

76. A gentleman divided his fortune among his sons, giving A. £9 as often as B. £5, and C. £3 as often as B. £7. C.'s dividend was £$1537\frac{5}{8}$. To what did the whole estate amount? *Ans.* £11583 8*s.* 10*d.*

77. If $\frac{5}{7}$ of $\frac{3}{8}$ of $\frac{4}{5}$ of a ship be worth $\frac{2}{9}$ of $\frac{7}{8}$ of $\frac{12}{13}$ of the cargo, valued at £1000, what was the value of both ship and cargo? *Ans.* £1837 12*s.* $1\frac{75}{117}$*d.*

78. Three men purchase a lot of land in company. A. paid $\frac{2}{5}$, B. $\frac{3}{7}$ of the whole, and C. paid £256. How much did A. and B. pay, and what part of the land had C.?

Ans. {
A. paid £597 6*s.* 8*d.*
B. paid £640.
C.'s share, $\frac{6}{35}$.

79. A gay fellow soon got the better of $\frac{2}{7}$ of his fortune. He then gave £1500 for a commission, and his profusion con-

tinued until he had but £450 left, which he found to be just $\frac{3}{8}$ of his money, after purchasing his commission. What was his fortune at first? *Ans.* £3780.

80. A. and B. are on opposite sides of a circular field, 268 rods in circumference. They start both at the same time to go round it, and go the same way. A. goes 22 rods in 2 minutes; B. goes 34 rods in 3 minutes. How many times will B. go round the field before he will overtake A.?

Ans. 17 times.

81. How high above the earth must a man be raised to see $\frac{1}{3}$ of its surface?* *Ans.* One diameter high.

82. The girt of a vessel round the outside of the hoop is 22 inches, and the hoop is 1 inch thick. What is the true girt of the vessel? *Ans.* $15\frac{5}{7}$.

83. The hour and minute hand of a watch are exactly together at 12 o'clock. When are they next together?

Ans. 1 h. 5 m. $27\frac{3}{11}$ s.

84. Three men trade in company till they gain £120. The sums put in were in such proportion, that as often as A. had £5 of the gain, B. had £7; and as often as B. had £4, C. had £6. What was each man's share of the gain?

Ans. A.'s, £26 13*s.* 4*d.*
B.'s, £37 6*s.* 8*d.*
C.'s, £56.

85. A. and B. cleared by an adventure at sea 45 guineas, which was £35 per cent. upon money adventured. With this gain they agreed to purchase a genteel horse and carriage, which they were to use in proportion to their sums adventured, which was found to be 11 to A. as often as 8 to B. What money did each adventure?

Ans. A. £104 4*s.* $2\frac{10}{19}$*d.*
B. £75 15*s.* $9\frac{9}{19}$*d.*

86. A military officer drew up his soldiers in rank and file, having the number in rank and file equal. On being reinforced with three times his first number of men, he placed them all in the same form, and then the number in rank and in file was just double what it was at first. He was again reinforced with three times his whole number of men; and, after placing the whole in the same form as at first, his number in rank and in file was 40 men each. How many men had he at first?

Ans. 100 men.

---

* This question is for the student in Geometry.

87. A general disposing his army into a square battalion, found he had 231 over and above, but increasing each side with one soldier, he wanted 44 to fill up the square. Of how many men did his army consist? *Ans.* 19000.

88. There are 3 horses belonging to different men, employed to draw a load of salt from Boston to Lowell for $9.50. A.'s and B.'s horses are supposed to do $\frac{7}{8}$ of the work, A.'s and C.'s $\frac{7}{12}$, B.'s and C.'s $\frac{17}{24}$. They are to be paid proportionally. What is each man's share of the gain?

Ans. { A.'s, 3.288\frac{6}{13}$. B.'s, 4.384$\frac{8}{13}$. C.'s, 1.826$\frac{12}{13}$.

89. A., B., and C. are to share £100, in the proportion of $\frac{1}{3}$, $\frac{1}{4}$, and $\frac{1}{5}$, respectively; but C. dying, it is required to divide the whole sum properly between the other two.

Ans. { A.'s share, £57 2*s.* 10$\frac{1}{4}$*d.* B.'s share, £42 17*s.* 1$\frac{5}{7}$*d.*

90. There is an island 50 miles in circumference, and 3 men start together to travel the same way round it. A. travels 7 miles a day, B. 8, and C. 9. When will they all come together again, and how far will each travel?

91. A man died leaving $1000 to be divided between his two sons, one 14 and the other 18 years of age, in such a manner that the share of each being let, at 6 per cent. interest, should amount to the same sum when they should arrive at the age of 21. What did each receive?

Ans. { The eldest, $546.153+. The youngest, $453.846+.

92. A hare starts 12 rods before a greyhound, but is not perceived by him till she has been up 45 seconds. She scuds away at the rate of 10 miles an hour, and the dog, on view, makes after, at the rate of 16 miles an hour. How long will the course hold, and what space will be run over from the spot where the dog started?

Ans. { 97$\frac{1}{2}$ seconds. 2288 feet.

93. There is a circular field, surrounded by a rail-fence 10 rails high, each one rod in length, and the number in the fence equals the number of acres in the field? What is the area of the field? *Ans.* 201062.4 acres.

94. How much greater is the circle described by the head of a man 6 feet high, than by his feet, in the revolution of the earth on its axis?

95. In an orchard, $\frac{1}{2}$ the trees bear apples, $\frac{1}{4}$ pears, $\frac{1}{6}$ plums,

and 50 of them produce nothing. How many trees are there in all? *Ans.* 600.

96. Sound moves at the rate of 1142 feet in a second. If the time between the lightning and the thunder be 30 seconds, what is the distance of the explosion? *Ans.* 6.488+ miles.

97. In a thunder-storm I observed by my watch that it was 6 seconds between the lightning and the thunder. At what distance was the explosion? *Ans.* 6852 feet.

98. Tubes may be made of gold, weighing not more than at the rate of $\frac{1}{1625}$ of a grain per foot. What would be the weight of such a tube, which would extend across the Atlantic, from Boston to London, the distance being 3000 miles?

Ans. 1 lb. 8 oz. 6 pwt. $3\frac{9}{13}$ grs.

99. Suppose one of those meteors called fireballs, to move parallel to the earth's surface, and 50 miles above it, at the rate of 20 miles per second, in what time would it move round the earth?

The earth's diameter being 7964 miles, the diameter of the orbit will be $7964+\overline{50\times 2}=8064$, and $8064\times 3.1416=25333.8624$, its circumference. Then $25333.8624\div 20=1266.693120=21'\ 6''\ 41'''\ 35''''\ 13'''''\ 55''''''$, the *Ans.*

100. In giving directions for making a chaise, the length of the shafts between the axletree and back-band being settled at 9 feet, a dispute arose whereabout on the shafts the centre of the body should be fixed. The chaise-maker advised to place it 30 inches before the axletree; others supposed that 20 inches would be a sufficient incumbrance for the horse. Now, supposing 2 passengers to weigh 3 cwt., and the body of the chaise $\frac{3}{4}$ cwt. more, what will the horse, in both these cases, bear in addition to his harness? *Ans.* $\begin{cases} 116\frac{2}{3}. \\ 77\frac{7}{9}. \end{cases}$

101. A piece of square timber is 10 feet long, each side of the greater base 9 inches, and each side of the less 6 inches. How much must be cut off from the less end to contain a solid foot? *Ans.* 3.392 feet.

102. A carpenter put a curb of oak round a well: the inner diameter of the curb was $3\frac{1}{2}$ feet, and its breadth $7\frac{1}{4}$ inches. What was the expense of it, at 8*d.* per square foot?

Ans. 5*s.* $2\frac{1}{4}$*d.*

BOOK-KEEPING.

ALL mercantile transactions consist in exchanging articles of trade, either for money or its equivalent. A systematic record of such transactions is called Book-keeping. Every person should possess sufficient knowledge of this science to keep such record of his business as will at any time exhibit a true state of his affairs.

The person who purchases goods, or receives any thing of me, is debtor to me; and he who pays me money, or delivers any thing to me, is creditor.

The following is a plain and simple method of keeping accounts without a Day-book, and will be found sufficient for the purposes of farmers and mechanics, where their business is such, that charges are made only at considerable intervals; but in all cases where some three or four, or more charges are made daily, a Day-book should be regularly kept.

1837.	James Trueworthy, Dr.	$	C	1837.	James Trueworthy, Cr.	$	C
Jan 1,	To 4 cords of wood, at $3,	12	00	Jan. 1,	By 3 gals. molasses, at 40 cts.	1	20
Jan. 9,	To 1 load of hay, at $20, -	20	00	March 1,	By 1 lb. starch, at 12 cts.		12
Feb. 2,	To 20 bush. oats, at 50 cts.	10	00	April 2,	By 50 lbs. coffee, at 14 cts.	7	00
June 1,	To 15 lbs. veal, at 5 cts. -		75	June 13,	By 12 lbs. sugar, at 9 cts.	1	08
July 6,	To pasturing horse 3 weeks, at 50 cts. -	1	50	" 20,	By 3 yds. broadcloth, at $5	15	00
				Aug. 19,	By cash to balance, - -	19	85
		44	25			44	25

BOOK-KEEPING BY SINGLE ENTRY.

By Single Entry two principal books are required, the *Day*, or *Waste Book*, and *Ledger*.

The Day-book should be ruled with two columns on the right hand, for dollars and cents, and one column on the left for noting the month, day, and year when the charges were made.

All charges should be made at or near the time when they bear date, or when they purport to have been made. Where an individual is charged with articles delivered to him, he is a debtor for the amount of them, and against his name on the Day-book, " Dr." is written to designate him as debtor. When such individual delivers articles to be allowed him on his account, he is said to be a creditor, and " Cr." is written against his name on the Day-book, to designate that he is credited with such articles.

THE LEDGER.

The Ledger is a book to which the accounts entered on the Day-book are transferred from time to time, in order that each man's whole account may appear by itself. All articles for which he is debtor are usually entered on the left-hand side of the page, and all articles for which he is creditor are entered on the right-hand side of the page.

The date of the charge, and the amount in dollars and cents, are entered in the same manner as on the Day-book; but where there are a number of charges of different articles made under one date on the Day-book, they are usually entered upon the Ledger as " Sundries," which is a general term for a variety of articles; and the total amount of all such articles charged on any one day, is carried out against such entry in dollars and cents.

When a charge is posted from the Day-book to the Ledger, a bracket, or other mark, is made on the left of the charge on the Day-book, to denote that it has been transferred to the Ledger.

When the account is to be settled, the sum requisite to balance the same is ascertained, and cash, or a note, is given, which is entered upon the book, and the account is thus made equal. Traders and mechanics most usually give receipted bills of their accounts on settlement. In some instances, a settlement is written at the bottom of the account, as follows:— " Settled and balanced all accounts to this date," which is signed by the parties.

FORM OF THE DAY-BOOK.

1837.			
Jan. 1,	John Newton, Dr.		
	To 4 gallons molasses, at 40 cts.	1	60
	15 lbs. sugar, at 10 cts.	1	50
	Joel Stephenson, Dr.		
	To 6 lbs. chocolate, at 1*s*.	1	00
	Jedediah Jones, Dr.		
	To 50 lbs. fish, at 4½ cts. .	2	25
	2 lbs. starch, at 14 cts. .		28
	6 lbs. coffee, at 15 cts. .		90
" 3,	John Silvers, Dr.		
	To 38 lbs. rice, at 5 cts. . .	1	90
	2 lbs. cocoa, at 28 cts. .		56
	Joel Mason, Dr.		
	To 50 lbs. iron, at 7 cts. .	3	50
	1 barrel of flour, ...	7	50
" 4.	Timothy Styles, Dr.		
	To 6 gals. of oil, at 6*s*. . . .	6	00
	50 lbs. fish, at 5½ cts. . .	2	25
" 5,	John Silvers, Dr.		
	To 40 lbs. of raisins, at 14 cts.	5	60
	John Newton, Dr.		
	To 1 gal. oil	1	00
	8 lbs. coffee, at 15 cts. .	1	20
	Cr. By cash, 10*s*. 6*d*.	1	75
	Joel Mason, Dr.		
	To 3 bush. of salt, at 6*s*. .	3	00
	John Newton, Dr.		
	To 2 lbs. saleratus, at 11 cts.		22
" 10,	John Newton, Cr.		
	By 40 lbs. dried apples, . .	2	00
	Dr. To 2 lbs. Y. H. tea, at 3*s*. 6*d*.	1	16
	To 2 lbs. raisins, at 2*s*. . .		34
" 11,	Jedediah Jones, Dr.		
	To 2½ gal. oil, at 6*s*.	2	50
	Joel Mason, Cr.		
	By 5 bush. apples, 2*s*. . .	1	67
	1¼ cords wood, at 15*s*. .	3	75
	Cash to balance ac't, . .	10	44
	John Silvers, Cr.		
	By 30 bush. potatoes, at 1*s*. 6*d*.	7	50
	100 lbs. cheese, ...	10	00
" 12,	Jedediah Jones, Cr.		
	By cash rec'd of I. Hardy.	3	00
	2 days' labor, at 4*s*. 6*d*.	1	50
	Cash to balance ac't, . .	3	33

1837.			
Jan. 6,	John Silvers, Dr.		
	To 4 yds. broadcloth, at $4.50	18	00
	Cr. By 8 bush. of rye, at 4*s*. 6*d*.	6	00
	By 15 lbs. dried apples, at 5 cts.		75
	40 lbs. cheese, at 10 cts.	4	00
" 7,	John Newton, Dr.		
	To 3 yds. flannel, at 3*s*. . . .	1	50
	Joel Stephenson, Dr.		
	To 1 bush. corn, at 9*s*. . . .	1	50
	1 fur hat	4	00
" 8,	John Newton, Cr.		
	By 1 ton of hay	15	00
	Dr. To 1 bbl. flour	7	50
	To 50 lbs. sugar	5	00
	Jedediah Jones, Dr.		
	To 38 lbs. rice, at 5 cts. . .	1	90
	Joel Mason, Dr.		
	To 1 lb. Y. H. tea, at 3*s*. . .		50
	1 bush salt	1	00
	1 lb. cocoa		28
	John Silvers, Dr.		
	To 1 pr. of boots	3	50
	1 pr. shoes	2	00
	Timothy Styles, Dr.		
	To 100 lbs. nails, at 6 cts. .	6	00
	Joel Stephenson, Dr.		
	To 40 lbs. butter, at 20 cts	8	00
	1 firkin, 3*s*.		50
" 12.	John Silvers, Cr.		
	By 2 days' work	2	00
	3 cords wood, at 15*s*. . . .	7	50
	Dr. To 1 lb. flour	0	10
" 13,	Joel Stephenson, Cr.		
	By 2 Cords bark, at 12*s*. . .	4	00
	Timothy Styles, Cr.		
	By 2 firkins butter, 80 lbs., at 1*s*.	13	34
	Cash		91
" 14,	John Newton, Dr.		
	To 100 lbs. fish, at 5 cts. .	5	00
	Cr. By 2 cords of wood, at 15*s*.	5	00
	By cash, to balance ac't.	2	27
	John Silvers, Dr.		
	To 100 lbs. iron, at 7 cts. .	7	00
" 15.	Joel Stephenson, Cr.		
	By 10 lbs. lard, at 1*s*.	1	67
	20 lbs. butter, at 20 cts.	4	00

FORM OF THE LEDGER.

Dr. **JOHN NEWTON,** **Cr.**

1837.				1837.			
Jan. 1.	To 4 gallons molasses, at 40 cts.	1	60	Jan. 5.	By Cash	1	75
	" 15 lbs. sugar, at 10 cts.	1	50	" 8.	By 1 ton of hay	15	00
" 5.	" sundries	2	42	" 10.	By 40 lbs. dried apples, ..	2	00
" 7.	" 3 yds. flannel, at 3s. ..	1	50	" 14.	By 2 cords of wood, at 15s.	5	00
" 8	" sundries	12	50	" "	Cash to balance accounts.	2	27
" 10.	" sundries	1	50			$26	02
" 14.	" 100 lbs. fish, at 5 cts..	5	00				
		$26	02				

Dr. **JOEL STEPHENSON,** **Cr.**

1837.				1837.			
Jan. 1.	To 6 lbs. chocolate at 1s..	1	00	Jan. 13	By 2 cords of wood, at 12s.	4	00
" 7.	" sundries	5	50				

Dr. **JEDEDIAH JONES,** **Cr**

1837.				1837.			
Jan. 1.	To sundries	3	43	Jan. 12	By sundries	4	50

Dr. **JOHN SILVERS,** **Cr.**

1837.				1837.			
Jan. 3.	To sundries	2	46	Jan. 6.	By sundries	10	76

Dr. **JOEL MASON,** **Cr.**

Dr. **TIMOTHY STYLES,** **Cr.**

The student should be required to complete posting the above accounts from the Day-book.

To each Ledger there is an Alphabet, or Index of Names, consisting of a small book with the letters of the alphabet pasted upon the leaves, so that each man's name can be readily found by turning to the leaf marked by the first letter of his sirname.

INDEX TO THE LEDGER.

	PAGE
A	
Allen James	1
B	
Bacon Samuel	2
Brown Leonard	3
C	
Canning George	3
D	
Dalton Levi	4
E	
Edgell Oliver	4
F	
Fulton Curtis	5
Freeman Nathan	6
G	
Grosvenor Jacob	7
H	
Hoit William	8
Horsley Henry	9
J	
Jenkins Abraham	9
Jones Jedediah	10
K	
Knowles Michael	11
L	
Leeds William	12
M	
Mason Joel	13
M'Farland Asa	13
N	
Newton John	14
Nelson John	15
O	
Odlin Woodbridge	16
P	
Pratt Thomas	17
Putney Andrew	18
R	
Ramsay Amos	18
Rollins Zenas	20
S	
Stephenson Joel	21
Silvers John	22
Styles Timothy	23
T	
Trueworthy James	24
W	
Wyndham Augustus	25

Every person engaged in business should take an inventory of his notes and accounts, his stock in trade, and other property, and of the debts owed by him, once or twice a year. By comparing this with former inventories, he will know his gain or loss, from time to time.

Inventory of Stock in Trade, Notes, Accounts, and other Property, taken from the foregoing example, July 1, 1837.

	$	cts.
200 lbs. coffee, at 1*s.*	$33	34
2 kegs tobacco, 100 lbs., at 20 cts.	20	00
5 bales cotton, 1500 lbs., 10 cts.	150	00
3 boxes sugar, 1200 lbs., 9 cts.	108	00
4 casks nails, 1600 lbs., 6 cts.	96	00
40 bushels corn, at 6*s.*	40	00
6 tons hay, at $10	60	00
400 lbs. iron, at 6 cts.	24	00
90 gals. Molasses, at 40 cts.	36	00
Am't carried up	$567	34
Am't brought up	$567	34
Note against James Allen, for $50 and int., dated Jan. 1, 1836	53	00
Note against Thomas Pratt for $300, dated July 1, 1836, and interest	309	00
Bal. on ac't against John Silvers	5	81
do. against Joel Stephenson	5	33
1 horse	75	00
1 chaise	125	00
Homestead	1000	00
	2140	48
Inventory taken Jan. 1, 1837	1950	75
Net gain in 6 months	189	73

NOTES, RECEIPTS, ETC.

NOTES.

(1.)

Orford, August 10, 1837.

For value received, I promise to pay to John True, or order, seventy-five dollars fifty cents, on demand, with interest.

JOSEPH DENMAN.

Attest: JOEL TRUSTY.

(2.)

Concord, July 4, 1837.

For value received, I promise to pay to James Doughty, or bearer, ten dollars thirty-four cents, six months after date.

JOHN MORSE.

(3.)

[By two persons.]

New Haven, Oct. 6, 1837.

For value received, we jointly and severally promise to pay to S. T., or order, seventeen dollars and eighty-eight cents, on demand, with interest.

ALONZO FONTAINELLE,
JAMES WHITEHEAD.

Attest: TIMOTHY TRUSTY.

For Bank Notes, see "Discount," page 222.

RECEIPTS.

(1.)

Norwich, June 7, 1837. Received of Mr. Nicholas Jewett, five dollars, in full of all accounts.

HENRY SLOCUM.

(2.)

RECEIPT FOR AN ENDORSEMENT ON A NOTE.

New York, Sept. 9, 1837. Received of Mr. John Hadley, (by the hand of James True,) twenty-five dollars fifty cents, which is endorsed on his note of May 6, 1836.

PETER TRUSTY.

(3.)

RECEIPT FOR MONEY RECEIVED ON ACCOUNT.

Mount Holly, Dec. 9, 1837. Received of Mr. John Van Dyke, twenty dollars on account. THOMAS BEAN.

(4.)

RECEIPT FOR INTEREST DUE ON A BOND.

Received, this fourteenth day of May, of Mr. S. W., the sum of six dollars, in full of one year's interest of 100 dollars due to me on the 16th day of April last, on bond from the said S. W.

By me, C. B.

ORDERS.

(1.)

Mr. Joel M'Knight: Sir—

For value received, pay to O. S., ten dollars, and place the same to my account. SUEL RYNO.

Hooksett, Sept. 7, 1837.

(2.)

Lowell, Oct. 10, 1837.

SIR:—For value received, pay S. O. twenty cents, and this, with your receipt, shall be your discharge from me.

To Mr. Daniel Holden. JUSTUS PRUDES.

FORM OF A BOND.

Know all men by these Presents, That I, [we] A. B., of C., in the county of R., and state of New Hampshire, gentleman, [and C. D., of, &c.] am [are] held and firmly bound to E. F., of said C., yeoman, [and G. H. of, &c.,] in the sum of one hundred dollars, to be paid to the said E. F. [and G. H., or either of them,] or his [their] certain attorney, executors, administrators, or assigns, to which payment, well and truly to be made, I [we jointly and severally] bind myself, [ourselves,] my [our] heirs, executors, and administrators, firmly by these presents.

Sealed with my [our] seal [s], and dated the tenth day of May, A. D. 1837.

With a condition to pay money.

The condition of this Obligation is such, that if the said A. B., [C. D., &c., or either of them,] his [their] heirs, executors, and administrators, do and shall well and truly pay, or cause to be paid, to the said E. F., [G. H., or either of them,] his [their] executors, administrators, or assigns, the sum of fifty dollars and interest, on or before the tenth day of May next; then this obligation to be void, otherwise in force.

[Signed and sealed as the preceding forms.]

GENERAL FORM OF AN AGREEMENT.

Articles of Agreement, indented, made, and concluded the tenth day of May, in the year of our Lord one thousand eight hundred and thirty-seven, (or A. D. 1837,) between A. B., of C., in the county of M., yeoman, of the one part, and C. D., of said C., husbandman, on the other part.

The said A. B., for the consideration hereafter mentioned, doth hereby covenant and agree, that, etc.

And the said C. D. doth hereby covenant and agree, etc.

In testimony whereof they have hereunto interchangeably set their hands and seals, the day and year above [or first above] written.

[Signed and sealed as the preceding forms.]

FORM OF A WARRANT OF ATTORNEY.

TO CONFESS JUDGMENT.

To A. B., Esquire, of, etc., an attorney of —— court of ——, to be holden at ——, on the —— day of ——, or to any other attorney of said court.

This is to authorize you, or any of you, to appear for me, E. F., of, etc., in the said court, or any other subsequent term, at the suit of G. H., of, etc., and by *non sum informatus*, *nil dicit*, or otherwise, confess judgment against me unto him, the said G. H., in an action of debt for one hundred dollars, and costs of suit; and for your, or either of your so doing, this shall be your warrant.

In witness whereof, I have hereunto set my hand and seal, this tenth day of June.

BRIEF FORM OF A RELEASE.

May 20, 1837. I, A. B., do hereby release to E. E. all suits, promises, covenants, and demands, which I have or can claim against him.

[Signed, sealed, and witnessed, as other instruments.]

OBS.—By a release of all demands are barred all rights and titles to lands, warrants, debts, duties, actions, judgments, and executions, and all contracts except those which are to be performed on a future contingency. By a release of all covenants and promises, are released all such covenants and contracts as are not released by the word demand, etc.

www.ingramcontent.com/pod-product-compliance
Lightning Source LLC
LaVergne TN
LVHW020220110826
845151LV00003B/774

* 9 7 8 1 4 2 5 5 3 2 5 2 9 *